POLYMER SCIENCE AND TECHNOLOGY
Volume 10

POLYMER ALLOYS
Blends, Blocks, Grafts, and
Interpenetrating Networks

POLYMER SCIENCE AND TECHNOLOGY

POLYMER SCIENCE AND TECHNOLOGY
Volume 10

POLYMER ALLOYS

Blends, Blocks, Grafts, and
Interpenetrating Networks

Edited by

Daniel Klempner
and Kurt C. Frisch

Polymer Institute
University of Detroit

SPRINGER SCIENCE+BUSINESS MEDIA, LLC

Library of Congress Cataloging in Publication Data

Main entry under title:

Polymer alloys.

(Polymer science and technology; v. 10)
"Based on a symposium sponsored by the Division of Organic Coatings and
Plastics Chemistry at the 173rd meeting of the American Chemical Society held in
New Orleans, March 1977."
Includes index.
1. Polymers and polymerization — Congresses. I. Klempner, Daniel. II. Frisch,
Kurt Charles, 1918-   III. American Chemical Society. Division of Organic
Coatings and Plastics Technology. IV. Series.

QD380.P636   668   77-21559

ISBN 978-1-4684-0876-8   ISBN 978-1-4684-0874-4 (eBook)
DOI 10.1007/978-1-4684-0874-4

Based on a symposium sponsored by the Division of Organic Coatings and
Plastics Chemistry at the 173rd meeting of the American Chemical Society
held in New Orleans, March, 1977

© 1977 Springer Science+Business Media New York
Originally published by Plenum Press, New York 1977
Softcover reprint of the hardcover 1st edition 1977

A Division of Plenum Publishing Corporation
227 West 17th Street, New York, N.Y. 10011

PREFACE

Alloy is a term commonly associated with metals and implies a composite which may be single phase (solid solution) or heterophase. Whichever the case, metallic alloys generally exist because they exhibit improved properties over the base metal. There are numerous types of metallic alloys, including interstitial solid solutions, substitutional solid solutions, and multiphase combinations of these with intermetallic compounds, valency compounds, electron compounds, etc. A similar situation exists with polymers. There are numerous types of composites, or "alloys" of polymers in existence today with new ones being created continuously. Polyblends are simple physical mixtures of the constituent polymers with no covalent bonds occuring between them. As with metals, these may be homogeneous (single phase) solid solutions or heterogeneous (multiple phase) mixtures. With polymers, the latter case is by far the most prevalent situation due to the thermodynamic incompatibility of most polymers. This is due to the relatively small gain in entropy upon mixing the polymers due to contiguity restrictions imposed by their large chain length. If the component polymers in the blend are crosslinked, a relatively new situation arises. The resulting material is known as an inter-penetrating polymer network(IPN). IPN's represent a mode of blending two or more crosslinked polymers to produce a mixture in which phase separation is not as extensive as would be otherwise, due to "permanent" entanglements occuring between the networks. If covalent bonds exist between the component polymers, a graft copolymer results. The term graft is employed since these copolymers(and they are true copolymers since the polymers are covalently bonded to each other) are generally produced by "grafting" a growing chain of one polymer onto a chain of another polymer. If the differing monomer units are arranged linearly in segments covalently bonded

to one another along the chain, a block copolymer results. Phase
separation generally occurs with the latter types of polymer composites, although it is not as extensive as in the case of polyblends.

As with metals these "polymer alloys" generally exhibit enhanced and in many cases synergistic properties. Indeed, this is the
major impetus for their formation. Determination of the morphology
of these composites has presented a unique challenge to polymer
scientists and engineers. This book presents a compilation of the
most recent efforts of both American and foreign scientists in one
of the most active areas of polymer research today, the study of
the preparation, characterization, and properties of polymer alloys.
The major portion of the chapters in this book was presented at a
symposium sponsored by the Division of Organic Coatings and Plastics
Chemistry at the 173rd meeting of the American Chemical Society,
held in New Orleans in March, 1977.

The editors wish to express their gratitude to the numerous
authors who contributed to this book and to the University of
Detroit for its encouragement of this effort.

                                        Daniel Klempner

                                        Kurt C. Frisch
Polymer Institute

University of Detroit

Contents

CONTENTS

ADVANCED POLYMER ALLOYS CONTAINING POLYPHOSPHONATE COMPONENTS

I. CABASSO*, J. JAGUR-GRODZINSKI, AND D. VOFSI
*Gulf South Research Institute, P. O. Box 26518
New Orleans, Louisiana 70186, and the Weizmann
Institute of Science, Rehovot, Israel

INTRODUCTION

Polymer alloys consist of two (or more) polymers which are compatible at the molecular level. The virtue of such materials is that they assimilate the properties of each of the polymeric components into a new, distinct product. Such an alloy displays a glass transition temperature, $Tg°$, intermediate between those of the two polymer components, and can be tailored into a product with desired properties, e.g., hydrophilic/hydrophobic balance.

Following Gibb's expression $\Delta G^m = \Delta H^m - T\Delta S^m$, the blending of two different polymer species into a homogeneous alloy is dependent on the heat and entropy of mixing. To achieve homogeneity at the molecular level, the condition $\Delta G^m \leq 0$ must be fulfilled; however, the gain in the free energy of mixing due to the entropy, ΔS^m, is negligible when long-chain polymers are involved. Since mixing of organic molecules is usually an endothermic process, polymer blends would, in general, resemble a heterogeneous mixture with distinct phase (or microphase) separation. Exceptions are mainly the result of specific interactions between sites of attraction, e.g., hydrogen bonding. In such cases, the differentiation between "true" compatibility and microphase separation is frequently obscure, and often depends upon the method by which the polymers are blended. From the practical point of view, compatibility is achieved when the alloy displays homogeneity with regard to some desired properties, e.g., single Tg, thermal conductivity, etc.

1

POLYMERS CONTAINING PHOSPHORYL ESTER GROUPS (PPN)

In recent years the synthesis of several polymers containing pendent phosphonate ester (1) and phosphate ester (2) groups
has been reported. These polymers are derivatives of
commercial polystyrene, poly-[2,6-dimethyl-1,4-phenylene
oxide] (PPO), and laboratory-prepared copolymers of styrene-
vinylidene chloride. The polyphosphonate (PPN) derivatives are
of the following types:

PPN-I (PSP)   PPN-II (CSVCP)

R = CH_3-, CH_3CH_2-

PPN-III (PPOP)   PPN-IV (PPOBrP)

The synthesis of these polymers has been described in detail
in previous publications (1). In short, chloromethylation of
polystyrene (and its vinylidene chloride copolymers) and bromina-
tion of the PPO methyl groups are followed by an Arbuzov phosphonylation with trialkyl phosphite: $(MeO)_3P$ or $(EtO)_3P$. It has been
shown that these PPN derivatives are capable of alloying various
polymers with diverse chain units (1-6).

The incorporation of pendant phosphoryl ester groups onto a
nonpolar polymer chain imparts to the latter a wide range of
dipole-to-dipole and hydrogen bonding interaction capabilities.
This can be demonstrated by illustrating the solubility latitude
of such a polymer in Hansen's two-dimensional solubility
parameter diagram (7), as shown in Figure 1. The plane is

defined by δ_p and δ_h coordinates, fractions of the Hildebrand solubility parameter, which reflect the polar and hydrogen bonding interactions. (The complete correlation is given by the equation

$$\delta^2 = \delta^2_d + \delta^2_p + \delta^2_h$$

where δ_d represents the dispersion forces.)

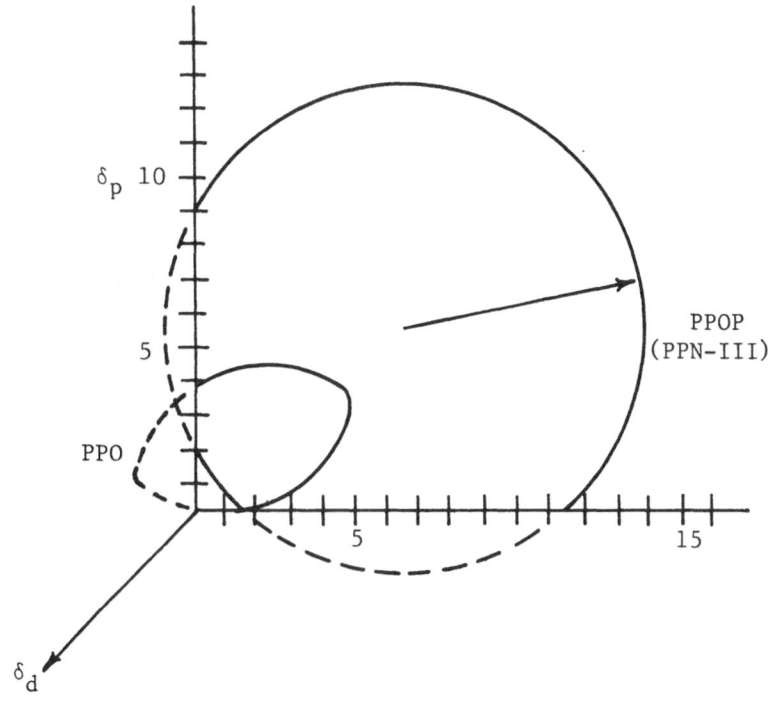

Figure 1.   Solubility parameter areas of PPO and its polyphosphonate derivative, PPOP (PPN-III), as projected on the δ_p vs δ_h solubility grid.

In Figure 1, the solubility parameter diagrams of PPO and PPOP (PPN-III) are shown. The encircled areas encompass all the organic compounds that are solvents for the designated polymers; every organic solvent is represented as a point on the δ_p–δ_h plane according to its solubility parameter value (7). The

solubility parameter areas do not provide a single δ value for a
polymer; rather, they provide a range of "intimate" interaction.
Thus, the solubility area of the PPN in the figure, an empirical
reflection of the polymer interactions, is rather excessive.
This is in accord with the fact that the basic oxygen in a
phosphonate ester is capable of forming strong hydrogen bonds
with water (7-8 kcal/mole) and displays a dipole moment of 3-4
D(Debyes). The hydrophilicity of the PPN film is a primary
function of the phosphonate content of the PPN. Some examples
are shown below:

PPN	P(%)	Water uptake (wt%)
I	8.0	40
II	8.6	315
III	9.5	38
IV	10.5	20

The unphosphonylated parent polymer displays water uptake
of less than 1%. The high water uptake of PPN-II is due to the
strong plasticizing effect exerted by the water, causing collapse
of the glassy state of this polymer which displays a Tg of
46°C (1).

The PPN's are nonflammable and were found to be thermostable.
Of the above four species, PPN-I seems to be the most stable;
it thermally crosslinks at 330°C, and chain degradation starts
at 470°C (1).

Solvent Cast Polymer Alloys with PPNs

Solvent cast alloys from PPN and another polymer were
usually prepared by dissolving 1-15 wt% of each polymer separately
but in the same solvent. The two solutions were mixed to give a
clear binary polymer solution. (Incompatibility in solution
results in a cloudy solution which often subsequently separates
into two distinct phases.) The clear solutions were cast on a
glass or Teflon plate with a "doctor blade" to give uniform
layers, 0.1-2 mm thick. The solvent then was allowed to evaporate
slowly to give a dense film. If no macrophase separation could
be detected in the cast product, physical measurements were
taken to characterize the alloy and to confirm homogeneity on
the microscale. Glass transition temperature was measured on a
Perkin-Elmer differential scanning calorimeter DSC-1B and on a
thermomechanical analyzer TMS-1. Some samples were examined with
a scanning electron microscope down to a resolution of 200Å. In

some cases, PPN-IV, which was synthesized (1) to contain 42% Br, was used to increase resolution. To enhance the scattering effect imparted by the presence of Br atoms, some of the film samples were exposed to solutions containing 1M $UO_2(NO_3)_2$ and 3M $NaNO_3$ at pH 2; they absorbed uranyl ions by virtue of the phosphonate ligand (8). The secondary electron technique was applied for the inspection of domains in such films.

The following polymers were found to be capable (some only in part) of forming a homogeneous alloy with PPN when cast from a mutual solvent.

```
Cellulose Acetate------------------------CA
Cellulose Triacetate--------------------CTA
Hydroxyethyl Cellulose------------------HEC
Nitrocellulose--------------------------NC
Cellulose Acetate Butyrate--------------CAB
Cellulose Acetate Propionate------------CAP
Polystyrene and its Copolymers----------PS
Polyphenylene Oxide---------------------PPO
Polyacrylonitrile-----------------------PAC
Unsaturated Polyesters------------------PEU
```

In addition, various ion-exchange polymers (e.g., polystyrene quaternary salts) proved to form homogeneous alloys with the PPN; these are presently under study.

When alloyed with PPN, the above polymers can be redissolved by a mutual solvent to give a coherent solution from which each component can be isolated. On the other hand, leaching the alloy with a solvent which is suitable for one of the components but is a nonsolvent for the second polymer does not lead to a complete extraction of the soluble component. The intimate interactions and dispersion of the polymer molecules result in entanglement of the polymeric chains from which the soluble component can no longer be removed. This characteristic may serve as a compatibility criterion but is a less clear-cut requirement for an alloy. Clearly, one cannot expect this to hold over the whole range of compositions, and an arbitrary limit has to be set for this test. In the present work, the standard test involved the immersion for 1 hour into a boiling solvent of a 100-μm-thick 200-mg film strip which contains 50% by weight of each component. The chosen solvent would, under identical conditions, dissolve films comprising only one of the two components of the alloy within a few minutes. A loss of weight smaller than 0.5% was regarded acceptable.

As was mentioned above, the homogenity of a system is often regarded as a continuum property. Such a definition often fits into the macroscale category, where microphase separation is overlooked. However, blends in which the dispersion of the polymeric components approaches the molecular level are considered to possess specific interactions, and/or to display mutual overlap in their solubility areas (6) as shown in Figure 2 for PPN-III/CA. (This relationship indicates that for such polymer combinations, the Flory-Huggins interaction parameter $\chi \to 0$.) Almost all the cellulose derivatives listed above exhibit some hydrogen bonding capabilities through their hydroxyl groups (in most cases a residual primary hydroxyl group). It is believed that hydrogen bonding interaction prevails between the phosphoryl oxygen (which is an electron donor) and the rather isolated hydroxyl of the substituted cellulose derivatives. Where such specific interactions do not exist, e.g., PPN/polystyrene, compatibility prevails for a limited weight fraction and depends upon the degree of phosphonylation of the PPNs. Elaboration on this aspect is given elsewhere (6).

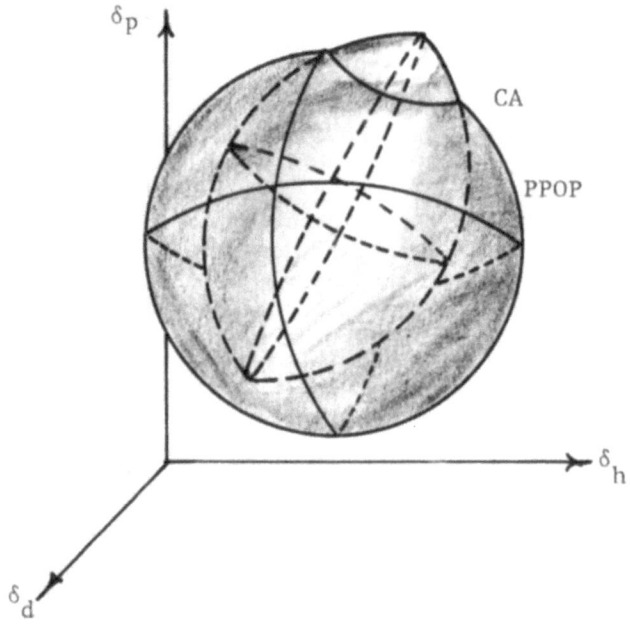

Figure 2. Three-dimensional solubility parameter space shows
          substantial overlap between PPOP (PPN-III) and
          cellulose acetate (CA), a compatible polymer couple.

The PPN alloys reported herein were studied with regard to a unique property that enables them to selectively absorb organic liquids which are considered solvents of only one of the alloy polymeric components. This property makes the alloys effective membranes for separating organic liquid mixtures. Also, in view of their high phosphorous and halogen content, the alloys have been examined for possible commercial importance as nonleachable fire retarding additives.

PPN Alloys with Cellulosic Derivatives

The cellulose derivatives listed above are compatible through the entire range of weight fraction with PPNs that contain 5-6% P or more. [This excludes some grades of CTA that upon casting seem to undergo some microphase separation. They are less clear (transparent) due to a "milky shade" in their alloys.] It was found that the molecular weights of PPNs I-IV and of cellulose derivatives do not influence the alloy formation; for example, no phase or microphase separation was detected when PPNs of M_n > 100,000 and high molecular weight CA (Eastman 394-45) were employed. These polymers display mutual overlap when they are projected on the two- or three-dimensional solubility parameter space as shown in Figure 2.

The glass transition temperatures of these alloys (1) follow the linear equation:

$$Tg(alloy) = W_1(Tg_1 - Tg_2) + Tg_2$$

where W_1 is the weight fraction of component 1. (This relationship is a convenient rearrangemant if $T_g = W_1 T_{g1} + W_2 T_{g2}$; the latter assumes a weight fraction linear average between the T_g's of the alloy components.)

Some glass transition temperatures for PPN, cellulose acetate and their alloys are shown in Table 1.

Table 1

Glass Transition Temperatures of PPNs and their Alloys
with Cellulose Acetate

PPN Designation	$Tg°(PPN)$	$Tg°(CA)$	$Tg°(PPN/CA\ 1{:}1\ w/w)$
I	47	163	110
II	96	163	130
III	118	163	142
IV	180	163	171

Tensile strength and flammability measurements were conducted on some compositions and are shown in Table 2 and Figure 3. All the films represented in this data were transparent, dense alloys.

Highly flammable nitrocellulose and polyacrylonitrile were spun with the PPNs to form alloy fibers which display self-extinguishing properties; this will be reported elsewhere (9).

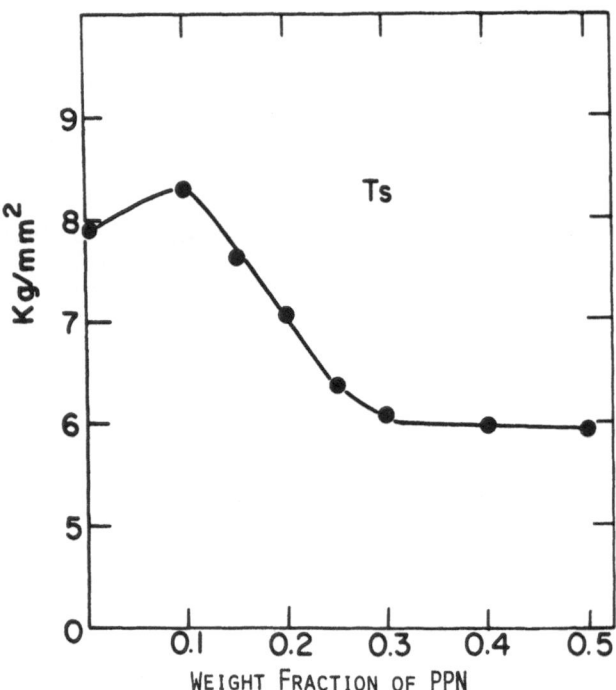

Figure 3.   Tensile strength curve of CA/PPN-IV alloys. (PPN-IV= 9% P, 42% Br, TS-3 kg/mm^2; CA=Eastman 394-45).

Table 2

Tensile Strength and Flammability of Polymeric Alloys

Component		Tensile Strength kg/mm^2	Flammability* (ASTM D) 1433-58
I	II		
Cellulose Acetate			
(Eastman 394-45)100%	–	8.0	burn – 3.5 cm/sec.
(Eastman 394-45) 80%	PPN-IV 20%	7.3	self-extinguishing (2-2.5 sec., 3-6 cm)
(Eastman 394-45) 75%	PPN-IV 25%	6.4	self-extinguishing (0-2 sec., 0-4 cm)
(Eastman 394-45) 70%	PPN-IV 30%	6.3	self-extinguishing (0-1.5 sec., 0-2 cm)
(Eastman 394-45) 60%	PPN-IV 40%	6.0	nonflammable
(Eastman 394-45) 50%	PPN-IV 50%	5.8	nonflammable
(Eastman 394-45) 50%	PPN-III 50%	5.9	nonflammable
(Eastman 394-45) 60%	PPN-I 40%	6.3	self-extinguishing (2-3 sec., 5-6 cm)
(Eastman 394-45) 50%	PPN-II 50%	5.7	self-extinguishing (1-2 sec., 2-4 cm)
Cellulose Triacetate			
(A-43.2%) 75%	PPN-IV 25%	4.75	self-extinguishing (0-2 sec., 0-2 cm)
(A-43.2%) 75%	PPN-III 25%	4.7	self-extinguishing (1-2 sec., 1-2 cm)

*Determined for 20 μm thick films, the following PPNs were employed

PPN-I = Cl-15.1%; P-9.2%   PPN-III = Br-12%; P-12%

PPN-II = Cl-1.1%; P-7.5%   PPN-IV = Br-42%; P-9%

    As was mentioned above, the exposure of an alloy to a nonmutual
solvent does not lead to a complete extraction of the soluble component. Some compositions remain intact in such a leaching
treatment; e.g., an alloy of PPN-I/CA 4/6 w/w retains its integrity
and composition when treated with benzene (a good solvent for
PPN-I). The solvent does not leach out the PPN from the matrix
but is progressively dissolved into the alloy as the PPN fraction
increases (Fig. 4). This quality of the alloys was exploited
successfully (2-5) in producing a semipermeable membrane which
can separate azeotropic liquid mixtures into their individual
components. Such a membrane separation capitalizes on the preferential absorption of the permeating liquid into the alloy membrane.

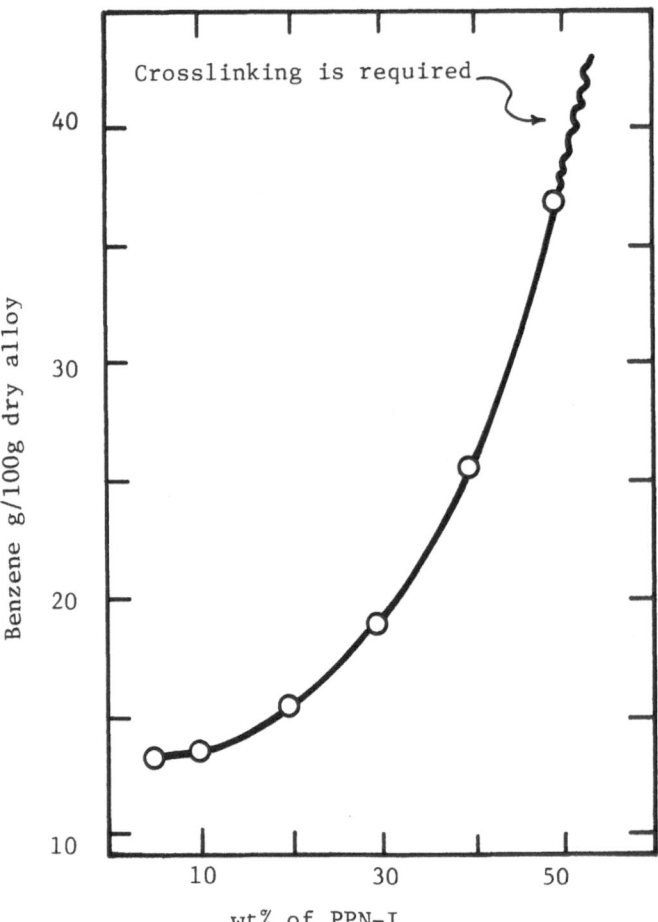

Figure 4.   Solubility of benzene in CA/PPN-I alloys equilibrated
            at 30°C.

Upon exposing the above alloy to the benzene/cyclohexane azeotrope
(55 parts benzene), the alloy membrane imbibes 17 wt% liquid, more
than 95% of which is benzene. As shown in Figure 4, the alloys
retain their integrity when immersed in benzene up to the composition ratio of PPN/CA=1/1 w/w. It has been (7) shown that an
increase in the PPN fraction would lead to a loss of this component
to the solvent; however, the alloy would elute PPN down to a
stable composition ratio. For cellulose acetate this composition
is 1/1; with cellulose acetate butyrate the composition is ∿6/4.

Alloy membrane separation of several organic liquid mixtures
has been reported (2-5). The separation was effected through pervaporation and "osmotic distillation" processes. The rationale
behind tailoring of an alloy membrane for such separations is
demonstrated in Figure 5.

Figure 5 illustrates the solubility parameter areas for an
alloy membrane. The components of this membrane are cellulose
acetate (Eastman 294-45) and PPN-II. The two areas of the
component polymers have a mutual overlap of approximately 60% in
the two-dimensional representation shown. In three-dimensional
representation, the overlap is somewhat greater. The PPN is
soluble in benzene and insoluble in cyclohexane. Cellulose
acetate is insoluble in both solvents. The coordinates for each
of the solvents are shown on the diagram.

Although a solvent for PPN, benzene does not destroy the
alloy membrane, since its coordinates are sufficiently distant
from the boundary of the cellulose acetate. By contrast, chloro-
form (δ_d= 8.7, δ_p=1.5, δ_h=2.8) destroys the membrane, since it
is a swelling liquid for the CA and a solvent for the PPN.
Cyclohexane, which forms an azeotrope with benzene, does not
affect the alloy at all and displays less than 1% solubility in
it. With respect to separation capability of such alloy membranes,
the separation factor is defined by the expression:

$$\alpha_{ab} = \frac{a}{A} \times \frac{B}{b} .$$

where A and B are the weight fractions of liquid components in
the feed mixture and a and b are the product components. The
separation of aromatics from their mixture with aliphatic liquids
proved to be very efficient employing membranes of the type
shown in Figure 5. Some permeation rates employing PPN-I/CAB
1/1 w/w membranes are shown in Figure 6.

Two other binary liquid mixtures are shown in Figure 5 to
illustrate separation efficiency: ethanol/heptane (E/H) and
styrene/ethylbenzene (S/E). The values of the fractional (Hansen's)
solubility parameters for the binary system are shown in the

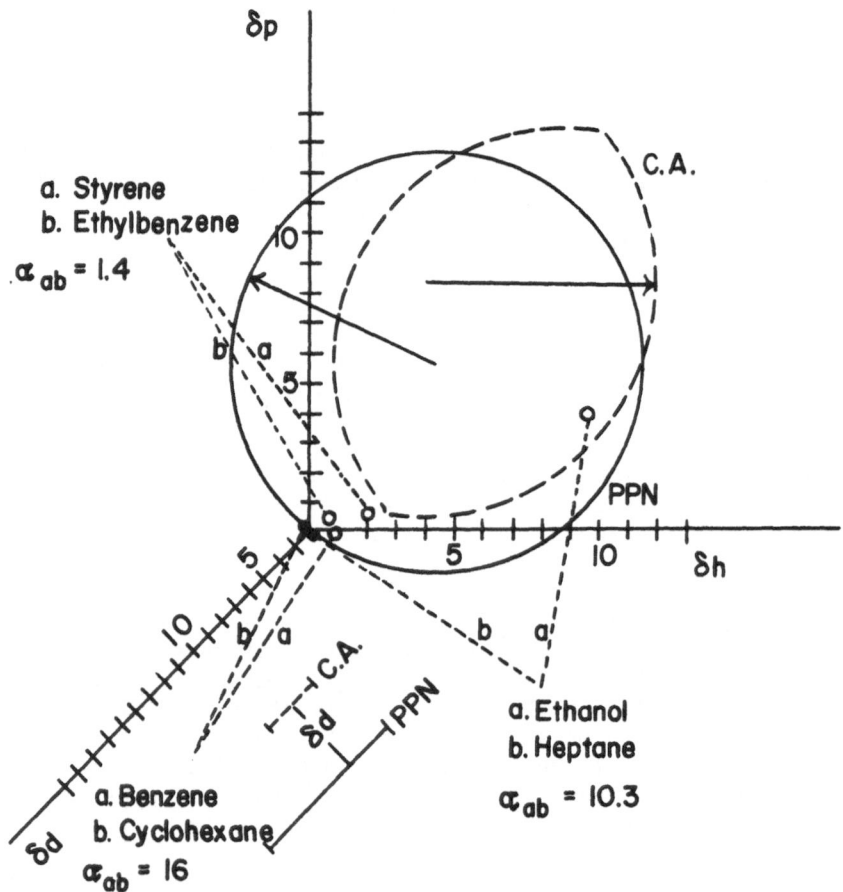

Figure 5.   Solubility parameter diagram of PPN/CA (1/1) alloy
             membrane, containing three binary azeotropic mixtures.
             (α_{ab} is the separation factor maintained by the
             membrane for each mixture using pervaporation
             technique).

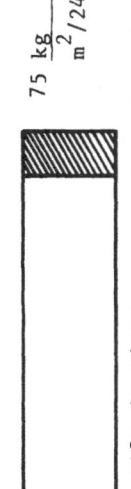

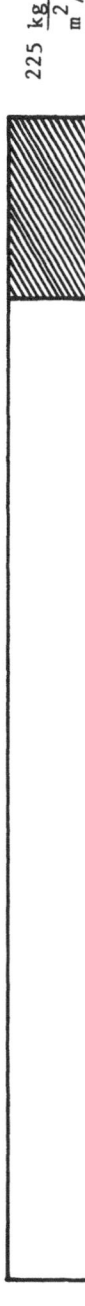

Figure 6. Separation of Aromatic/Aliphatic 1:1 w/w Mixture by Pervaporation Employing Alloy Membranes.

figure. In these examples, the location of the solubility
parameters of the liquid components indicates that the same
reasoning outlined previously (i.e., preferential absorption)
would apply. From these values, it is apparent that ethanol is
in the boundaries of the PPN solubility area as a solvent, and
near the boundaries of cellulose acetate as a swelling agent;
heptane is completely outside both areas. The observed separation
factor for an azeotropic mixture E/H – 48/52 was α_{ab} =10.3. This
is a high separation factor for a system which in ordinary
distillation yields an azeotrope.

The binary system containing styrene and ethylbenzene has
coordinates inside the PPN solubility area, but outside the
cellulose acetate area. At room temperature ethylbenzene has a
higher vapor pressure than does styrene, so that simple evaporation
from a surface would lead to greater concentrations of ethylbenzene
in the product gases. But as expected from the solubility
diagram, after applying membrane separation, the product was
richer in styrene, with a separation factor of 1.4 (observed)
for a feed composition of 1/1). Styrene is located closer to the
center of the alloy coordinate system and is thus expected to
have a greater interaction with the membrane, leading to better
solubility and permeation rates. The observed separation factor
is significantly smaller than the values for the azeotropes of
benzene/cyclohexane and ethanol/heptane. This, too, is predictable
on the basis of the small difference in the relative location of
the solutes in the solubility parameter space, and from the fact
that both ethylbenzene and styrene are within the space of one
component of an alloy membrane.

Using solubility parameter diagrams as shown in the figures,
polymers may be qualitatively selected for preparing alloy
membranes. The same rationale can be applied to predicting the
feasibility of separating certain mixtures by membrane permeation.

Intensive study of polymeric alloy membranes of this type,
including diffusion characteristics of various molecules in the
alloy, has been reported elsewhere (3-5).

Crosslinking of PPN/Cellulose Derivative Alloys

By exposing the alloys to some swelling liquids that display
high solubility in the alloy (>40 wt%), microphase separation
may result. The high solubilities of some chlorinated hydrocarbons
in PPN/cellulose derivative alloys represent such a case. The
highly absorbed liquid might decrease the softening point of the
alloy below its Tg and thus enhance solvent plasticization.

In such cases where the alloy is virtually composed of
three components, two polymers and a dissolved liquid, the
latter might compete on the potential sites of interaction.
Thus, for example, specific interaction between chloroform (proton donor) and PPN (proton acceptor) in an amorphous nonglassy
matrix may eventually produce micro domains of PPN-chloroform in
the alloy. This may jeopardize the integrity of the alloy and
sometimes impose limitations on the workable conditions of the
alloy similar to those shown in Figure 4, which illustrates
that the integrity of the alloy is jeopardized above a certain
composition ratio (e.g., PPN/CA 1/1 w/w).

A controlled crosslinking reaction between the alloy components can be easily carried out to form a durable, stabilized
alloy which is not susceptible to solvent leaching even above the
limiting alloy compositions.

The PPNs are produced from halomethylated polymers that are
phosphonylated via the Arbuzov reaction with alkylphosphites (1)
by sparing trace quantities of unreacted halomethyl groups in the
PPN chain. A crosslinking reaction between the polymer components
can take place either by heat or chemical treatment, or by exposure to light. The cellulose/PPN alloy derivatives are crosslinked
via interraction of the residual primary hydroxyl groups of the
cellulose with the halomethylated groups on the PPN to form an
ether-linkage. For example, by sparing 2-3% of the original
chloromethylated group on PPN-I, a crosslinked alloy of PPN-I/CA
4/1 remained intact when treated with benzene. It is important
to note that the crosslinked alloys retained the mechanical,
thermal and absorption properties of the uncrosslinked ones since
they were only slightly crosslinked.

Crosslinking reactions were carried out with CA, CAB, CAP
and derivatives of PPN-III and PPN-IV that contained residual
groups of

The trace bromine (halomethylated) in the polymer did not exceed
5% (Br). Such reactions occur spontaneously when the alloy is
exposed to sunlight for several days, heat treated for 1 hour at
120°C, or exposed to pyridine vapor. Crosslinked alloys of PPN-
III/CA, CAB, CAP (6/4 w/w) could be swelled by solvents such as
acetone and chloroform. The loss of weight by alloys treated
with such solvents did not exceed 2% of the nontreated alloy.

Some highly durable membranes were produced from alloys
which contained residual chloromethylated groups (2-3%) and
were exposed to heat treatment at 120°C for 30 minutes. Such
alloy membranes were tested under severe conditions, at 80-100°C
for 2000 hours with a feed solution of toluene/heptane 1/1, main-
taining a constant flux rate and separation factor α_{ab}=19 (9).

Alloys with Polystrene (PS) and Styrene Reinforced Polyester

Polystyrene is compatible with PPO in the entire weight
fraction. The blend PPO/PS displays a single Tg and can be cast
as a transparent film (10). The compatibility of these polymers
is an outcome of the low value of their Flory-Huggins interaction parameters, χ, and does not reflect any specific interaction
mechanism (this perpetuates the debate as to whether or not a
true alloy of PPO/PS prevails). Introducing a phosphonate group
to PPO, polystyrene (and its copolymers) limits their compatibility with the orginial parent polymers (1). The compatibility
of PPN-IV with polystryene is dependent upon the degree of
phosphonylation, the temperature of mixing, and weight fraction
of each polymer component. However, phosphonylated polystyrene
is compatible with phosphonylated PPO in the entire weight fraction
range. In addition, it was shown that the PPNs are compatible
among themselves (1).

Clear alloys of polystyrene and the entire family of PPNs
can be produced when styrene monomers are polymerized in the
presence of the PPNs. This is of a practical value since the
product is a self-extinguishing, transparent polystyrene alloy.

Styrene is a solvent for all the above-mentioned PPNs.
Alloys of PS/PPN were prepared by bulk polymerization of the
styrene to give a transparent resin or mold with desired shape
and properties. For example, 25 g of PPN-IV were dissolved in
75 g styrene and 0.5 g of photosensitizer (Triagonal-14) was
added. The solution was brought to 40°C and exposed to radiation
from a 160-watt mercury lamp for 1 hour. A transparent, homogeneous (Tg=121°C), solid material was obtained. Its hardness,
measured with Barcoll-940, was 40, and its self-extinguishing time
was 3 seconds (measured according to ASTM D-635).

Several formulas of unsaturated polyester resin solutions
were blended with PPN and were subsequently cured to yield a
durable, nonflammable, transparent product. Unsaturated polyester
solutions usually consist of highly viscous, low molecular weight
(1000-5000) polymers (polymerized by step reaction from dibasic
acids--as maleic acid--and dihydric alcohol) that have been thinned
by vinyl monomers, primarily styrene. The unsaturated polyester
forms a homogeneous blend with the PPN. When styrene is used as

a diluent to such a blend, the solution can be cured to yield a product with the desired self-extinguishing property. For example, 3.5 g PPN-IV were mixed with 7.5 g of a commercial unsaturated resin (crystic LV-191), producing a transparent self-extinguishing laminate when cured by the above method. This mixture was also cured slowly at 50°C for 24 hours by adding 0.1 g benzoylperoxide to the mixture. Rapid curing of such systems is often used as a means to suspend and blend incompatible polymers in the solid state before phase separation can take place in the intermediate liquid phase. Slow curing was employed to let phase (or microphase) separation take place if the polymer components were not compatible. No indication of microphase separation could be detected.

The reinforced PPN/styrenated polyester alloy formulated above produced a transparent laminate. The hardness (Barcoll-940) was 46 and the self-extinguishing time was 3.5 seconds.

REFERENCES

1. I. Cabasso, J. Jagur-Grodzinski and D. Vofsi, J. Appl. Polym. Sci., 18 1969 (1974).
2. I. Cabasso, J. Jagur-Grodzinski and D. Vofsi, J. Poly. Sci. Chem. Ed., 12 1141 (1974).
3. I. Cabasso, J. Jagur-Grodzinski and D. Vofsi, Ger. Pat. 2.363.635, Isr. Pat. 39.426 (1973); U.S. Patent 4.008.191 (1977).
4. I. Cabasso, J. Jagur-Grodzinski, D. Vofsi, J. Appl. Sci. 18 2137 (1974).
5. I. Cabasso and I. Leon, "Liquid Mixture Separation by Flat-Sheet and Hollow Fiber Membranes" Boston AIChE Meeting, Manuscript #7527 (1975).
6. I. Cabasso, Am. Chem. Soc. "Organic Coating and Plastic Chemistry," preprint 38 (1977).
7. C. Hansen, and A. Beerbower, Kirk-Othmer Ency. of Chem. Tech., Supplement Vol., p. 889 (1971).
8. J. Kennedy, J. Appl. Chem., 9 26 (1959).
9. I. Cabasso, to be published.
10. A.R. Shultz and C.R. McCullough, J. Polym., Sci., A-2, 10 307 (1972).

AN APPROACH TO NEW POLYMERIC MATERIALS VIA BLOCKS, GRAFTS AND BLENDS

James F. Kenney

M&T Chemicals Inc.

Rahway, New Jersey 07065

INTRODUCTION

There are at least four ways to formulate a polymeric material with unique properties: (1) develop new monomers, (2) develop new methods and techniques of polymerization, (3) combine existing monomers in such a way that the resulting material has certain superior properties, and (4) combine existing polymers in unusual ways to achieve new and useful properties. Thermoplastic elastomers are ordered, block copolymers of the general structure A-B-A, where A is a thermoplastic polymer segment and B is an elastomeric polymer segment. The thermoplastic elastomers are composed of "hard" and "soft" segments which exhibit rigid properties and elastomeric properties at one and the same time. A two-phase system is formed, with the center segment phase constituting a continuous three-dimensional elastomeric network and the dispersed end segment phase serving as multi-junction points for the ends of the center segment. Choice of monomers, block length, the weight fractions of A and B, and the segment arrangement are crucial in achieving elastomeric performance. Block copolymers having segment arrangements such as A-B, or B-A-B do not exhibit the tensile behavior characteristic of thermoplastic elastomers, since for a continuous network to exist both ends of the elastomer segment must be immobilized in the non-elastomeric domains.

Thermoplastic elastomers, without vulcanization, have rubber-like properties similar to those of conventional rubber vulcanizates but flow as thermoplastics at temperatures above the glass transition of the end segment. Melt behavior with respect to shear and temperature, is similar to that of conventional thermoplastics.

Melt viscosities, however, are very much higher than those of either homopolymer of the same total molecular weight. High tensile strength, in the absence of reinforcing fillers or crystallization, is attributed to the immobile domains of the "hard" segment, dispersed in a continuous matrix of "soft" segment. A highly perfected network exists.

The commercially important thermoplastic elastomers, styrene-butadiene block copolymers, are prepared by anionic polymerization. The remarkable physical properties of polystyrene-polybutadiene-polystyrene block copolymers are attributed to the formation of a two-phase system. This system consists of a discrete phase of polystyrene segments, acting as multifunctional junction points for the continuous phase of the polybutadiene segments. The development of elastomeric properties requires the relative proportions of the two types of segments to be such that the elastomeric material forms a continuous phase. High resilience, high tensile strength, highly reversible elongation, and abrasion resistance are obtained. Novel products are manufactured by modern, rapid, thermoplastic processing techniques, such as extrusion and injection molding, and without any chemical vulcanization.

Graft copolymerization involves the polymerization of monomers onto a polymer backbone other than at the chain ends. Graft copolymers are of major industrial interest. The industrially important process is addition polymerization in emulsion of vinyl monomers initiated by free-radical catalysts. The kinetic steps in graft copolymerization of a vinyl monomer are initiation, propagation, chain transfer and termination. Chain transfer is the most important step in graft copolymerization. The practical importance of the chain transfer reaction in many industrial polymerization processes is enormous. In the chain transfer reaction, an actively growing chain molecule abstracts an atom from another molecule, transferring the site of activity and ending its own growth. Chain transfer can occur by transfer of the free-radical site to monomer, a chain transfer agent such as a mercaptan or a polymer chain. When the growing polymer radical reacts with a mercaptan (-SH) or with some other active center, such as C-Cl, the hydrogen or halogen atom is transferred, terminating the growing chain, and at the same time forming a new radical site on the transfer molecule. Chain propagation from these transfer sites results in grafted side chains which terminate as usual by combination or disproportionation.

The polymerization of a monomer in the presence of a preformed polymer may result in grafting of the two types of polymeric chains. In graft copolymers the backbone and side chains may both be homopolymeric or copolymeric. Grafting a glassy phase polymer to a rubbery phase polymer increases the interfacial strength or adhesion between the two phases, assuring uniform dispersion and retarding particle aggregation. Graft copolymerization stabilizes the resulting heterophase structure, thereby enhancing physical properties.

For many industrial products, a pure graft copolymer is not required to enhance polymer performance properties. The presence of only 5-10% grafting is sufficient in some products to improve the physical properties of a polymeric material. The yield of grafted copolymer is favored in general by high concentrations of the preformed transfer polymer and high concentrations of polymerizing radicals. The presence of a small amount of graft copolymer acts to reduce the tendency to phase separation.

In commercially important ABS polymer, the presence of graft copolymer prevents microphase separation of the polybutadiene rubbery component and styrene-acrylonitrile copolymer glassy component during processing and in the product when it is subjected to impact.

One of the most commercially significant areas for the development of new polymeric materials has been polymer blending. An important application for polyblends is impact improvement, in which small particles of a rubbery polymer are dispersed in the matrix of a glassy polymer. Another commercially important application of polyblending is improvement in the processability of relatively intractable polymers such as polyphenylene oxide, by blending with polystyrene in the production of Noryl.

Polyblends are physical mixtures of polymers and since the mixing of polymers is normally an endothermic process, heterogeneous polyblends are generally obtained. The degree of polymer incompatibility varies widely, however. Blending two polymers usually leads to a class of materials whose properties are due to the presence of two phases, with two glass transition temperatures and usually opaque. On the other hand, if the polymers comprising the mixture have a strong enough affinity for one another, they will be compatible and mutually soluble. Such mixtures form homogeneous polyblends, have a single phase, are characterized by a single glass transition temperature, and are usually transparent. For compatibility, the solubility parameter of the polymers must match very closely. The degree of compatibility is determined by the size and distribution of the phases, i.e., how finely one polymer is dispersed within the other.

The size of the particles or domains that constitute the dispersed phase can range from Angstrom-size to micron-size in multi-phase materials. The ultimate in compatibility, short of true solubility, is Angstrom-size particles of one polymer highly dispersed in another. Conversely, the dispersed phase can be so large and coarse that the polyblend has no mechanical strength. While high compatibility is desirable for ease of blending, some degree of incompatibility often leads to useful properties. A completely homogeneous polyblend tends to average the properties of the two polymers comprising the polyblend, in direct proportion to the volume fraction of each polymer blended, whereas a multiphase polyblend often provides a superior balance of useful properties. In some cases, synergistic effects occur, where one or more properties are far superior to those of any of the polymers comprising the polyblend.

Several important families of polyblends have been commercialized, but the potential for new polymeric materials via polyblending remains very large. Properly exploited, polyblending promises a versatile route to superior combinations of useful properties from existing polymers.

Monomer units in copolymers can be arranged in a random fashion, grafts, blocks and in an alternating fashion. Methods available for producing these copolymeric structures are well documented in the literature and will not be discussed further. The effects of monomer unit arrangement on the physical and mechanical properties of copolymers have been correlated in order to assess the effects of structural arrangement of a polymer on its properties (1).

DISCUSSION

Styrene (S) - Butadiene (BD)

The first block copolymers of practical significance were SBS block copolymers. Block copolymers representing SB, SBS, BSB and SBSB sequence arrangements were synthesized by using n-butyl lithium catalyst in benzene at room temperature (2). Figure 1 shows that each structure exhibits two dynamic loss peaks E" at -80° and 110°C corresponding to polybutadiene and polystyrene, respectively. Thus, a heterogeneous two-phase system is present. The stress-strain behavior is markedly affected by changes in the sequence arrangements as shown in Figure 2. SBS and SBSB are tough, as evidenced by their high elongation at break, while SB and BSB are as brittle as polystyrene. A polyblend of the same composition is also brittle. In SBS, the elongation at break increases and yield stress decreases with increasing BD sequence length. SBS exhibits the highest impact strength of the sequence arrangements but does not exceed high impact polystyrene. SBS and SBSB show crazing or stress whitening under tensile stress as observed in ABS polymer.

Heat distortion temperatures decrease with increasing BD ratio in
SBS. The flow properties indicate easy processability, as the
flow curves lie between that of polystyrene and high impact polystyrene. It is well known that two-phase systems of different
refractive indices are opaque. The S-BD block-copolymers described
here are transparent in contrast to the opacity of a polyblend and
high impact polystyrene of the same composition. Transparency in
these block copolymers is due to the small phase size. Electron
micrographs of SBS show spherical PBD particles of about $300A^0$
dispersed in the PS matrix. On the other hand, SB and BSB sequence
arrangements exhibit PBD chains linked together to form irregularly
shaped rod-like structures.

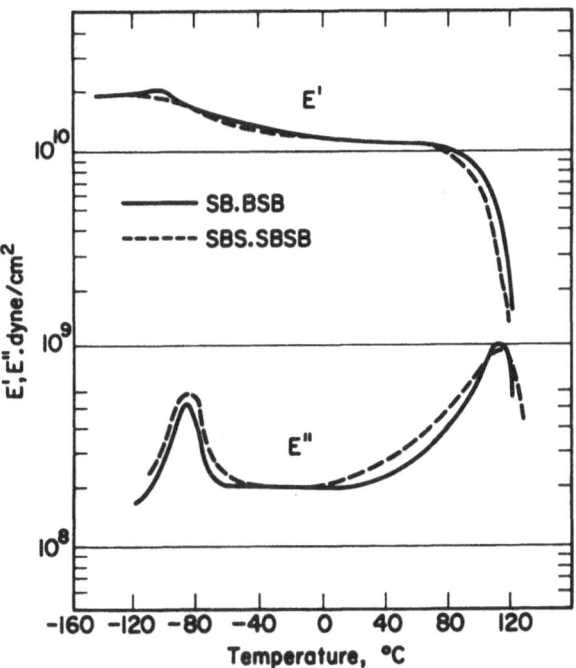

Figure 1. Temperature dependencies of dynamic modulus (E') and
dynamic loss (E") for styrene-butadiene block copolymer arrangements. Frequency 110 c/s (2).

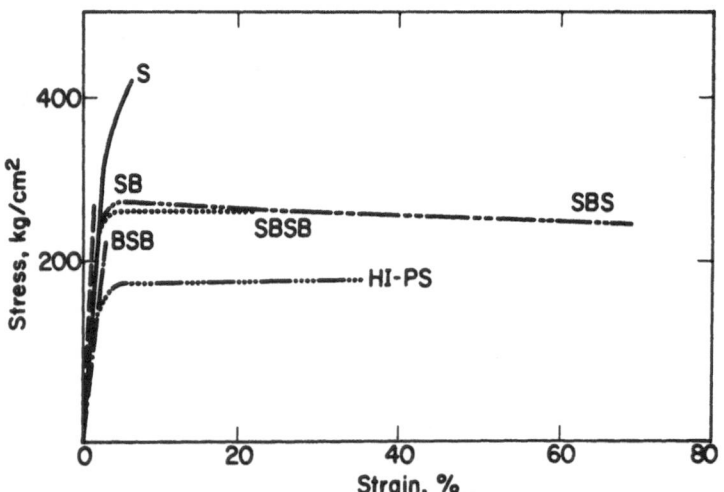

Figure 2. Stress-strain properties for styrene-butadiene block copolymer arrangements (2).

Acrylonitrile-Butadiene-Styrene (ABS)

The addition of rubber to a glassy polymer increases toughness without substantial loss in modulus and heat distortion temperature. Figure 3 shows the loss modulus E" and impact strength with temperature for graft and blend ABS containing the same 20 weight -% rubber (3). The E" - temperature curves for the graft and blend exhibit two loss peaks at -80°C and 110°C corresponding to polybutadiene and styrene-arcylonitrile copolymer phases, respectively. However, the impact strength of the graft ABS in the useful temperature range is far superior to the polyblend of the same composition. Grafting the SAN glassy phase to the PBD rubbery phase increases the interfacial strength or adhesion between the two phases, resulting in higher impact strength.

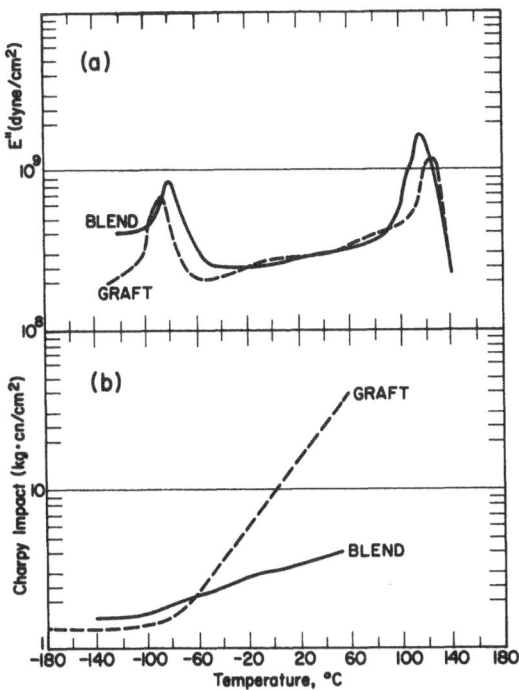

Figure 3. Temperature dependencies of dynamic loss modulus (E")
and Charpy impact strength for graft- and blend-ABS. (Rubber
contents are the same: 20 wt/%. (3).

    Figure 4 shows a typical tensile strength-elongation curve of
the two-phase graft copolymer. It exhibits a yield point below
the ultimate tensile strength of the glass, followed by ductile
deformation to failure. It is the deformation after yield that
results in the high impact strength of the graft copolymer. This
ductile deformation is accompanied by "stress-whitening." It
commences at the yield point and as the tensile strain increases
beyond yield, the area of stress-whitening expands and deepens
with increase in strain. A blend of polybutadiene and SAN
copolymer does not possess the properties of the graft copolymer,
i.e., both a high modulus and high impact strength together.

    During high shear melt processing, both blends and graft
copolymers may suffer a very large decrease in impact strength.
The two-phase graft copolymers tend to aggregate and exhibit poor
dispersibility under high shear conditions during processing,

resulting in poor impact strength. The loss in properties on
processing can be traced to the inability of the rubber phase
particles to preserve the desired size. The rubber particles not
being crosslinked deform and aggregate during processing. Cross-
linking the graft copolymer retards the deformation and aggregation
process. Crosslinking of rubber and glass phases concurrent with
grafting leads to better preservation of morphology and improved
mechanical properties. The graft copolymer must be sufficiently
tightly crosslinked to resist deformation.

    Crosslinking the rubber phase as well as the graft glass
phase improves melt processability and reduces the amount of
elastic strain energy which causes melt fracture or extrudate
roughness and die swell. On processing, the crosslinked graft
copolymer exhibits lower die swell, and higher extrusion rates are
possible.

    Crosslinking increases the yield stress which tells us
how much stress a fabricated piece can withstand before undergoing
a deformation, or how well it can absorb an impact without fracture.
A crosslinked graft copolymer exhibits a high yield stress, low
die swell, reduced melt fracture and high impact strength.

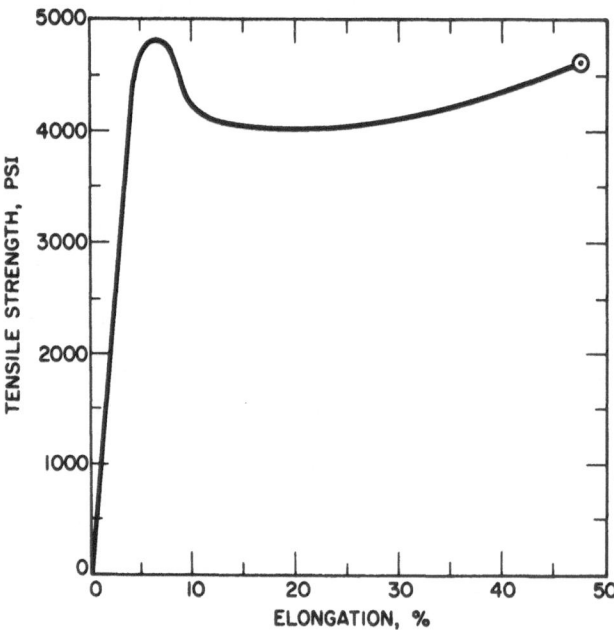

Figure 4. Typical tensile strength-elongation curve for rigid ABS.

Alpha-Methylstyrene-Methacrylonitrile (AMS-MAN)

High molecular weight AMS-MAN copolymer has low impact strength exhibiting an Izod impact strength of only 0.3 ft-lb/in. notch (4). The need for improving the impact strength is apparent. The impact strength of the copolymer was improved by copolymerizing AMS and MAN in a preformed butadiene-methacrylonitrile (BD-MAN) copolymer emulsion (4). The resulting graft copolymer had substantially improved impact strength over the unmodified copolymer on grafting upon 20-30 wt/% rubber. The effect of percent BD-MAN rubber in the copolymer on notched Izod impact strength and on distortion temperature under load (DTUL) is shown in Figure 5. The impact modified copolymer possessed an Izod impact strength of 6 ft-lb/in. notch and a DTUL of 115°C. Figures 6 and 7 show the effect of mercaptan chain-transfer agent used in the graft copolymerization on impact strength and DTUL at a rubber content of 30%. It is clear that tert-dodecylmercaptan is more efficient than n-dodecyl-mercaptan in enhancing the impact strength. Increasing tert-dodecylmercaptan concentration apparently enhances grafting of AMS-MAN glass phase via chain transfer processes onto the poly-butadiene copolymer substrate rubber.

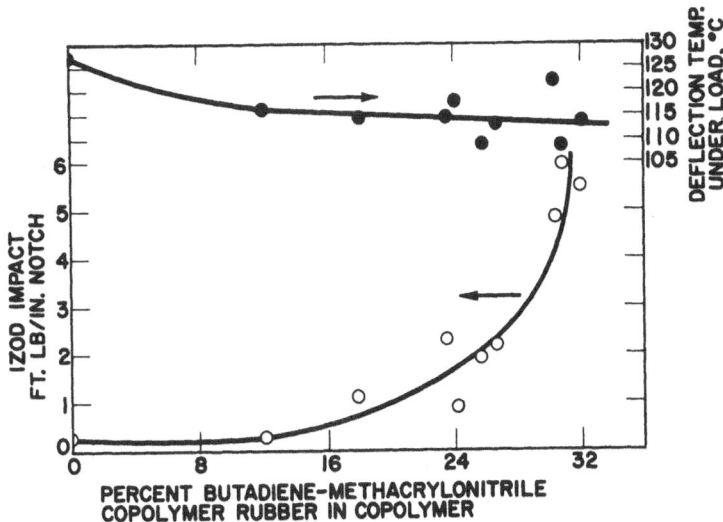

Figure 5. Effect of rubber content on impact strength and deflection temperature under load.

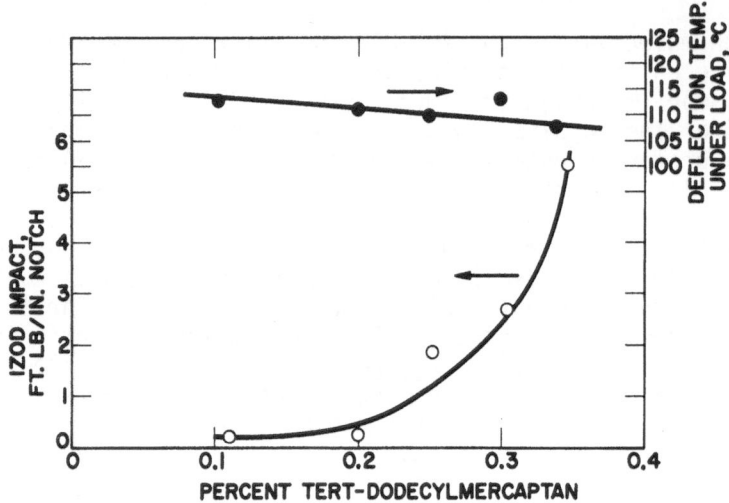

Figure 6. Effect of tert-dodecylmercaptan on impact strength and deflection temperature under load.

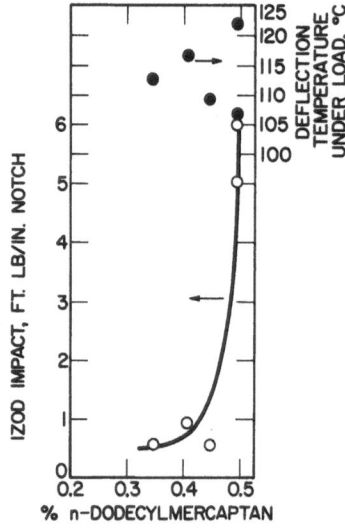

Figure 7. Effect of n-dodecylmercaptan on impact strength and deflection temperature under load.

Reaction (1) shows the reaction of growing glass phase polymer with mercaptan, resulting in termination of the growing chain and generation of the transfer radical RS·

$$M\overset{\cdot}{_X} \ + \ RSH \ \longrightarrow \ M_X\text{-}H \ + \ RS\cdot \qquad (1)$$

Reaction (2) shows the actual chain transfer step where radical RS· abstracts hydrogen from substrate polybutadiene rubber, forming a graft site on the rubber.

$$RS\cdot \ + \ \text{-C-C=C-C-} \ \longrightarrow \ RSH \ + \ \text{-C-C=C-}\overset{\cdot}{\underset{}{C}}\text{-} \qquad (2)$$

Reaction (3) shows formation of the graft chain by glass phase monomer addition to the graft site on the polybutadiene rubber.

$$\text{-C-C=C-}\overset{}{\underset{\cdot}{C}}\text{-} \ + \ M \ \longrightarrow \ \text{-C-C=C-}\underset{\underset{\cdot}{M}}{\overset{|}{C}}\text{-} \qquad (3)$$

Figure 6 shows that 0.35% tert-dodecylmercaptan in the glass phase monomers results in an Izod impact strength of 6 ft lb/in. notch and a DTUL of 115°C.

A two-phase system is readily apparent from Figure 8, with two glass transition temperatures -- one for the butadiene-methacrylonitrile (62/38) copolymer rubber at -10°C and another for the glassy copolymer at around 145°C. The impact modified copolymer is opaque to translucent whereas the unmodified copolymer is transparent. The opaqueness is due, of course, to a difference in refractive index of the rubber phase and the glassy copolymer phase. Substantial impact improvement of the copolymer could not be obtained by mixing rubber and copolymer latexes, melt blending of rubber and copolymer, or melt blending with ABS.

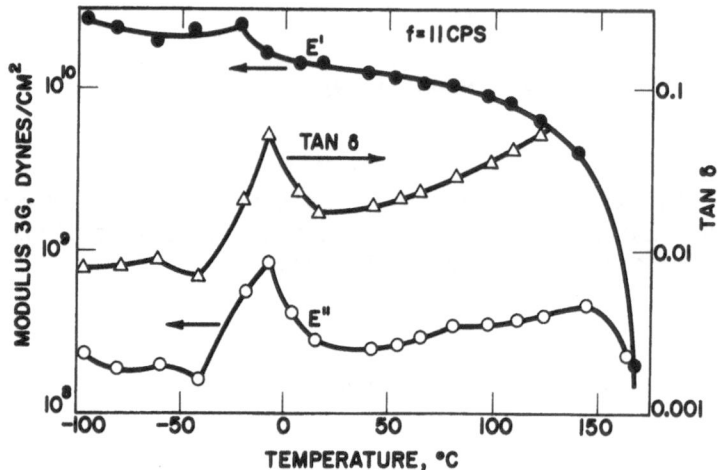

<u>Figure 8.</u> Modulus-temperature and tan δ temperature curves.

Poly(vinyl chloride (PVC)-Alpha-Methylstyrene-Methacrylonitrile-Ethyl Acrylate (AMS-MAN-EA) Polyblend

The preparation of high molecular weight AMS-MAN-EA terpolymer has been described previously (5). The composition of the terpolymer is 58 wt/% AMS, 40 wt/% MAN, and 2 wt/% EA. Polyblends of AMS-MAN-EA and PVC were prepared by melt blending on a two-roll mill the terpolymer and PVC with 2% Thermolite 31 in the melt at 180-200°C (6). A compression molded plaque of the polyblend was transparent. A film of the polyblend cast from tetrahydrofuran solution was also transparent.

Table I shows the dependence of deflection temperature under load (DTUL) on composition of the polyblend.

Table I

Dependence of DTUL of PVC-AMS-MAN-EA Polyblend on Composition

Wt Fraction PVC	DTUL, °C
0	129
0.20	110
0.35	104
0.50	95
0.65	90
0.80	81
1.0	75

Figure 9 shows the relationship between DTUL and composition of the polyblend. The broken line was calculated from the well-known Fox equation for random copolymers.

$$\frac{1}{DTUL} = \frac{W_1}{DTUL_1} + \frac{W_2}{DTUL_2} \qquad (4)$$

where W_1 and W_2 are the weight fractions of each of the components of the polyblend. DTUL is substituted for T_g.

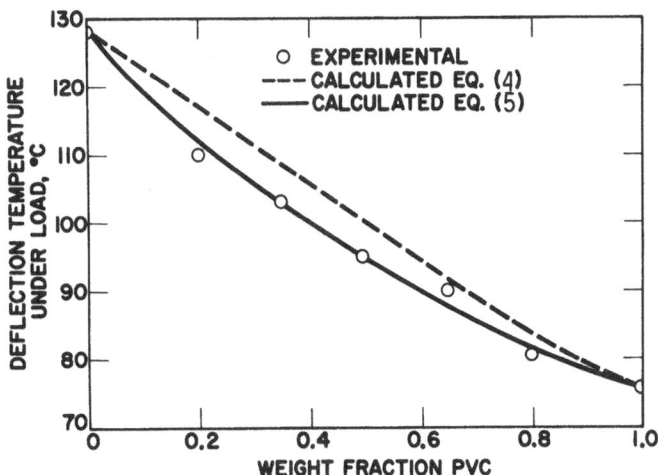

Figure 9. Dependence of deflection temperature under load on composition.

The polyblend approximately obeys the Fox equation for random copolymers, exhibiting a DTUL about 5°C below that calculated. This behavior is not unusual even for random copolymers since volume additivity of the components is seldom ideal. Many copolymer systems which do not obey the Fox equation have been found to obey the following form of the Gordon-Taylor relationship.

$$DTUL = \frac{DTUL_1\ W_1\ +\ K\ DTUL_2\ W_2}{W_1\ +\ KW_2} \qquad (5)$$

where K is a factor which takes into consideration the thermal expansion coefficients of the polymer in both the liquid and glassy states. If eq. (5) is rewritten in the form

$$DTUL = DTUL_2 - \frac{1}{K}\ (DTUL-DTUL_1)\ \frac{W_1}{W_2} \qquad (6)$$

then it is seen that a plot of DTUL against $(DTUL - DTUL_1)\ W_1/W_2$ will have a slope $-1/K$ and intercept $DTUL_2$. This plot, shown in Figure 10 is based on experimental data for AMS-MAN-EA/PVC poly-blends of different composition; a value of K = 0.60 is obtained. Figure 9 shows that there is good agreement between the experimental points and the theoretical line calculated from eq. (5) for K = 0.60. Thus, the properties of this polyblend are similar in properties that one would expect from a random copolymer of the components of the polyblend.

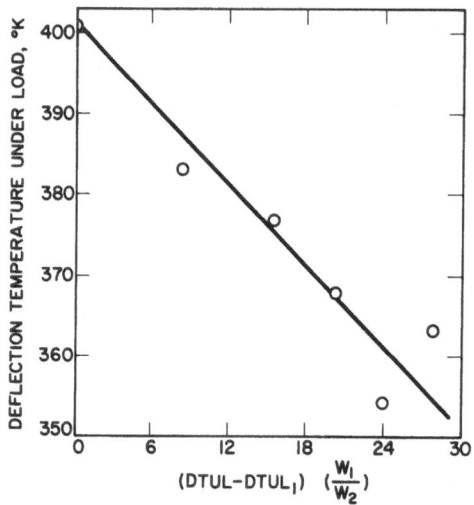

Figure 10.   Determination of K factor.

The modulus-temperature and tan δ-temperature curves of a
50/50 wt/% AMS-MAN-EA/PVC polyblend is shown in <u>Figure 11</u>.   It
can be seen that a single T_g at approximately 100°C exists.   This
unusual feature of this polyblend is unexpected and suggests
homogeneity.   The single T_g suggests either a single-phase system
or that if two phases are present they must be extremely small.

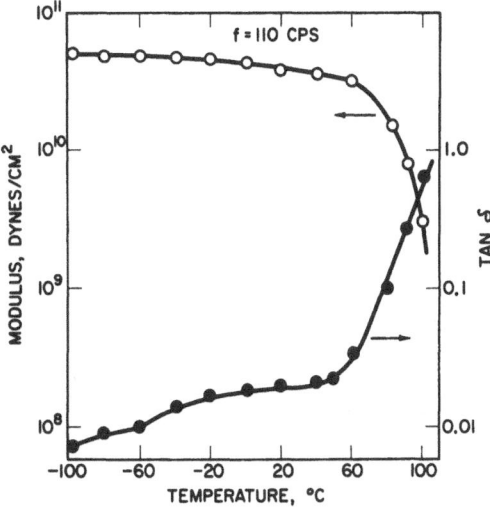

<u>Figure 11</u>.   Modulus-temperature and tan δ-temperature curves for
50/50 wt/% polyblend.

The two polymers making up the polyblend are mutually soluble.
The solubility parameter (δ) of PVC is 9.6.   The solubility
parameters of the terpolymer was calculated from eq. (7) to be
also 9.6.

$$\delta_{mixture} = V_1\delta_1 + V_2\delta_2 + V_3\delta_3 \quad (7)$$

V_1, V_2 and V_3 are volume fractions of the components of the
terpolymer.   Since the solubility parameters of the two components
are identical, they exhibit rarely observed polymer-polymer
compatibility.   The refractive index of the polyblend was found
to be 1.546 at 23°C.   The properties of a polyblend prepared by
blending on a mill 50 wt/% PVC and 50 wt/% AMS-MAN-EA terpolymer
are shown in <u>Table II</u>.

Table II

Properties of AMS-MAN-EA/PVC (50/50 wt/%) Polyblend

Deflection temperature under load, $^{\circ}$C	95
Izod impact, ft lb/in. notch at 23°C	0.53
Elongation, %	5
Stress, psi	10,180
Tensile modulus, psi	214,000
Flexural strength, psi	17,375
Flexural modulus, psi	500,600

Preliminary processing behavior of the AMS-MAN-EA/PVC polyblend was obtained on the Brabender Plasticorder. Figure 12 shows torque versus time curves for AMS-MAN-EA and PVC homopolymers and also polyblends containing 10 and 50 wt/% PVC. Curve A of the AMS-MAN-EA terpolymer is broad, indicating high damping and high melt viscosity; a material with difficult processability. Curve B of PVC is typical of a processable polymer. Curves C and D of the polyblends are narrow and indicate good processability. Curve C was unexpected in that a small amount (10%) of PVC would reduce the high melt viscosity of the AMS-MAN-EA terpolymer such a great extent. This unusual feature suggests that PVC is a good processing aid for the AMS-MAN-EA terpolymer.

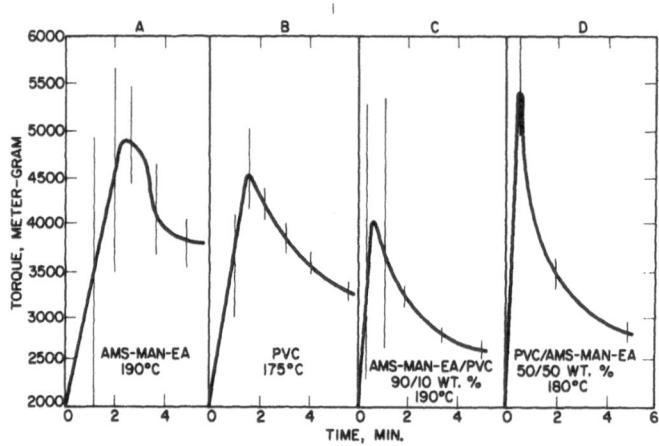

Figure 12. Brabender torque-time curves for homopolymers and polyblends.

Table III summarizes the flux properties of AMS-MAN-EA terpolymer, PVC, and polyblends of AMS-MAN-EA and PVC evaluated in the Brabender Plasticorder. There is substantial reduction in 5-min torque at 190°C from 3500 m-g for AMS-MAN-EA terpolymer to 2750 m-g for a 90/10 AMS-MAN-EA/PVC polyblend, indicating the improved flow properties on blending a small amount of PVC into the terpolymer. The same torque value could be maintained at lower temperatures on increasing PVC in the polyblend.

Table III

Brabender Rheometer Torque Characteristics of
AMS-MAN-EA/PVC Polyblends at 32 rpm

PVC, wt/%	Torque at 5 min, m-g	Wall temp, °C	Stock temp, °C
0	3500	190	192
10	2750	190	191
30	2500	185	191
40	3500	180	191
50	3500	180	187
75	3500	176	188
95	2950	175	187
100	3400	175	183

Poly(vinyl chloride) (PVC) -
Methyl Methacrylate-Butadiene-Styrene (MBS) Polyblend

MBS graft copolymers were prepared in emulsion by copolymerizing styrene with butadiene to form a first stage rubbery
copolymer and then copolymerizing styrene with methyl methacrylate
to form a second stage glassy copolymer. The second stage glassy
copolymer phase is thereby chemically grafted to the preformed
first stage rubbery phase. The MBS graft copolymers were blended
with PVC on a two roll mill and the milled sheet compression
molded. The properties of the PVC-MBS polyblend are highly dependent on the composition of the graft copolymer. PVC-MBS
polyblend properties are dependent upon the amount of tert-docecyl-
mercaptan chain transfer agent employed in the second stage graft
copolymerization and the styrene/methyl methacrylate ratio used in
the second stage graft copolymerization.

The effect of tert-dodecylmercaptan employed in the second
stage graft copolymerization on Izod impact strength of the PVC-
MBS polyblend is shown in Table IV. Tert-dodecylmercaptan
markedly enhances the impact strength of the polyblend by increasing
grafting of the glassy copolymer to the rubbery copolymer. The
effect of increased grafting on the properties of two-phase polymeric
materials has been pointed out in a previous section.

Table IV

Effect of Tert-Dodecylmercaptan on
Impact Strength of PVC-MBS Polyblend

Tert-Dodecylmercaptan, %	Izod Impact, ft lb/in. notch
0	12
0.05	18
0.06	20
0.4	25

The effects of the styrene/methyl methacrylate ratio employed
in the second stage graft copolymerization on PVC-MBS polyblend
properties are shown in Table V. Izod impact strength, light
transmission, haze and fluorescence properties of the polyblend are
markedly affected by the styrene/methyl methacrylate ratio in the
glassy phase of the MBS graft copolymer. The impact strength and
light transmission are at a maximum, while haze is at a minimum
at 46-58 wt/% styrene. The polyblend exhibits yellow fluorescence
at low styrene content and blue fluorescence at high styrene content.
No fluorescence is observed in the range 46-58 wt/% styrene in the
glassy phase.

Table V

Effects of St/MMA Ratio in Glassy Phase of
MBS Graft Copolymer on Properties of PVC-MBS Polyblend

Styrene, Wt/%	Methyl Methacrylate, %	Izod Impact, ft lb/in. notch	Light Transmission, %	Haze, Percent	Flourescense
0	100	1	-	-	Yellow
20	80	3	78	11	Yellow
46	54	21	82	11	None
50	50	25	81	11	None
52	48	20	83	8	None
58	42	20	82	11	None
64	36	15	83	9	Blue
70	30	5	80	16	Blue
100	0	1	61	75	Blue

CONCLUSION

Block copolymers, graft copolymers and polyblends offer opportunities to make polymeric materials having unique "built-in" properties to satisfy end-use performance requirements. The commercial importance of these materials has not become clearly recognized.

REFERENCES

1. J. F. Kenney, "Recent Advances in Polymer Blends, Grafts, and Blocks," Ed. L. H. Sperling, Plenum Press, 1974, p. 117.

2. M. Matsuo, Polymer 9, 425 (1968).

3. M. Matsuo, A. Udea and Y. Kondo, Polymer Engineering and Science 10, 253 (1970).

4. J. F. Kenney and P. J. Patel, J. Appl. Polym. Sci., 20, 449 (1976).

5. J. F. Kenney and P. J. Patel, J. Polym. Sci., 14, 105 (1976).

6. J. F. Kenney, J. Polym. Sci., 14, 123 (1976).

THE CRYSTALLIZATION BEHAVIOUR OF BLENDS OF LOW DENSITY

POLYETHYLENE AND POLY(ε-CAPROLACTONE)

D. Krevor and P.J. Phillips[1]

Department of Chemical Engineering
State University of New York at Buffalo
Buffalo, New York   14214

The term alloy, often used to describe blends of polymers, originated in metallurgy and is generally associated with a mixture of two or more metals. Structurally, there are two classes of alloys: (a) solid solutions, in which the atoms of one element may replace, in a random manner, the atoms in the crystal of the second and (b) multiphase alloys, in which one phase may be pure element and the second an intermetallic compound of specific composition (1-3).

In both classes of alloys, the phases referred to are crystalline. By analogy, therefore, only blends of polymers in which co-crystallinity has been detected should, logically, be referred to as alloys. It is precisely the modification of the properties of the crystals and of the interactions between different crystal phases through the alloying procedure, which has been responsible for practical advances in metals technology over the centuries. The authors are not aware of any reference in the literature to co-crystallization in a blend of two polymers.

The solid solution type of co-crystal should be the easiest to make since it does not require specific interactions between the polymers or specific mole ratios of the two polymers. It does, however,

[1]To whom correspondence should be addressed; present address: Department of Materials Science and Engineering, University of Utah, Salt Lake City, Utah   84112.

require that the crystalline space lattices be the same and that
the unit cell parameters be very close. Two polymers whose unit
cells are orthorhombic are polyethylene and poly(ϵ-caprolactone).

Poly(ϵ-caprolactone), PCL, is one of the most commonly used
polymers in blending studies, because of its apparent compatibility
with many polymers (4). This is presumably due to a combination
of low molecular weight (up to 40,000) and a relatively innocuous
chemical structure $[-(CH_2)_5 - CO_2 -]$. PCL can be blended with low
density polyethylene, LDPE, at concentrations of up to 30% without
any major changes in physical properties.

The crystalline structure of PCL has been determined by
Bittinger et al. (5). Unit cell parameters of PCL together with
those of polyethylene are given in Table 1.

The a and b dimensions which reflect the lateral packing of
the linear zig-zag chains differ by 1.28% and 0.88%, respectively.
This is very small and indicates a high potential for co-crystallization. In fact, as in copolymers of ethylene and carbon monoxide
(6,7) the base plane lattice is relatively unchanged. For
comparison, copper and nickel which form a complete series of
solid solutions have atomic radii of 1.278Å and 1.246Å, respectively,
i.e. a difference of 2.6%.

The major difference in the unit cell parameters of the two
polymers we are considering here is clearly the c dimension. This
parameter does not really represent a critical packing parameter
since it is reflecting the periodicity of the chemical structure in
an extended molecule. In the case of polyethylene (see figure 1)
this represents a straight line separation between any CH_2 group
and its second nearest neighbour. For poly(ϵ-caprolactone) it
represents double the length of an extended mer, and corresponds to
a total of 14 chain bonds. A chain of polyethylene containing the

Table 1

Unit Cell Parameters (Å)

	a	b	c
PCL	7.496	4.974	17.297
PE	7.40	4.93	2.534

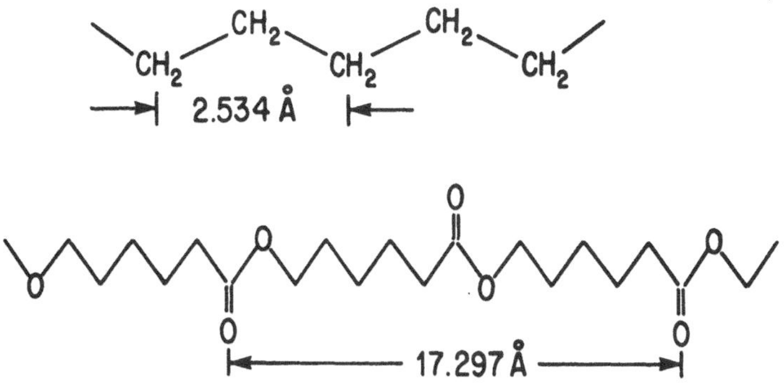

Figure 1. The Extended Form of Polyethylene and Poly(ϵ-caprolactone) showing the origin of the c unit cell parameter.

same number of bonds would be 17.738Å long. This corresponds to an actual difference of 2.49% and is still comparable to the difference in atomic radii of copper and nickel. However, over a long chain the discrepancy would become significant. For an average crystal of LDPE of, say, 150Å thickness the PCL chain would contain an additional three single bonds. The difference in length is caused solely by the different bond lengths and angles associated with an ester group. It is well established (8) that long chain ketones or esters can form solid solutions with paraffin wax. The polyethylene - poly(ϵ-caprolactone) blend would therefore be a reasonable system for the formation of a mixed crystal.

The molecular weight of the PCL is likely to be of some importance, since, in contrast to the paraffin wax systems it will be necessary for chain folding to occur in order for the PCL to be incorporated into a polyethylene crystal. It is a relatively simple matter to calculate the number of traverses of a 200Å thick crystalline lamella necessary to incorporate a PCL molecule of any particular molecular weight. Such figures are presented in Table 2. It is anticipated that PCL-150 and PCL-300 should be relatively easy to incorporate, but that PCL-700 might cause some problems. Since the melting points of the two polymers are not too different, being 72°C and 120°C (those of copper and nickel are 1083°C and 1453°C), successful co-crystallization experiments would be expected.

Table 2

PCL	150	300	700
M.W.	7,000	15,000	40,000
No. of traverses of a 200Å lamella	2	5	13

EXPERIMENTAL

Three samples of PCL were supplied by Union Carbide which exhibited weight average molecular weights of approximately 7,000, 15,000 and 40,000. Although a number of blends were prepared, the research has concentrated on a 9:1 weight ratio blend of PE and PCL. This blend is within the range known to give good physical and mechanical properties.

The blends were produced by first dissolving the polyethylene in p-xylene at 120°C and then adding the poly(ϵ-caprolactone). Precipitation was carried out by dripping the hot solution into a mixture of methanol and cardice at -20°C. Careful examination under a microscope failed to detect the characteristic white flakes of PCL crystals on the surfaces of the particles. It could, therefore, be assumed that the blending operation was successful. Similar testing was carried out after each preparation. The solubility of the PCL, being much greater than that of the PE, was such as to result in separation during precipitation under less drastic conditions.

After thorough drying, specimens were melted between a glass slide and cover plate and heat treated in a hot-stage mounted on a Reichert microscope. Quenched specimens were produced by melting at 150°C, followed by a rapid quench to room temperature. Annealed specimens were first melted at 150°C, crystallized for 3 hours at 105°C, for 1 hour at 70°C and then quenched to room temperature. Other melt temperatures have been used.

The specimens so produced were studied using hot-stage microscopy to observe the spherulitic superstructures and to determine transition temperatures. Films were separated from the slides, placed in aluminum sample pans and differential scanning calorimetry (DSC) carried out on them.

Some specimens were heat treated in the differential scanning calorimeter using the same procedure for heat treatment of thin films. The materials so produced were also used in thermal analysis experiments.

Dielectric relaxation studies were carried out on specimens prepared using the heat treatments described earlier for thin films, but were in the form of plates approximately 1/16 inch thick. The experiments were carried out using a three terminal submersible cell over the frequency range 10^3 to 10^6 Hz and the temperature range -60 to +60°C.

RESULTS AND DISCUSSION

Specimens Crystallized from 150°C Melt

DSC curves of quenched specimens generally showed only the two expected melting points of PCL and PE at approximately 60° and 115°C, respectively. They do, however, generally show a shoulder on the low temperature side of the polyethylene peak (figure 2a).

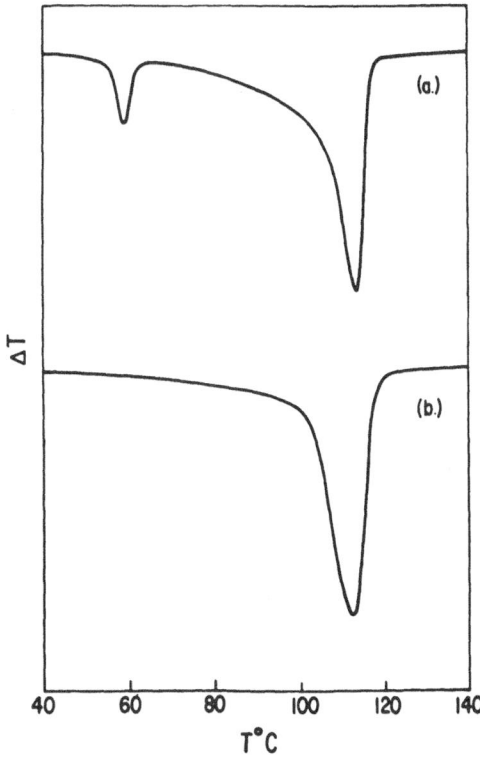

Figure 2. DSC traces of (a) a quenched specimen of a 9:1 blend of LDPE and PCL-300 (b) a quenched specimen of LDPE, both at heating rates of 10°C/min.

The polyethylene endotherm narrowed considerably compared to that of quenched low density polyethylene (figure 2b). The breadth is 6° at the half point compared to 10° for LDPE. These changes are not reflected in spherulitic texture of the blend. Hot-stage microscopy shows single spherulites of PCL which melt at ca 60^OC isolated in a matrix of PE spherulites, which melted at 115^OC.

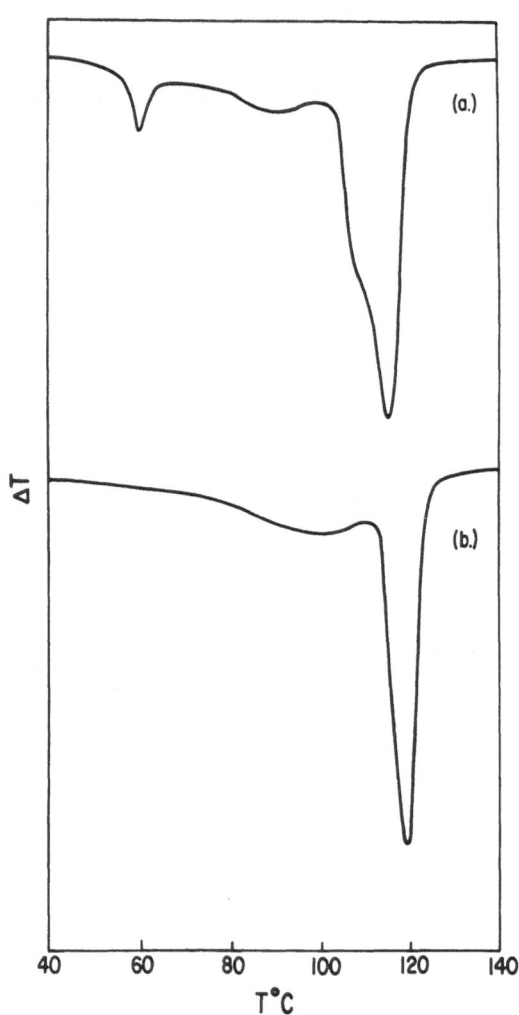

Figure 3. DSC traces of (a) an annealed specimen of a 9:1 blend of LDPE and PCL-700 and (b) an annealed specimen of LDPE, both at heating rates of 10^OC/min.

For annealed materials, the behaviour is very different and complex. In addition to the main PE and PCL peaks, a broad diffuse peak was observed at 90°C and an additional sharp peak (or shoulder) observed at about 105°C (figure 3a). Crystallization of LDPE for extended periods of time (i.e. ca. 24 hours) at 105°C reproduces the 90°C peak, but is unable to duplicate the 105°C peak (figure 3b). The 105°C peak is highly dependent on the heating rate used in the DSC experiment; the slower the heating rate the better resolved the peak (figure 4). The phenomenon is observed in thin films of blends of LDPE and is independent of the molecular weight of PCL being used. Hot stage microscopy showed, as in quenched specimens, that only two types of spherulite exist. The PCL is found, again, as isolated spherulites within a general matrix of LDPE spherulites. The 90°C DSC peak corresponds to a general darkening of the LDPE spherulites without any decrease in their size. The 105°C and 115°C peaks appear as part of a general melting of the spherulites. Melting begins both in the vicinity of spherulite boundaries and radially within the spherulites.

The 90°C peak in annealed LDPE is believed to be due to low molecular weight material which is able to fractionate during slow crystallization. If its origin is the same in the blend, then

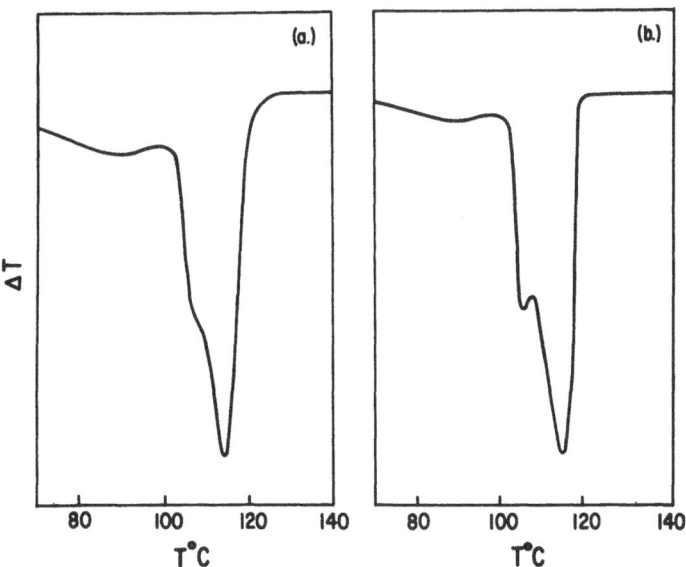

Figure 4. DSC traces of an annealed specimen of a 9:1 blend of LDPE and PCL-300 at (a) 10°C/min and (b) 5°C/min.

fractionation is assisted by the presence of PCL, the process taking only a tenth of the time. Either or both of the two new peaks could, however, be caused by a co-crystal since such a species should exhibit an intermediate melting point. Also of interest is the observation that the melting point of LDPE is 3-4° higher in the unblended state. This could indicate the incorporation of PCL molecules into the PE crystals. Alternatively, it could be a manifestation of the already molten PCL acting as a solvent and thus depressing the melting point.

Specimens Crystallized from Other Melt Temperatures

The highest temperature achieved in the molten state has been found to have a major effect on the endotherms observed and hence on the morphological structure. This is in contrast to the highest crystallization temperature since it has been found that the use of 110°C or 105°C has no effect on the DSC traces.

The use of a melt temperature of 170°C followed by the standard method of preparation of annealed specimens results in curves similar to those reported for 150°C. The new peaks do however appear less intense.

A melt temperature of 125°C, i.e., only slightly above the melting point, produces the most interesting results. An example of this data is shown in figure 5. Curve (a) is similar to those already referred to (any differences are due to the use of a DuPont Thermal Analyzer in earlier figures and the use of Perkin-Elmer DSC IB in this figure). Curve (b) was obtained by stopping the scan for curve (a) at 125°C., performing the standard preparation, and then rescanning. Curve (c) was obtained by stopping the scan for curve (b) at 125°C., cooling to room temperature, and then rescanning.

Clearly, curve (c) shows the absence of a PCL melting peak. The very long low temperature shoulder of the polyethylene peak extends to the PCL melting point. It would therefore appear that the PCL has been incorporated into the, still predominantly, poly-ethylene crystallites.

The shape of curve (b) is intriguing and appears to be indicative of a recrystallization phenomenon. The start of the curve corresponds to normal melting behaviour. A few degrees before the normal maximum occurs, a recrystallization occurs causing the 'negative' peak. This material then melts causing the second 'sharp' peak. The curve is therefore a composite of several peaks.

Scanning at other heating rates has confirmed this behaviour. At 10°C/min. the phenomenon does not occur. At 2.5°C/min. the first

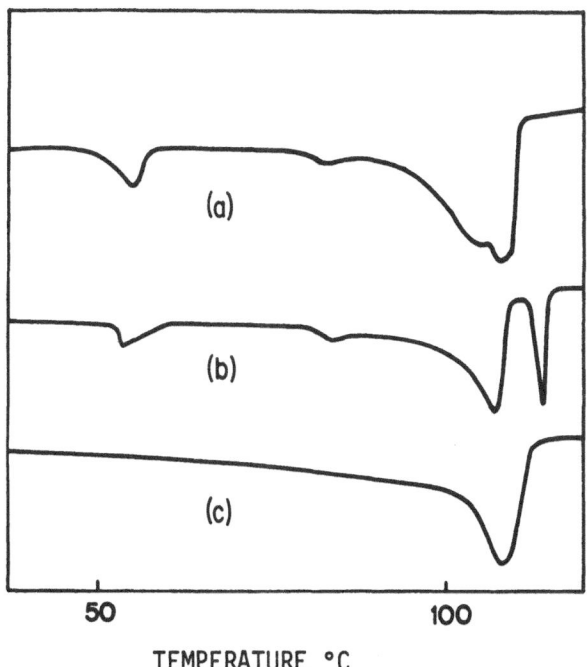

Figure 5. D.S.C. traces of a 9:1 blend of LDPE and PCL-300 (a) as previously reported (b) melt temperature of 125°C followed by annealing (c) specimen of (b) melted to 125°C and cooled to room temperature.

peak is smaller, the second sharper peak being intensified. This behaviour occurs for blends of PCL-150 and PCL-300, but not for PCL-700

Dielectric Relaxation Studies

The dielectric relaxation behaviour of polyethylene is well characterized, consisting of α, β, and γ peaks (9). These dispersions correspond to a librational mode within the PE crystal, the glass transition and mixed local modes, respectively. The librational mode is found at ca. 60°C; the glass transition at about -20°C. Dielectric spectra of PCL have not yet been reported in the literature. Our data show a major dispersion at ca. -50°C corresponding to the glass transition. There is no evidence of a librational mode similar to the α peak of PE although a peak does occur which is associated with the melting point of the PCL. Dielectric relaxation studies of copolymers of ethylene and carbon monoxide (6) showed that the temperature of the α-peak was directly proportional to the carbonyl content. The presence of PCL molecules in the polyethylene crystals would therefore be expected to show up as an α peak. Unfortunately, the melting peak of the PCL appears as a dielectric dispersion in the same temperature region and considerably complicates the analysis. A shoulder does however exist. Further experimentation is expected to permit resolution of the peak.

CONCLUSIONS

A number of interesting effects have been observed in crystallization studies of low density polyethylene with poly(ϵ-caprolactone). The compatibility of the two polymers in the melt seems to control the resultant crystallization behaviour. The polymers appear to phase separate in the melt as temperature increases. Maximum melt compatibility occurs close to the melting point of polyethylene. Crystallization proceeding from a compatible melt appears to result in co-crystallization through a mechanism not presently understood. It is, however, clear that only a fraction of the polyethylene is involved in a co-crystallization process. This may be lower molecular weight material. Any co-crystals formed are present in the spherulites of polyethylene.

REFERENCES

(1) L.H. Van Vlack, "Elements of Materials Science and Engineering," Addison Wesley (1975).

(2) A.G. Guy, "Essentials of Materials Science," McGraw-Hill (1976).

(3) K.M. Ralls, T.H. Courtney and J. Wulff, "Introduction to Materials Science and Engineering," Wiley (1976).

(4) Union Carbide Inc., (Bound Brook, NJ) Product Information Sheet, No. F-42501.

(5) H. Bittinger, R.H. Marchessault and W.D. Neigisch, Acta Cryst., B26, 1923 (1970).

(6) P.J. Phillips, G.L. Wilkes, B.W. Delf and R.S. Stein, J. Polym. Sci., A-2, 9, 499 (1971).

(7) Y. Chatani, T. Tukizawa and S. Murahashi, J. Polym. Sci., 62, 527 (1962).

(8) C.A.F. Tuijnman, Polymer, 4, 315 (1963).

(9) N.G. McCrum, B.E. Read and G. Williams, "Anelastic and Dielectric Effects in Polymeric Solids," Wiley, New York (1967).

ELASTOMER BLENDS - IMPROVED STRENGTH PROPERTIES OF UNCURED

RUBBER COMPOUNDS

Emil M. Friedman, Richard G. Bauer, and Diego C. Rubio

The Goodyear Tire & Rubber Company, Research Division

142 Goodyear Blvd., Akron, Ohio 44316

The green strength of natural rubber (NR) has long been known to be superior to that of synthetic cis-1,4-polyisoprene (SYNPI). This paper first addresses the source of this difference. It then describes methods to improve the green strength of natural or synthetic polyisoprene (PI) compounds.

SOURCE OF DIFFERENCE

One school of thought holds that the greater stereoregularity of NR promotes more and faster stress crystallization, and that the crystallites reinforce the rubber. However, it is not immediately obvious that the difference in cis-1,4 content[1] (98% vs 97%) (1) is large enough to create an appreciable difference. Brock and Hackathorn suggested (1) that when a 3,4 unit interrupts a string of cis-1,4 units, the next string may have the opposite sense (head-tail vs. tail-head). The reversal of chain sense inhibits chain folding crystallization,and a single 3,4 unit can therefore prevent many other units from crystallizing. Thus, 3,4 units exert an effect upon crystallization at -25°C out of proportion to their concentration (1). Gregg and Macey (2) extended this idea and suggested that the same phenomenon creates an out of proportion dependence of stress crystallizability at room temperature upon stereoregularity. Neither paper, however, tries to quantify this idea. We will do this, and will also show that direct head-head and tail-tail addition of cis, 1-4 units is also capable of ruin-ing the stress crystallizability of SYNPI.

[1]The cis-1,4 contents used should be considered illustrative, as various workers have reported various numbers. However, the general results obtained here are unchanged.

Quantifying the Effect of Stereoregularity

What is needed is an expression for the weight percent of perfect sequences at least L units long--where L represents the minimum length needed to permit crystallization. Let the probability that a randomly chosen monomer unit does not cause an "error" (i.e., a reversal of the head-tail sequence) be called the "correctness", p. Call the probability that a given sequence of "correct" units is exactly L units long, "P(=L)". Then,

$$P(=L) = p^{L-1} (1-p), \qquad\qquad L = 1,2,\ldots \qquad (1)$$

The probability that a randomly chosen "correct" unit is part of a sequence L "correct" units long is, therefore

$$W_c(=L) = \frac{L\, P(=L)}{\Sigma\ \ L\ P(=L)}$$

$$= Lp^{L-1} (1-p)^2, \qquad\qquad L = 1,2,\ldots \qquad (2)$$

Since the probability is p that a randomly chosen unit is a "correct" unit, the probability that a randomly chosen unit is part of a sequence of "correct" units exactly L units long is

$$W(=L) = \begin{cases} Lp^L (1-p)^2 & \text{for } L = 1,2,\ldots \\ (1-p) & \text{for } L = 0 \end{cases} \qquad (3)$$

(The case L=0 represents the random unit being an "error" unit.) Finally, the probability that a randomly chosen unit is a part of a sequence of "correct" units at least L units long (i.e., the weight fraction of perfect sequences at least L units long) is

$$W(\geq L) = \sum_{n=L}^{\infty} W(=n) = \sum_{n=L}^{\infty} np^n (1-p)^2, \qquad L = 1,2,\ldots \qquad (4)$$

Evaluating the sum gives

$$W(\geq L) = Lp^L - (L-1)\, p^{L+1}, \qquad\qquad L = 1,2,\ldots \qquad (5)$$

---

[2]This treatment assumes that the errors are randomly distributed. Because of their small number, a tendency towards alternation will have almost no effect and blockiness will be unlikely. Therefore, the assumption is good. (See appendix.)

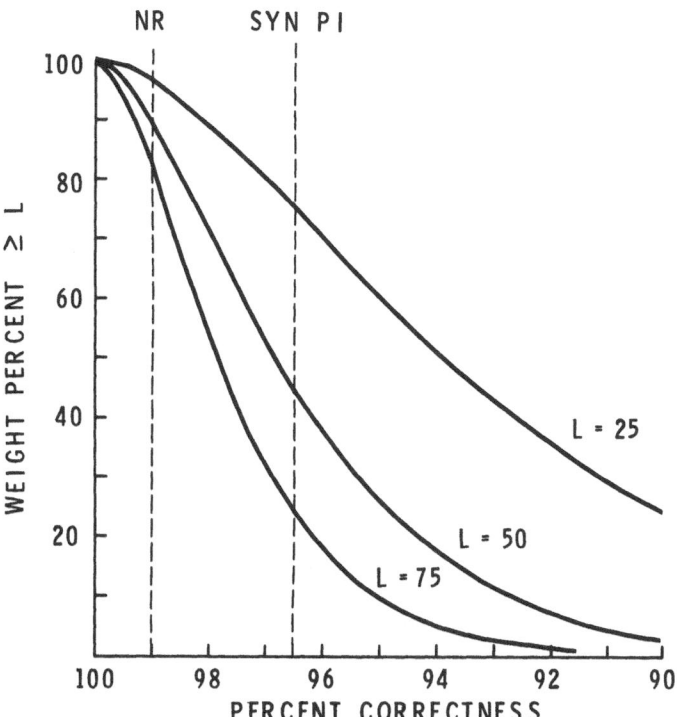

Figure 1. Weight percent "correct" sequences at least L units
           long vs. percent correctness (for various L).

        Figure 1 shows W($\geq$L) as a function of "percent correctness"
(px100%) for various values of L. (The rapidity with which W($\geq$L)
decreases as p decreases is even greater than a first glance shows
since the horizontal scale is 10 times the vertical.) This rapidity
means that small differences in correctness can lead to large
differences in W($\geq$L). While our thinking is not sufficiently
advanced to permit us to <u>directly</u> relate the weight fraction of
long perfect sequences, W($\geq$L), to the weight fraction of crystallites formed during stretching, the reasonable presumption that
they are intimately and monotonically related does permit us to
draw some rather significant conclusions.

                    Application to NR and SYNPI

        Applying eqn. (5) to NR and SYNPI requires first knowing the
"% correctness" and, second, knowing how long a sequence is needed
to permit crystallization (L). The first question is answered by
inspecting the data of Hackathorn and Brock (3, 1). For NR they

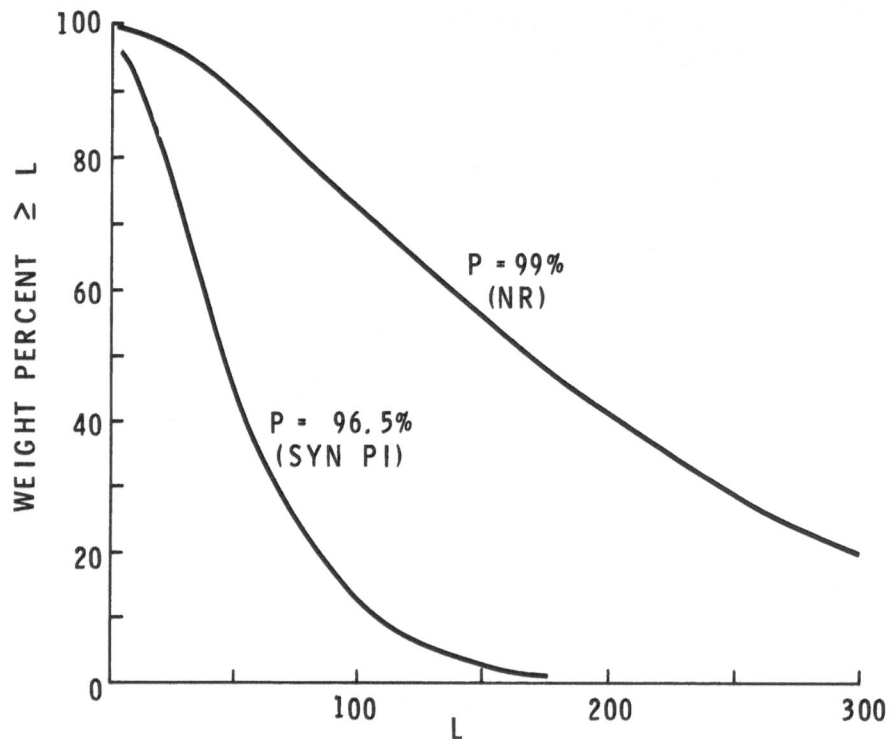

Figure 2. Weight percent "correct" sequences at least L units long
           vs. L (for NR and SYNPI).

found 98% cis,1-4, all of which was head-tail. If half of the
other 2% reversed the chain sense and half did not, then half were
"error" units and half were merely isolated errors which did not
propagate. Thus, the "% correctness" was 99%. (Their data
represents the worst possible case in terms of proving our point.)
For their Ziegler polyisoprene they found 97% cis-1,4 of which 95%
was head-tail, 1% was head-head and 1% was tail-tail. They found
3% 3,4; half of this we assume to be "error" units and the other
isolated errors. The "% correctness" was then 96.5%. These
"correctness" values are indicated in Figure 1 by vertical dashed
lines.

     Figure 2 shows $W(\geq L)$ as a function of what value we choose
for L for the "correctness" values estimated for NR and SYNPI
(i.e., the integral weight percent distribution of "correct"
sequence lengths).

     The best value to use for L is not easily arrived at. For
chain folding crystallization, one could hardly use less than 60.

For a fringed micelle, the number might be smaller. However, it is clear from Figure 2 that the differences in W(>L) between NR and SYNPI is significant whatever value of L (within reason) we choose. Furthermore, it must be remembered that in order to form a fringed micelle and, probably, in order for a chain-folded crystallite to be an effective reinforcing domain, a number of chains must come together. Therefore, the weight fraction of reinforcing crystallites will be proportional not to W($\geq$L) but to a higher power of W($\geq$L). As a result, the effect of small differences in "correctness" will be amplified still further.

The above analysis shows that the "small" difference in stereoregularity between NR and SYNPI produces a large difference in the weight fraction of long sequences and thus is capable of accounting for large differences in stress crystallizability and green strength.

Alternate Source - Ionic Groups

Smirnov, et. al, (4) have claimed that ionic groups (hydroxyl, carboxyl, and/or carbonyl) attached to the PI backbone in NR are the main source of its superior green strength. They measured the 300% modulus and tensile strength of NR and SYNPI gum stocks with and without adding ZnO and obtained:

	NR		SYNPI	
	No ZnO	With ZnO	No ZnO	With ZnO
300% Mod. (MPa)	0.32	0.43	0.24	0.23
T.S. (MPa)	0.47	1.51	0.25	0.21
Elongation (%)	480	690	1100	1280

The tremendous increase in the green tensile strength of NR upon adding ZnO was attributed to the ionic groups on the backbone reacting with the Zn^{++}.

We repeated their experiment using Test Recipe A (Table 1, next page) which is more representative of commonly used tire carcass compounds. We compared the recipe shown to one with 3.5 phr GPF black substituted for the ZnO. The stress-strain curves are shown in Figure 3. Again, ZnO increased the modulus of the NR compound but did not affect the SYNPI stress-strain curve at all. However, the quantitive effect was much smaller than Smirnov (4) observed and the decrease in ultimate elongation caused the tensile

E.M. FRIEDMAN ET AL.

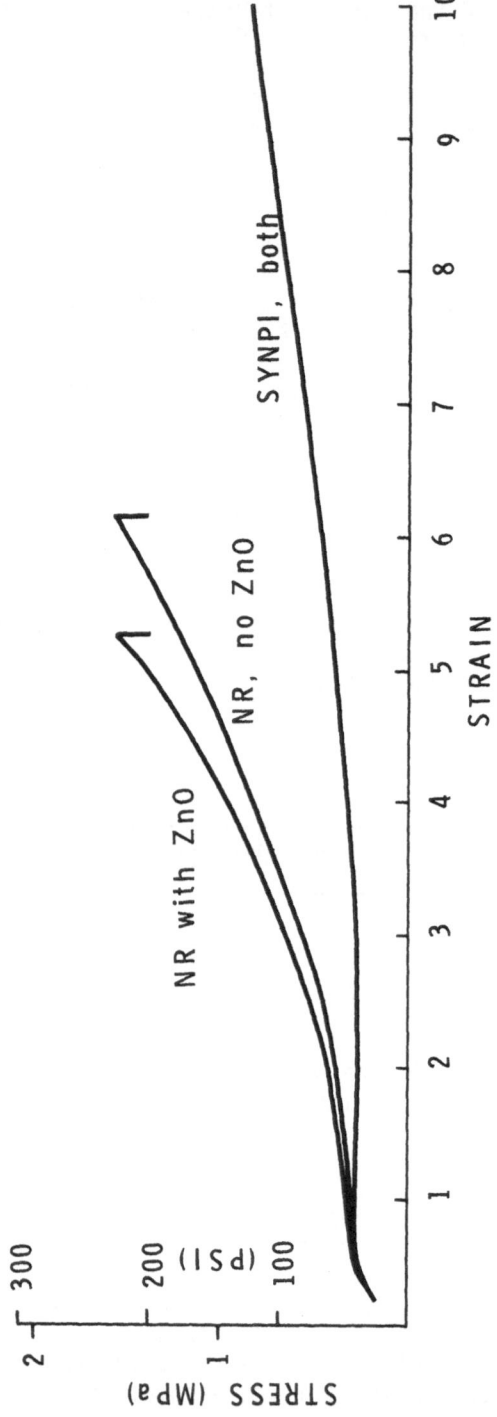

Figure 3. Typical stress-strain curves for a green PI compound with and without ZnO.

Table 1. Compounding Formulas

	Compound A+		Compound B+
NR** or SYNPI**	111.25	or ⎰ 70 pbw	70 pbw
SBR or ioSBR	0	⎱ 41.25	30
Stearic Acid	2.0		1.5
ZnO	3.5		3.5
SBR Black	-		15
FEF Black	-		25
GPF Black	50		-
CIRCO Light Rubber Process Oil (Sun Oil)*	-		3.0
CUMAR P-25 (Neville Chemical)*	-		2.0
SHELLFLEX 212 (Shell)*	4.0		-
AGERITE STALITE S (Vanderbilt)*	-		2.5
WINGSTAY 100 (Goodyear)*	1.0		-
Curatives			
Sulfur	2.5		2.1
AMAX (Vanderbilt)*	-		0.7
DPG DUSTLESS (Cyanamide)*	-		0.25
NOBS NO. 1 (Cyanamide)*	0.8		-

  * Tradename (Supplier).
 ** NR was #1 Ribbed Smoked Sheet; SYNPI was NATSYN 200 (Goodyear).
  + Adapted from the Vanderbilt Rubber Handbook (N.Y., 1968), p. 457.

----

strength to be unchanged. Furthermore, as with Smirnov's data, NR had a significantly higher green strength than SYNPI even without ZnO. This can be interpreted as indicating that while ionic groups may contribute to NR's green strength, some other effect (stress crystallization?) is much more important.

One objection to using the effect of Zn^{++} on green strength to test whether ionic groups account for NR's green strength is that a component may be present in NR which performs the same function as ZnO. However, Gregg and Macey (2) found that removal of the nonhydrocarbon fraction from NR did not destroy its green strength. Therefore, the objection is not valid.

OTHER GREEN STRENGTH IMPROVEMENT METHODS

If the rubber tree uses a template synthesis (as in DNA bio-synthesis) then most of the errors will be nonpropagating and isolated and the "correctness" will be even greater than 99%. To duplicate this feature in synthetic rubber would then require a propagation mechanism analogous to biosynthesis--i.e., quite

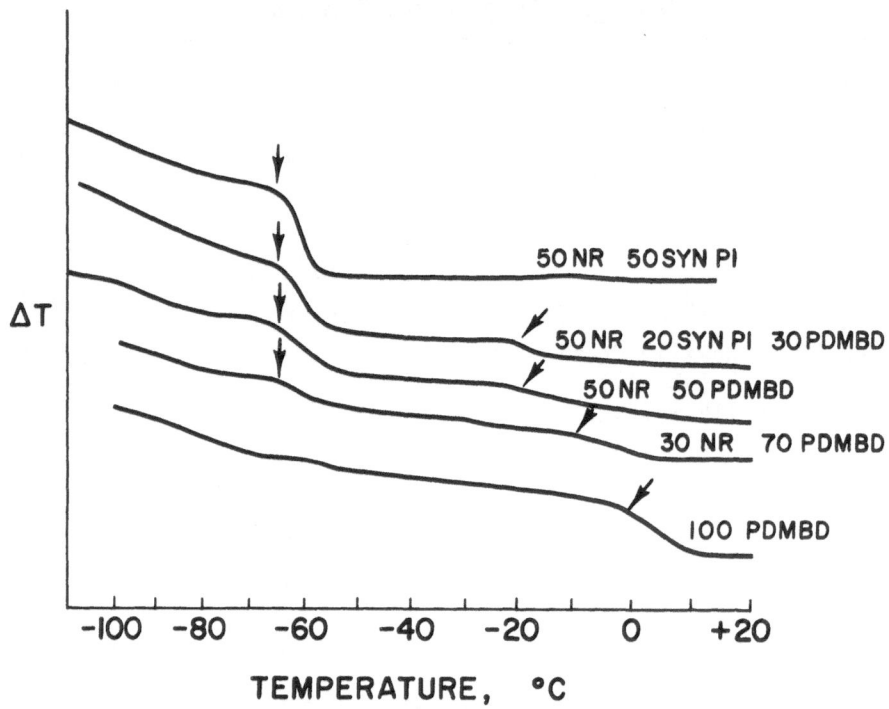

ΔT

50 NR 50 SYN PI

50 NR 20 SYN PI 30 PDMBD

50 NR 50 PDMBD

30 NR 70 PDMBD

100 PDMBD

TEMPERATURE, °C

Figure 4. DTA of PI/PBMBD Blends.

different from that believed to occur with modern stereo diene
polymerization catalysts. Since such a method seems unlikely to
be cheaper than growing rubber trees, other approaches were
studied.

    Crystallites reinforce natural rubber by acting as hard
domains which form under stress and to which the elastomeric chains
are attached. Any hard domain to which the elastomeric chains are
attached should perform the same function.

PDMBD/PI Blends[3]

    One very surprising way to get such domains is to prepare
blends of polydimethylbutadiene (PDMBD) with PI. Figure 4 shows a
set of DTA curves (DuPont Model 900 DTA) for a series of blends
containing various amounts of PDMBD. As the PDMBD content is

---

3 Patent applied for.

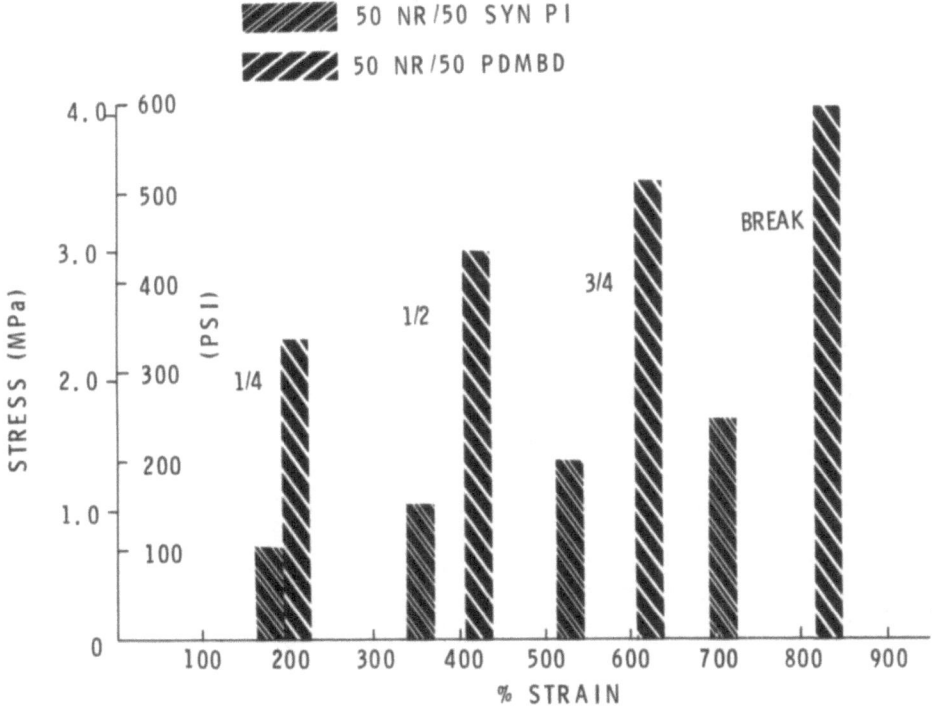

Figure 5.   Stress-strain data for green PDMBD blend and control.

increased, the polyisoprene (PI) transition at -65°C decreases in
amplitude but does not shift in position. The PDMBD transition at
-1°C, however, does appear to shift as the PI content is increased.
Apparently the blend contains PI domains and PI/PDMBD domains. The
latter have a higher Tg than the former and since the polymers are
sufficiently compatible to permit PI in the PI/PDMBD domains, PI
chains can traverse the domain boundaries. Thus, we have domains
which may perform the same function as the crystallites do in NR,
and we can reasonably expect a green strength enhancement.[4]

Stress-strain data for a 50:50 NR/SYNPI compound containing 60
parts carbon black is compared to that of an analogous 50:50
NR/PDMBD compound in Figure 5. Not only does the latter have much
better green strength, but its compound Mooney is unchanged
(46 vs. 44) and its cured modulus is slightly lower (16 vs. 18-1/2
MPa at 300%, 30 in./min., .389 in. dumbbell).

4   This section may lead the reader to believe that we predicted the
    green strength enhancement from the DTA. In actuality, we found
    the green strength enhancement and then looked for a reason.

Experimental. PDMBD was prepared via emulsion polymerization using a standard redox recipe with KOH/fatty acid (pH 10.5) emulsification and salt-acid coagulation. PDMBD compounds and controls were mixed in a Banbury using standard procedures.

Curatives were added on a cold mill. Samples for green tensile tests were cut directly from the mill and 0.5 in. strips pulled at 20 in./min. with an initial jaw separation of one inch.

Pendant Ionic Groups[5]

Previous workers (5, 6) have attached ionic groups to SYNPI backbones. Such groups (at the correct concentration) can react with counterions, and if they melt at processing temperatures, should serve the same purpose as crystallites. However, complicated post-polymerization reactions were required since polar groups kill typical stereo diene polymerization catalysts. A simpler method would be to prepare an ionically reinforced polymer which is compatible with PI. Since emulsion poly(butadiene-co-styrene) (SBR) is often a component of tire compounds and since polar groups do not seriously alter free radical polymerizations, carboxylated SBR's (ioSBR) were prepared by terpolymerizing styrene, butadiene, and methacrylic acid (MAA).

Figure 6 shows a series of typical stress-strain curves for a green 70:30 SYNPI:ioSBR compound (Test Recipe A, see Table 1) using various salt-acid coagulated oil-extended ioSBR's. In each case the mole % butadiene was held constant. The code "64H101" indicates that the ML4 at 100°C of the oil-extended ioSBR was 64, the level of MAA was "high" (3 phm charged), and the ML4 at 100°C of the compound was 101. "M" indicated 1.8 phm MAA charged; "L" indicates 0.75 phm. The Mooneys of the controls are indicated in an analogous fashion with the polymer Mooney being that of the SBR. For example, the Mooney of the SBR1 control was 44 while that of the SYNPI/SBR1 compound was 75.

All the SYNPI/ioSBR compounds have green strengths superior to the SYNPI/SBR control. Some even appear superior or equivalent to the NR/SBR control which is commonly used in tire compounds. Compounds 49M85 and 20H71 are likely compromises between high green strength and sufficiently low Mooney, to permit replacing NR with SYNPI in carcass compounds.

If the objective is to obtain much better green strength than the NR/SBR control, replacing the SBR with ioSBR is an excellent

---

5 Patent applied for.

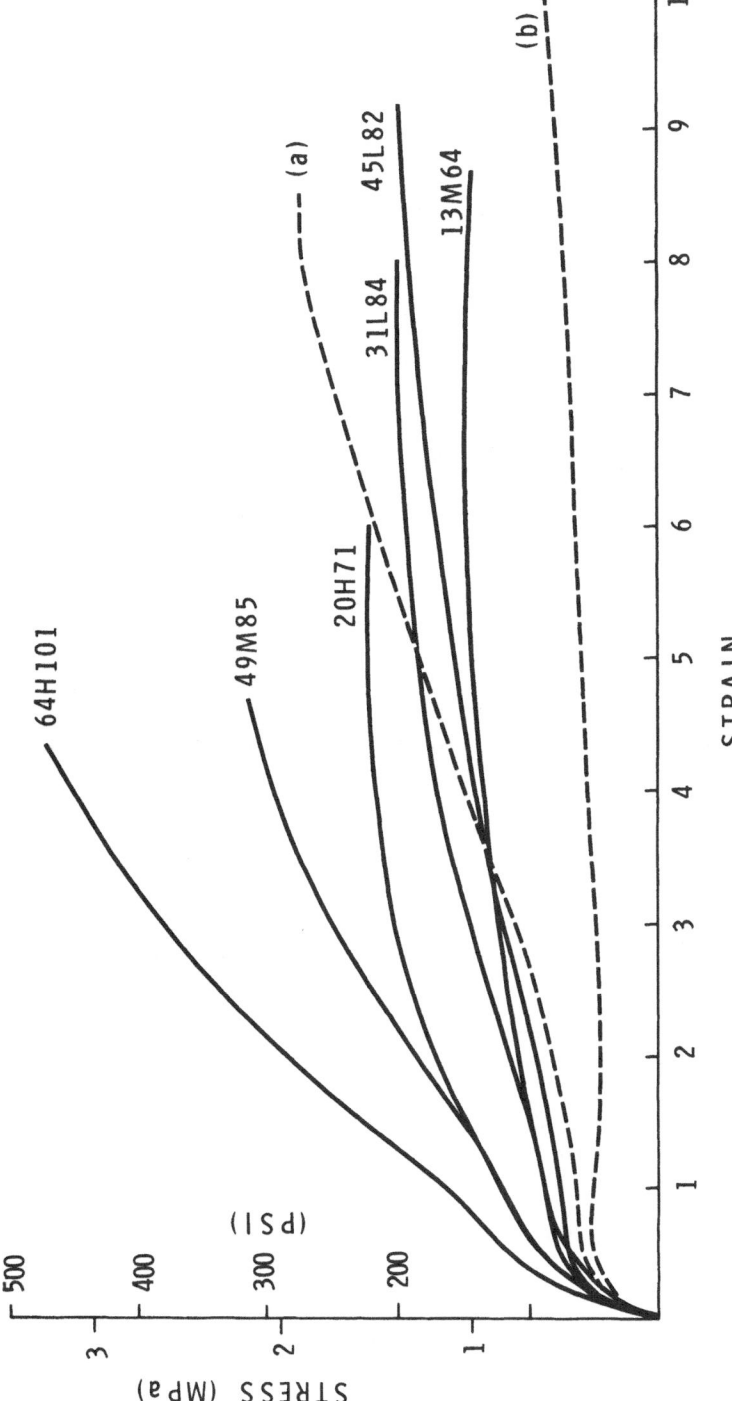

Figure 6.   Typical stress-strain curves for green SYNPI/ioSBR compounds and controls. Codes explained in text. Curve a: 44 NR/SBR1 55, b: 44 SYNPI/SBR1 75.

method. Figure 7 shows a set of typical stress-strain curves for green NR/ioSBR compounded as before.

The green strength improvements mentioned above were obtained without materially increasing the cured moduli. Figure 8 shows the improved green tensile properties found in a previous set of experiments. Table 2 shows that the cured moduli were not materially increased. The cure rate data in Table 3 shows that the experimental compounds were not scorchy.

It should be mentioned that the ZnO in the compounding formula (normally considered part of the cure system), is of great importance in achieving high green strength. We substituted 3.5 parts of GPF black for the ZnO and also tried using an additional 3.5 parts ZnO while decreasing the black accordingly. The result is illustrated in Figure 9. Presumably, the ZnO reacts with the carboxylic acid groups to form carboxylate ions which interact either with the Zn^{++} ions or with ZnO particles having a resultant excess of Zn^{++}. Of course, some other (polyvalent?) metal would probably do as well.

The method of preparing the dumbbells for green tensile measurements is also important--at least for the "L" series of ioSBR's. We normally cold mill out rough sheets, mold smooth sheets from them at 149°C for 30 min. and then cool them under pressure. If, instead we mill out reasonably smooth sheets at 110°C, the green tensiles are always lower[6]. However, for the SYNPI/ioSBR they are much lower--so much so that the green strength advantages are almost entirely lost (See Figure 10). Presumably, the molding process anneals the ionic crosslinks into a much more effective network.

The coagulation method used is of some minor importance. Figure 11 shows the results of green tensile measurements on five pairs of compounds as well as the raw and compound Mooneys. Barium chloride coagulation always gives a higher raw Mooney and higher green moduli and tensile strength than salt-acid coagulation; however, the compound Mooney is not affected.

Test Recipe B, not containing extender oil, was also tried. (See Table 1). In this case the SYNPI/ioSBR compound much more closely matched the green tensile properties of the corresponding NR/SBR compound. (Figure 12). Further investigation is currently under way in this regard.

---

6 It should be mentioned that whether samples are milled or molded into sheets for Mooney measurements does not affect their compound Mooney.

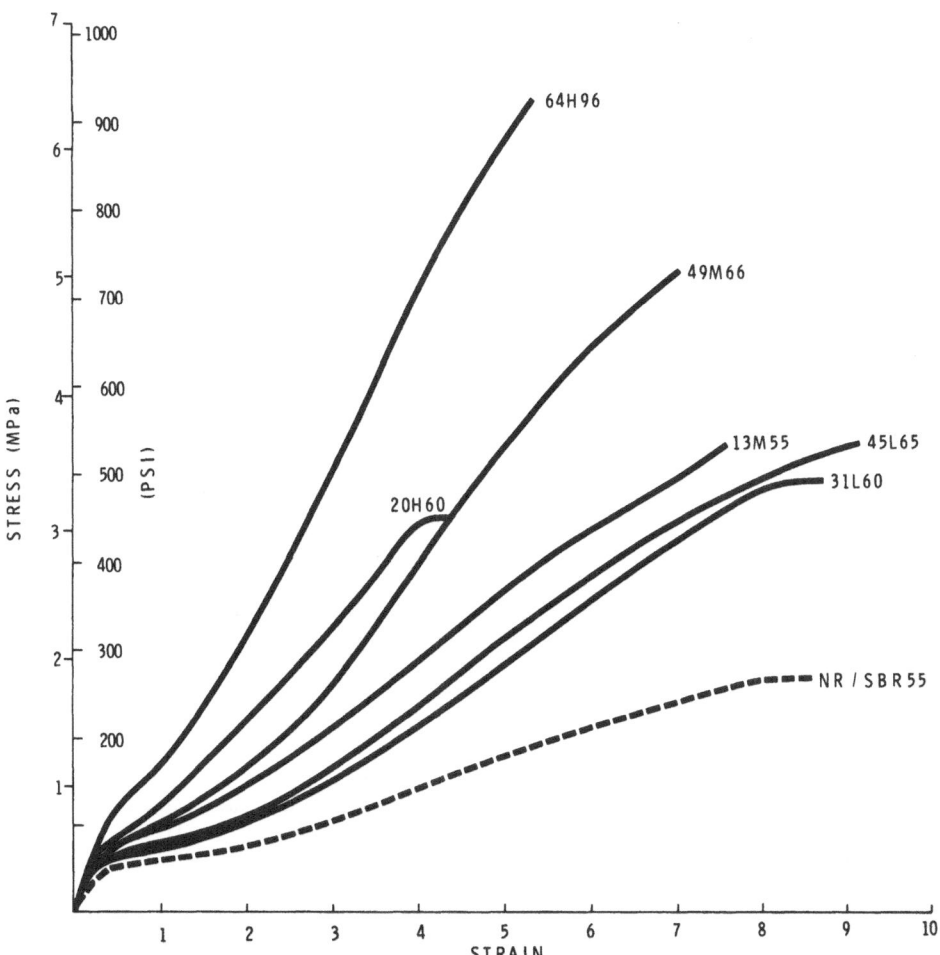

Figure 7. Typical stress-strain curves for green NR/ioSBR
compounds and control.

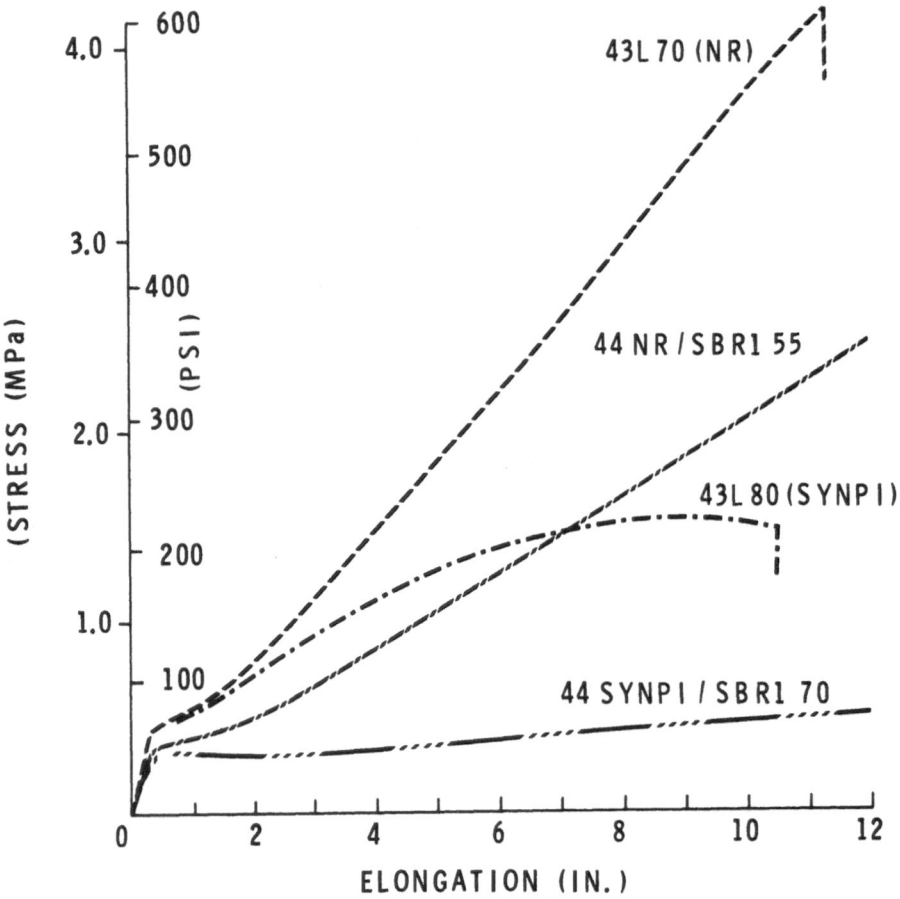

Figure 8.   Typical stress-strain curves for green PI/43L
           compounds and controls. Note change in horizontal
           scale.

Table 2.   Room Temperature Cured Tensile Properties

	Stress (MPa) at Elongation of (in.) of				Breaking	
	1.00	3.00	6.00	9.00	Stress	Elongation
70RSS/41.25 43L	2.2	7.0	17.9		26.9	7.80 in.
70RSS/41.25 SBR1	2.2	6.9	17.2	31.0	32.4	9.30
70 Natsyn/41.25 43L	2.3	7.2	17.9		29.0	8.40
70 Natsyn/41.25 SBR1	1.9	5.7	14.3	26.2	30.3	9.80

NOTE: For 0.1 in. dumbbells a 1 inch elongation roughly
corresponds to a 60% strain.

Table 3.   Cure Rate

Rubber	$t_{90\%}$	t_2	Rheometer Min.	Torque Max.
70RSS/41.25 43L	22.3	7.4	10.3	45.7
70RSS/41.25 SBR1	24.2	7.8	9.2	54.2
70 Natsyn/41.25 43L	22.2	8.4	13.8	45.3
70 Natsyn/41.25 SBR1	26.8	8.8	12.2	54.0

(Monsanto rheometer, micro die.)

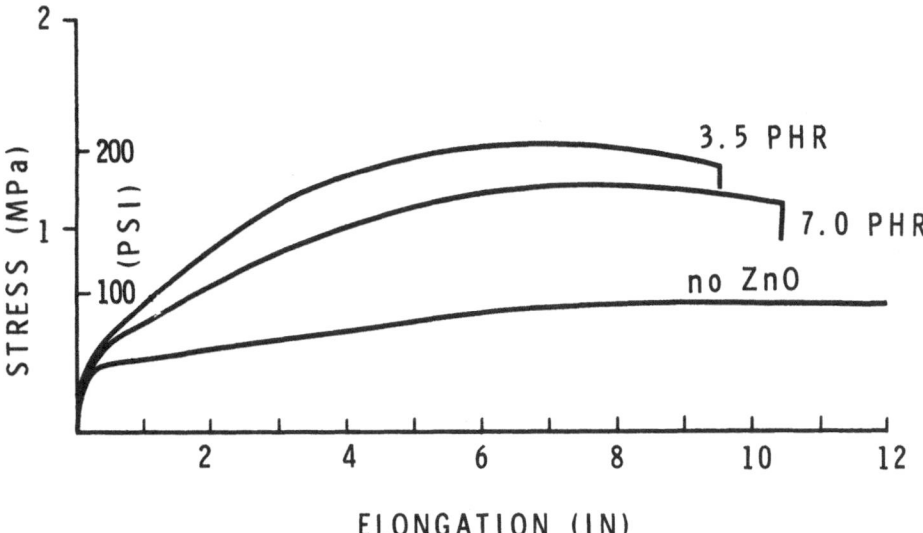

Figure 9.   Effect of ZnO on green SYNPI/43L.

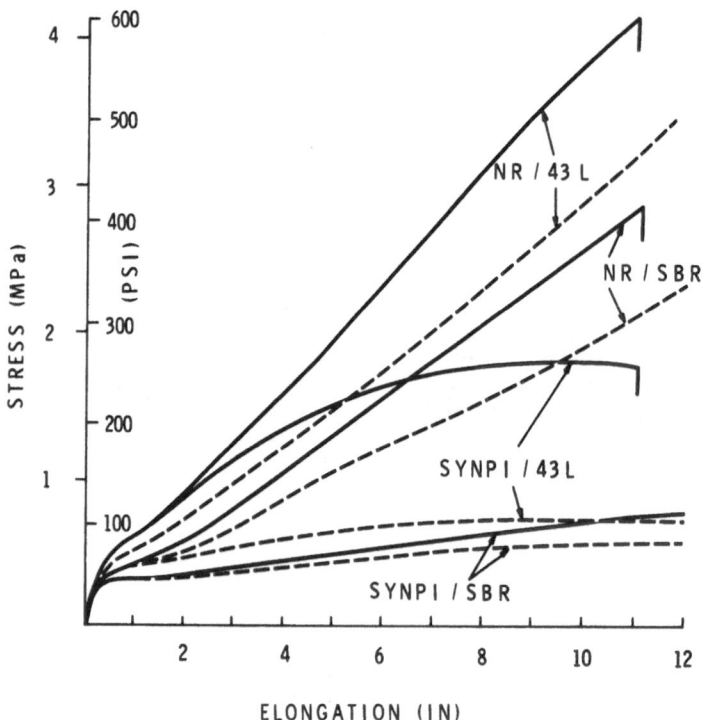

Figure 10. Effect of molding. Solid curves: milled and molded.
Broken curves: milled only.

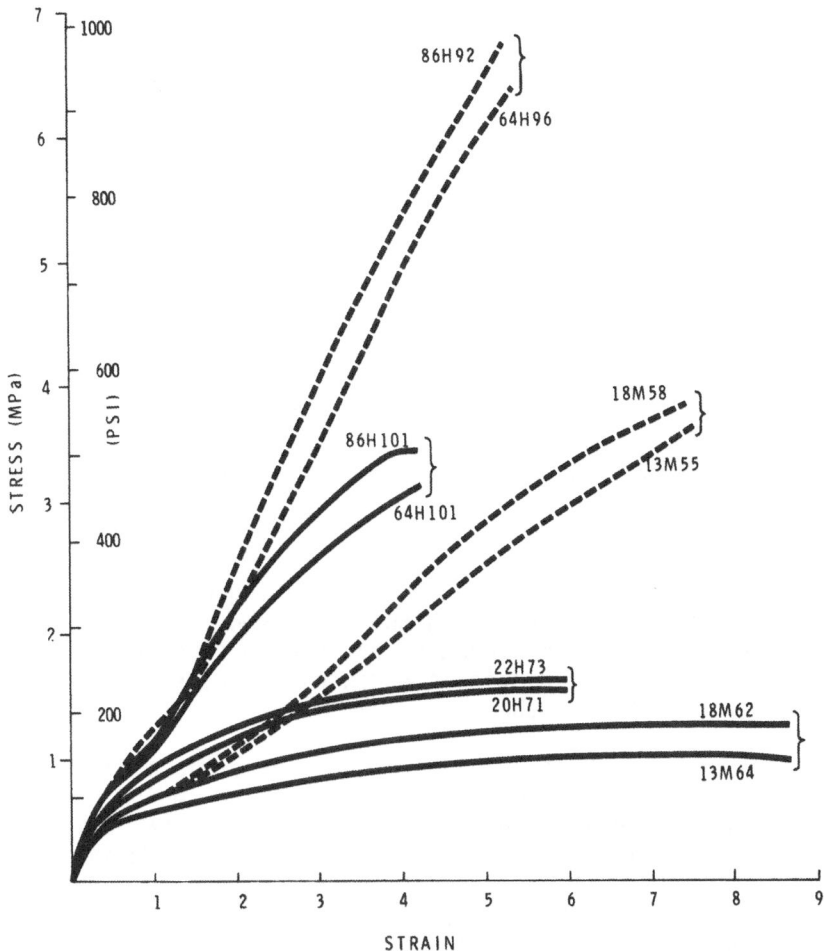

Figure 11. Effect of coagulation method. Brackets indicate that
crumbs were from the same latex. Upper curve of each
pair: BaCl₂ coagulation; lower curve: salt-acid.
Solid curves: SYNPI/ioSBR compounds; dashed
curves: NR/ioSBR compounds.

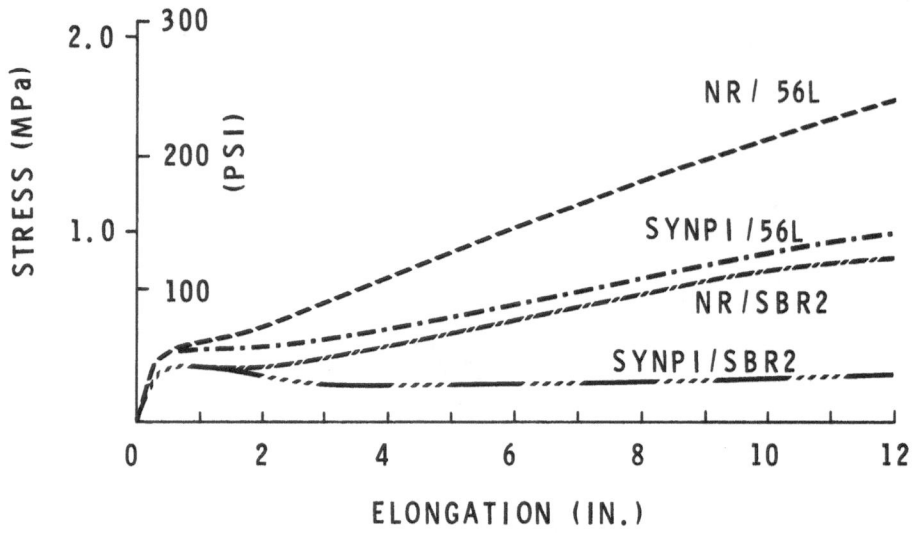

Figure 12. Typical stress-strain curves for green non-oil-extended
            ioSBR (56L) and SBR2 control in Compound B.

It is not yet known whether we are dealing with direct ionic
crosslinks involving one zinc ion and two carboxylate ions or
whether there is a domain structure based upon clusters of ions
including, perhaps, ZnO particles containing excess Zn^{++} formed
during neutralization of the carboxylic acid.

Experimental. SBR and ioSBR were prepared via emulsion
polymerization using a standard redox recipe and sodium dodecyl-
benzene sulfonate emulsification. SBR1 and SBR2 were commercial
samples of PLIOFLEX* 1778Y and 1502, respectively. NR and SYNPI
were commercial samples of #1 ribbed smoked sheet and NATSYN 200*.
The non-oil-extended ioSBR was alum coagulated. The oil extended
ioSBR's were extended with 37.5 parts SHELLFLEX 371* and then salt-
acid coagulated. Polymer compositions are shown in Table 4.

Compounding formulations are shown in Table 1. All compounds
were mixed in a Brabender for 5 min. at 141°C. Curatives were
added to samples for cured tensile tests on a cold mill; curatives
were not added to samples for green tensile tests. Nonproductive
stock was molded into 0.08 in. sheets for 30 min. at 149°C,
productive for the appropriate $t_{90\%}$ at 300°F. 0.1 in. dumbbells
were cut and pulled on an Instron at 20 in./min. at room temperature.

---

* PLIOFLEX and NATSYN are Goodyear trademarks; SHELLFLEX 371 is
  a Shell Chemical Co. trademark.

Table 4, Polymer Composition

Polymer	Wt % Charge			Pred mole % Comp			S/B*		mE MAA/100 gm+		
	Bd	St	MAA	Bd	St	MAA	Pred	Meas	Pred	Meas	Meas/Pred.
64H	79.5	15.5	3.0	88.5	8.0	3.4			58		
20H	79.5	15.5	3.0	88.5	8.0	3.4			58		
25H	75.6	21.4	3.0	86.4	10.1	3.5	10.5	10.3	57	39	0.68
49M	79.0	19.2	1.8	89.0	8.9	2.1			35		
13M	78.1	20.1	1.8	88.5	9.4	2.1			35		
45L	77.3	22.0	0.75	88.7	10.4	0.9	10.5	10.7	14	10.3	0.71
31L	77.3	22.0	0.75	88.7	10.4	0.9	10.5		14		
56L	74.4	24.8	0.75	87.1	12.0	0.9	12.1	13	14	10.9	0.75
44N	77.9	22.1	0	89.5	10.5	0	10.5	10.4	0	None	
67N	75.0	25.0	0	87.9	12.1	0	12.1	13	0		
SBR1	Predictions			88.5	11.5	0	11.5	11.8	0		
SBR2	Based on specs.			86.2	13.8	0	13.8	15.0	0		

* S/B is (moles St)/(moles St + moles Bd) in polymer -- measured via NMR.

+ mE MAA/100 gm. is milliequivalents MAA per 100 gm. polymer -- measured via titration. Predicted compositions were based upon the following reactivity ratios which we estimated based upon the best information available (reference 7):

Bd-St: 1,4, 0.6; Bd-MAA: 0.2, 0.5; St-MAA: 0.15, 0.7.

A few tests were run at 41°C and showed qualitatively similar behavior although all moduli were approximately halved.

CONCLUSIONS

This paper describes various methods of achieving high green strength in PI compounds. Two utilize domains through the boundary of which the PI chains pass. In one case (NR) the domains are PI crystallites; in the other we speculate that the domains are a PI/PDMBD blend. Another method involves pendant ionic groups on the SBR component of a PI/SBR blend. In this case domains (whether they be single Zn^{++} ions or larger) to which the SBR chains are ionically bound, are responsible for the enhancement of green strength, even though the SBR comprises less than 18% of the compound.

ACKNOWLEDGEMENTS

We thank R M Scriver for compounding and testing the PDMBD. We thank The Goodyear Tire and Rubber Company for permission to publish this work.

APPENDIX

This appendix relates to "Quantifying the Effect of Stereo-regularity".

Bernoullian statistics were implicity assumed -- i.e., the probability that a monomer causes an error is independent of whether the previous monomer caused an error -- as in the Wall (8) treatment of copolymerization. However, a Markov chain is more likely -- i.e., the previous monomer does have an effect -- as in the Mayo and Lewis (9) treatment. Even so, since the previous monomer is rarely an error unit, the Bernoullian treatment is a good approximation. This is shown below.

The PI chain is treated as a copolymer of correct units (C) and error units (E). Define K so that the probability a randomly chosen linkage is a C-E linkage is

$$[CE] = Kp(1-p) \tag{A1}$$

where p is defined as before.

Using the same reasoning as Harwood and Ritchey (10)

$$[EC] = [CE] = Kp(1-p) \tag{A2}$$

$$[CC] = p - Kp(1-p) \tag{A3}$$

$$[EE] = (1-p) - Kp(1-p) \tag{A4}$$

The probability that an error occurs <u>given</u> that the previous unit was not an error unit is

$$P_{CE} = \frac{[CE]}{[CE] + [CC]} = K(1-p) \tag{A5}$$

and similarly

$$P_{CC} = 1 - K(1-p) = p - (K-1)(1-p) \tag{A6}$$

$$P_{EC} = Kp \tag{A7}$$

$$P_{EE} = 1 - Kp \tag{A8}$$

Clearly, a Bernoullian distribution of error units occurs when K is unity, a blocky distribution when K is smaller, and an alternating distribution when K is larger.

Equation (1) becomes

$$P(=L) = P_{CC}^{L-1} (1-P_{CC}) \tag{A9}$$

and equation (5) becomes

$$W(\geq L) = \frac{p}{P_{CC}} \left[L P_{CC}^{L} - (L-1) P_{CC}^{L+1} \right] \tag{A10}$$

In other words, we use P_{CC} instead of p and then multiply by

$$\frac{p}{P_{CC}} .$$

In order to see how large an error is introduced by not making this correction we look at

$$S_E \equiv \frac{P_{EE}}{P_{CE}} = \frac{1-Kp}{K-Kp} \tag{A11}$$

Rearranging it

$$K = \frac{1}{(1-p)S_E + p} \qquad (A12)$$

Since S_E must be positive, $K < \dfrac{1}{p}$ and

$$P_{CC} \geq p - \frac{(1-p)^2}{p} \qquad (A13)$$

This expresses the fact that P_{CC} and the distribution of error units are hardly effected by even a large alternating tendency because the number of errors are small.

To find a maximum for P_{CC} we make the reasonable assumption that $S_E \lesssim 10.$* This gives

$$K = \frac{1}{1 + (S_E-1)(1-p)} \lesssim \frac{1}{1 + 9\,(1-p)} \qquad (A14)$$

and

$$P_{CC} = p + \frac{(S_E-1)(1-p)^2}{1 + (S_E-1)(1-p)} \lesssim p + \frac{9(1-p)^2}{1 + 9(1-p)} \qquad (A15)$$

Using relations (A13) and (A15) for the case $p = 0.965$ (SYNPI) gives

$$0.964 = p - .001 \leq P_{CC} \leq p + .008 = 0.973 \qquad (A16)$$

For the case $p = 0.99$ (NR) they give

$$0.99 = p - 10^{-4} \leq P_{CC} \lesssim p + .008 = 0.991 \qquad (A17)$$

Thus equations (1) and (5) give an adequate approximation to Markovian statistics if a small error range is permitted for p.

---

* This is less stringent than $\left(\dfrac{P_{EE}}{P_{EC}}\right) \cdot \left(\dfrac{P_{CC}}{P_{CE}}\right) \lesssim 10.$

REFERENCES

1. M. J. Brock and M. J. Hackathorn, Rubber Chem. Tech. **45**, 1303 (1972).

2. E. C. Gregg, Jr. and J. H. Macey, Rubber Chem. Tech. **46**, 47 (1973).

3. M. J. Hackathorn and M. J. Brock, Rubber Chem. Tech. **45**, 1295 (1972).

4. V. P. Smirnov, V. N. Reikh, L. S. Ivanova, and A. B. Kusov, Soviet Rubb. Tech. **28** (6), 5 (1969).

5. E. N. Marandzhera, A. S. Estrin, I. V. Garmonov, B. M. Bolkhovets, D. M. Rudkovskii, N. S. Imyanitor, and B. E. Kuvaev, Soviet Rubb. Tech. **30** (2), 51 (1971).

6. V. P. Smirnov, N. F. Kovalev, E. N. Marandzhera, I. V. Garmonov, and A. S. Estrin, Soviet Rubb. Tech. **30** (6), 3 (1971).

7. L. J. Young in Polymer Handbook, 2nd ed., ed. J.Brandrup and E. H. Immergut (Wiley, N.Y., 1975), pp. II-155, 156, 234, 235.

8. F. T. Wall, J. Am. Chem. Soc. **63**, 1862 (1941).

9. F. R. Mayo and F. M. Lewis, J. Am. Chem. Soc. **66**, 1594 (1944).

10. H. J. Harwood and W. M. Ritchey, Polym. Lett. **2**, 601 (1964).

PHYSICAL AND MECHANICAL PROPERTIES AND MORPHOLOGY OF

MIXTURES OF HIGH IMPACT POLYSTYRENE WITH HOMOPOLYSTYRENE

B. V. Kravchenko, V. D. Yenalyev

Donetsk State University

Donetsk 340055 (U.S.S.R.)

HIPS obtained by means of graft copolymerization of styrene with rubber was mixed with homopolystyrene in different apparatus: in industrial and laboratory extruders, and in plastic mixers of the Banbery type. While mixing in extruders, physical and mechanical properties of the mixture got worse to a lesser degree than by diluting HIPS with polystyrene: when homopolystyrene was added in a small quantity, physical and mechanical properties got even better.

HIPS mixed with homopolystyrene in a Banbery lowers physical and mechanical properties in proportion to homopolymer dilution.

RHEOPTICAL STUDIES OF BLENDS OF POLY(STYRENE-B-BUTADIENE-B-STYRENE)

AND POLYSTYRENE.

S. D. Hong[*] and M. Shen

Department of Chemical Engineering
University of California
Berkeley, California 94720
                  and
T. Russell and R. S. Stein
Department of Polymer Science of Engineering
    and Polymer Research Institute
University of Massachusetts
Amherst, Massachusetts 01003

ABSTRACT

Polyblends of poly(styrene-b-butadiene-b-styrene) and polystyrene were cast from a tetrahydrofuran/methyl ehtyl ketone mixture
using a spin caster. These samples were found to undergo a "strain-
induced plastic-rubber transition" upon deformation. Infrared
dichroism and birefringence studies both indicate the high orientation of the polybutadiene chains, but not of the polystyrene chains.
Thus most of the deformation after the "transition" has taken place
is due to the rubbery domains. Scanning electron micrographs show
some surface fracture upon stretching for those samples containing
high polystyrene content. Small angle x-ray scattering results
reveal that the morphology in the three coordinate directions was
different. Upon stretching, the polystyrene domains may have
broken up into smaller sub-domains, which could be responsible for
the observed plastic-rubber transition. Furthermore, there was a
decrease in the total scattering intensity for the stretched samples, indicating a change in the mean square fluctuation of electron density of the stretched sample. The small angle light scattering patterns appear to be consistent with the above proposed
interpretations.

* Jet Propulsion Laboratory, 4800 Oak Grove Drive
   Pasadena, California 91103

INTRODUCTION

Polymer blends in general have synergetic properties which the individual homopolymers do not have. High impact polystyrene and ABS block copolymers are excellent examples. Because polymers are generally found to be incompatible as a consequence of positive free energy of mixing, most polymer blends exhibit phase separation (1-3). Blends of incompatible homopolymer do not have good mechanical properties because of the lack of sufficient adhesion between the different phases (4-6). An alternate process to improve adhesion between different phases in a blend is to add a graft copolymer or block copolymer. It was found that the use of these copolymers as additives in homopolymer blends greatly improves the mechanical properties of the blends (4-9). The morphology and mechanical properties of these blends have been studied by a number of workers (10-17).

Upon deformation block copolymers and blends of copolymer and homopolymer often exhibit stress softening (18-20). Furthermore the deformed samples demonstrate a healing effect in that, upon removal of stress, properties of the original undeformed sample are restored (18,20). For example, when a sample of the blend of poly(styrene-b-butadiene-b-styrene) and polystyrene is strained beyond the yield point, it becomes rubbery and exhibits high elasticity rather than irreversible drawing (20). The stress-softening effect in pure block copolymer has previously been attributed to the breaking up of some sort of rigid structures (18,21, 22) and the healing effect to the reformation of the original domain structures (18). Inoue et al., (15), on the other hand, found that in the blends of poly(styrene-b-isoprene) with polystyrene, the sample underwent yielding upon stretching. In this work, a rheo-optical investigation of the strain-induced plastic-rubber transition is presented in an effort to further elucidate the mechanism of this interesting phenomenon.

EXPERIMENTAL

Poly(styrene-b-butadiene-b-styrene) block copolymer (SBS), designated as Kraton 1101, was received from Shell Chemical Company. The copolymer contains 28% polystyrene and 72% polybutadiene. The weight average molecular weight of the copolymer is 84,000 and the polydispersity index is 1.21. The styrene blocks of the copolymer have a number average molecular weight of 13,300. The polybutadiene (PB) blocks have 46% trans-1,4, 46% cis-1,4 and 8% vinyl structures. Polystyrene (PS) was supplied by Polysciences, Inc., and has a number average molecular weight of 41,000 and polydispersity index of 2.32.

Samples of the block copolymer and its blends with PS were cast in the form of sheets from a 10% solution of tetrahydrofuran and methyl ethyl ketone mixture (90/10 in volume ratio). A spin caster was used for casting. The cast films were heated in vacuo at 60°C until constant weight was reached, indicating that the residual solvent had been removed.

Small angle x-ray scattering (SAXS) measurements were made with a Kratky small angle x-ray camera equipped with a Kratky-Siemans x-ray tube operated at 40 Kv and 20 mA. The intensity was measured with a proportional detector in conjunction with a pulse-height analyzer. Slit collimation was used and no desmearing corrections were made for the data presented. The resolution of the apparatus was set at 1800 Å . Small angle light scattering patterns were recorded on polaroid films using a He-Ne laser (Spectra-Physics) as light source. Scanning electron micrographs were obtained on an ETEC scanning electron microscope.

Infrared dichroism measurements were carried out using a Perkin-Elmer Model 180 Infrared Spectrophotometer equipped with a Perkin-Elmer silver bromide gold wire grid polarizer. A stretching device capable of extending both ends of the film simultaneously was used, thus the same part of the film remained in the beam at all elongations. Birefringence measurements were made with a Babinet compensator equipped with a mercury lamp as light source.

RESULTS AND DISCUSSION

Infrared Dichroism and Birefringence

The SBS and its blends with 10%, 20%, 30% and 40% PS all showed similar dichroic ratio vs. elongation relationships, therefore only the results for the SBS and its blend with 20% PS will be presented. The assignments of the absorption bands chosen for this study are summarized in Table I. Since almost all absorption bands associated with PS displayed negligible dichroism, only the results from two bands will be given.

The orientation of the transition moment with respect to the sample stretching direction may be given as (27):

$$f_M = \frac{D - 1}{D + 2} \tag{1}$$

where the dichroic ratio $D = A_{//}/A_{\perp}$; $A_{//}$ and $A_{\perp}$ are absorbance with the incident infrared beam polarized parallel and perpendicular to the stretching direction, respectively. f_M can be used to estimate the orientation of chain axis of the molecule provided the angle between the chain axis and the transition moment is known (27). In the present case, however, this information is not

Table 1. Assignments of IR Absorption Bands

Frequency (cm^{-1})	Assignment
1602, 1493	The C=C skeletal in-phase vibration of the aromatic ring in polystyrene (23,24)
1410	CH$_2$ in-phase deformation of $-$CH=CH$_2$(23) or CH in-plane bending of cis $-$CH=CH$-$ (25)
1310	CH bending of cis 1,4 polybutadiene (26)
1243	Associated with polybutadiene (25)

available. Thus we will use f_M as a qualitative indication of molecular orientation.

Figs. 1 and 2 show f_M as a function of elongation for the SBS and its blend with 20% PS. One may note that the value for f_M for the absorption bands 1602 cm^{-1} and 1493 cm^{-1} (and other bands not shown) associated with vibration in PS chains are negligible. Although in this case the angle between the transition moment and the chain axis is close to 54.7°, the value of f_M is always very small even if the molecules have very high orientation (27). However, since in the present case the values of f_M for all the absorption bands associated with PS are very small, it may be reasonable to conclude that PS chains have negligible orientation. On the other hand, the absorption bands associated with the vibration of PB chains have noticeable orientation. The values of f_M for different absorption bands are different, because the angle between the transition moment and the chain axis for each chain may be different. Furthermore, it has been shown theoretically (28,29) and experimentally (30) that the orientations of chain segments in the amorphous regions are conformation-dependent. The chain segments of different conformations in the same parent chain may have different orientation.

Fig. 3 shows the birefringence of the SBS and the blend with 20% PS as a function of elongation. The samples have positive birefringences at all elongations. This is in contrast to the case of poly(styrene-b-isoprene) in which the birefringence measured in the necked regions was reported to be negative due to the orientation of PS (15).

The birefringence of a heterophase copolymer may be expressed in the form (31)

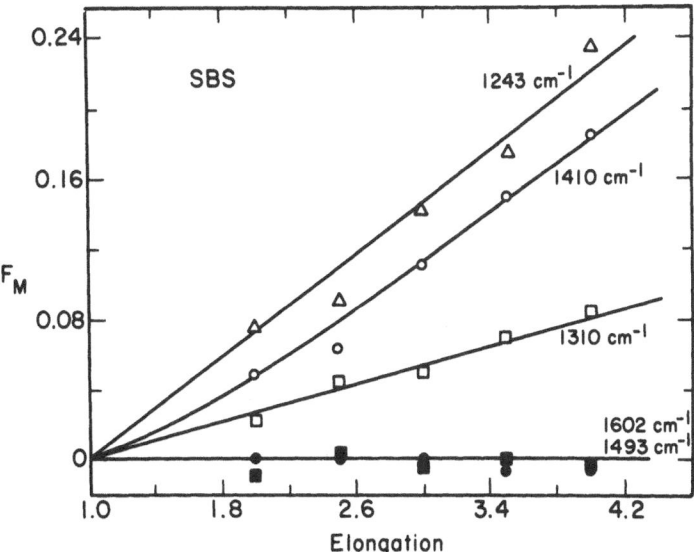

Figure 1.   Orientation functions of transition moment
            of various infrared absorption hands of
            polybutadiene and polystyrene components in SBS.

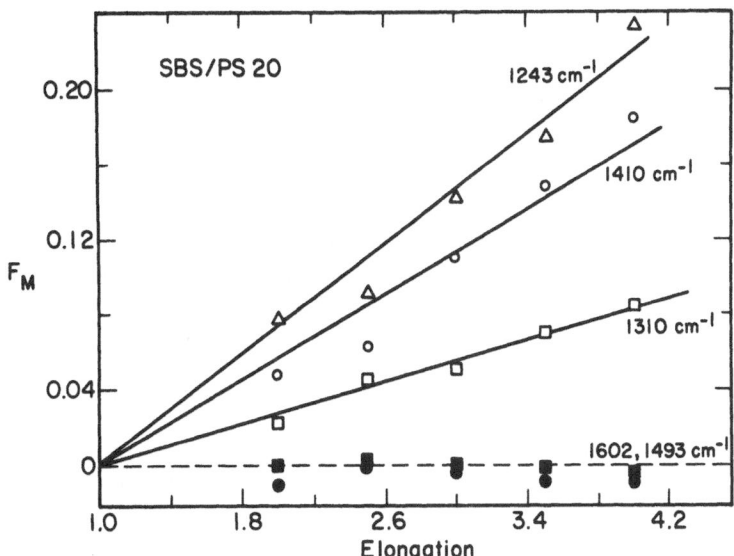

Figure 2.   Orientation function of transition moment
            of various infrared absorption hands of
            polybutadiene and polystyrene components in
            blend of SBS/20% PS

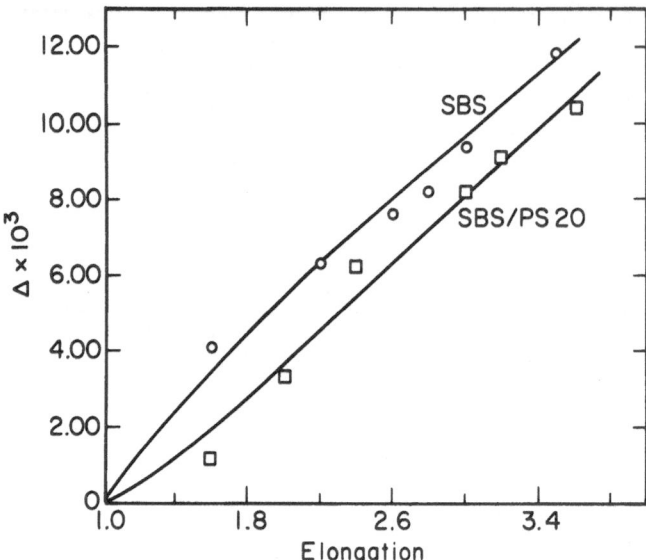

Figure 3. Birefringence vs. elongation for the
SBS (0) and SBS/20% PS (●)

$$\Delta = \phi_B C_B \sigma_B + \phi_S C_S \sigma_S + \Delta_f \qquad (2)$$

where S and B designate PS phase and PB phase separately, and ϕ_S and ϕ_B are volume fraction, C_S and C_B are stress-optical co-efficients and σ_B and σ_S are stresses. Δ_f is form birefringence. The value of form birefringence for a SBS block copolymer (20% styrene content) with the PS domains forming perfectly oriented rod structure was estimated to be on the order of 5×10^{-4} (32). In our case, the unstretched samples had negligible birefringence, indicating a negligible form birefringence. As the samples are stretched, the value of form birefringence may increase as a result of orientation of the optic axis of the domains. However, the value of the form birefringence is not expected to exceed 5×10^{-4} because the domains cannot achieve a high degree of orientation. It is more likely that Δ_f is very small in the stretched samples because electron micrographs did not show any appreciable orientation of the PS domains upon stretching (33). The value of C_S was estimated to be -5.2×10^3 Brewsters if PS is in the rubbery state and 10.2 Brewsters if it is in the glassy state (19), while C_B was estimated to be 3.1×10^3 Brewsters from the following equations (31)

$$C_B = X_c C_c + X_t C_t \qquad (3)$$

where X_c and X_t are the mole fractions of the cis and trans isomers in PB and C_c = 2700 Brewsters and C_t = 4100 Brewsters (15) are the respective stress optical coefficients. The contribution from the vinyl component is neglected.

Since significantly high values of positive birefringence were observed for the SBS and its blends after yield point, it seems to indicate that no appreciable orientation of PS chains has occurred. Otherwise one may expect the birefringence to be negative, as in the case of poly(styrene-b-isoprene) (15), because of the high negative value of C_S and the high value of σ_S in comparison with those of C_B and σ_B. This evidence seems to support the conclusion derived from infrared dichroism data discussed earlier.

From previous electron micrograph studies of SBS and its blends (11), the unstretched samples had a morphology of interconnecting PS domains and PB domains. The fact that PB chains were oriented while PS chains showed no orientation after yield point indicates that the continuity of polystyrene domains was disrupted to give smaller PS domains imbedded in polybutadiene matrix. Thus the PB phase must bear the bulk of the strain, resulting in chain orientation.

Small Angle X-Ray Scattering (SAXS)

The geometry of the SAXS scan is shown in Fig. 4, where the direction 1 is normal to the film surface, the direction 2 is parallel to the spin direction and the direction 3 perpendicular to both 1 and 2. The scattering profiles of the blend with 20% PS are shown in Fig. 5a for the unstretched sample, and in Fig. 5b for the stretched sample. The scattering profiles of the pure SBS are similar except for the different peak positions.

As can be seen from Fig. 5a, the scattering patterns are not the same in different directions, indicating an anisotropic morphology. This anisotropy may be a result of the relatively rapid rate of evaporation of the solvent in spin casting. It has been reported that the rate of evaporation of the solvent will greatly affect the morphology of the sample (34).

The unstretched sample shows a rather sharp scattering peak with the incident beam along the direction 1 (Fig. 5a). The same is true for SBS. These results indicate that the domains have higher regularity in the direction normal to the sample surface. There is a very intense scattering at lower angles when the incident beam is along the 2 direction (curve 2 of Fig. 5a), which appears to arise from total reflection, as shown in Fig. 6 in which the scattering profiles are seen to be dependent on the tilt angle.

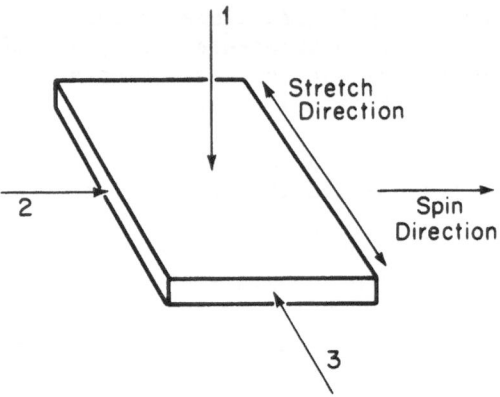

Figure 4. Geometry of SAXS scans. 1 is normal to
the sample surface. 2 is the spin
direction of the spin caster and 3 is
the stretching direction.

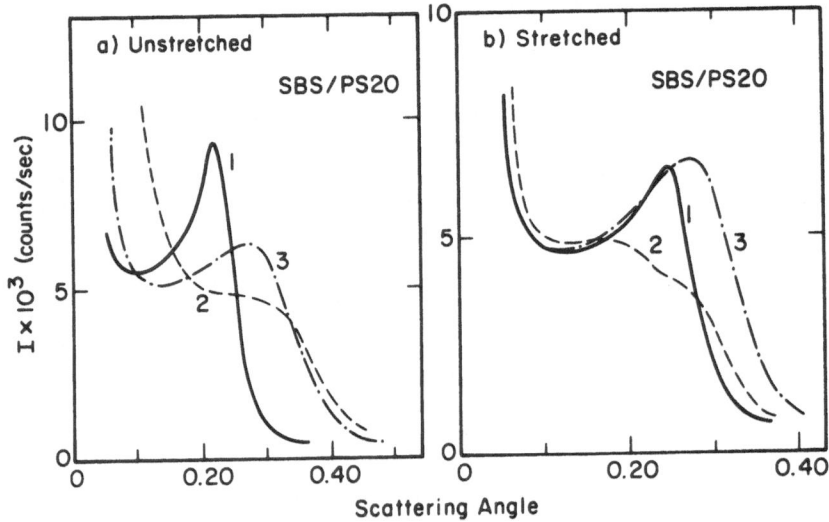

Figure 5. SAXS profiles for the blend of SBS/20% PS. (a) Sample
unstretched. (b) Sample was stretched until the
plastic-rubber transition completed. The stress
was then released. The residual strain was less
than 20%. Curves 1, 2 and 3 correspond to scanning
geometry shown in Fig. 4.

The unstretched sample of the specimen in the beam also has weak
second order scattering maxima in all three scattering profiles
while the blend with 20% PS has second order scattering only when
the incident beam is along the 1 direction.

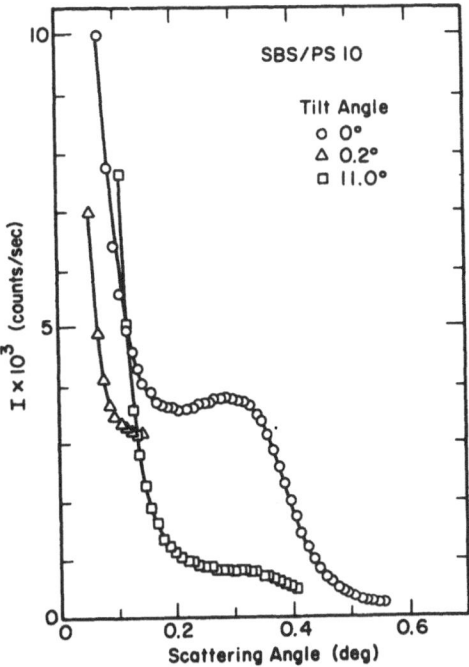

Figure 6.   Scattering profiles of curve 2 in Fig. 5
            with sample tilted at various angles.

Because the sample holder of the SAXS apparatus was not capable of holding an extended sample, the sample was first stretched
until the plastic-rubber transformation was completed and then
released. The residual strain was less than 20%. The scattering
profiles show drastic changes, as shown in Fig. 5b. The scattering
profiles become more diffuse and the first order scattering maxima
are shifted to higher angles. However, the second order scattering
maxima (not shown) are not present in the stretched sample.

While it may not be strictly valid, Bragg's law was used to
estimate the change in interdomain spacings from the peak positions
of the scattering curves. Since the residual strain remaining in
the stretched sample was less than 20%, changes in the interdomain
spacings from Bragg's law should serve as an approximate indication
of morphological changes upon deformation. The results are summarized in Table 2.

Table 2. Positions of Observed Small-Angle Maxima (Å)

Sample	Condition	Order of Maximum	Mode of View		
			1	2	3
SBS	Unstretched	1	366	306	309
		2	198	148	148
	Stretched	1	342	294	–
SBS/20% PS	Unstretched	1	443	355	355
		2	207	–	–
	Stretched	1	383	369	349
		2	177	–	–

There is a distinct decrease in the interdomain spacings with deformation as seen from the scattering curves for the case of the incident beam normal to the sample surface. The changes in the interdomain spacings when the incident beam is in the other two directions are not very certain because the scattering profiles are very diffuse. This decrease in interdomain spacings in conjunction with the more diffuse scattering profiles upon deformation seems to indicate that the polystyrene domains were disrupted into smaller domains. If the polystyrene domains maintained their integrity after deformation, one would not have expected such a pronounced change in interdomain spacings where less than 20% residual strain remained. This conclusion also agrees with recent electron micrograph studies of a different SBS sample prepared under the same condition (33).

The effect of annealing the stretched sample at ambient temperature is shown in Fig. 7. Curve 1 is for the unstretched sample, curve 2 for the sample which was stretched and scanned immediately and curve 3 for the stretched sample which was kept (unstretched) at ambient temperature for 4 days. It is seen that upon annealing the scattering curve became more similar to that of the original sample, indicating a partial restoration of the original morphology. This reformation of the original morphology may correspond to the reported healing process observed in the mechanical properties of block copolymers and polyblends (10,18).

Small Angle Light Scattering (SALS)

The V_V (parallel polarized) and H_V (cross polarized) scattering patterns of SBS at various elongations are shown in Fig. 8. The

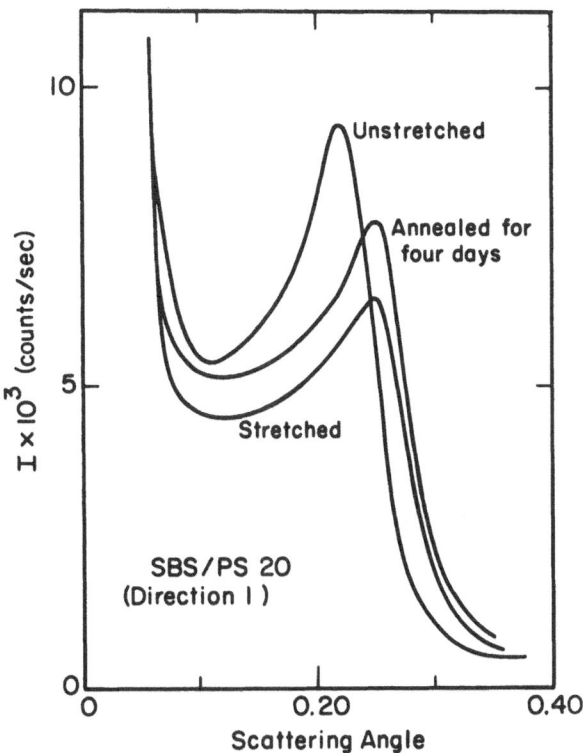

Figure 7. Scattering profiles for a SBS/20% PS sample
           with different healing history. Curve 1,
           unstretched sample. Curve 2, scanned
           immediately after stretching. Curve 3, four
           days after stretching.

intensity of H_v scattering is very weak compared to the V_v
scattering. The exposure times for the H_v patterns were much
longer than those for the corresponding V_v scattering patterns.
In Fig. 9 the scattering of a blend of SBS with 10% PS in the
necked and unnecked regions of the sample after elongation

    The unstretched samples of the pure block copolymer and its
blends have no detectable H_v scattering. The weak H_v scattering
of the stretched samples may be attributed to form birefringence
and orientational birefringence arising from the extension of
polymer chains, especially the PB chains. As discussed earlier,
the form birefringence of the unstretched samples is probably very
small but its magnitude will increase as the samples are extended.

    The fact that V_v scattering is much stronger than H_v scattering
suggests that correlated density fluctuation in the samples

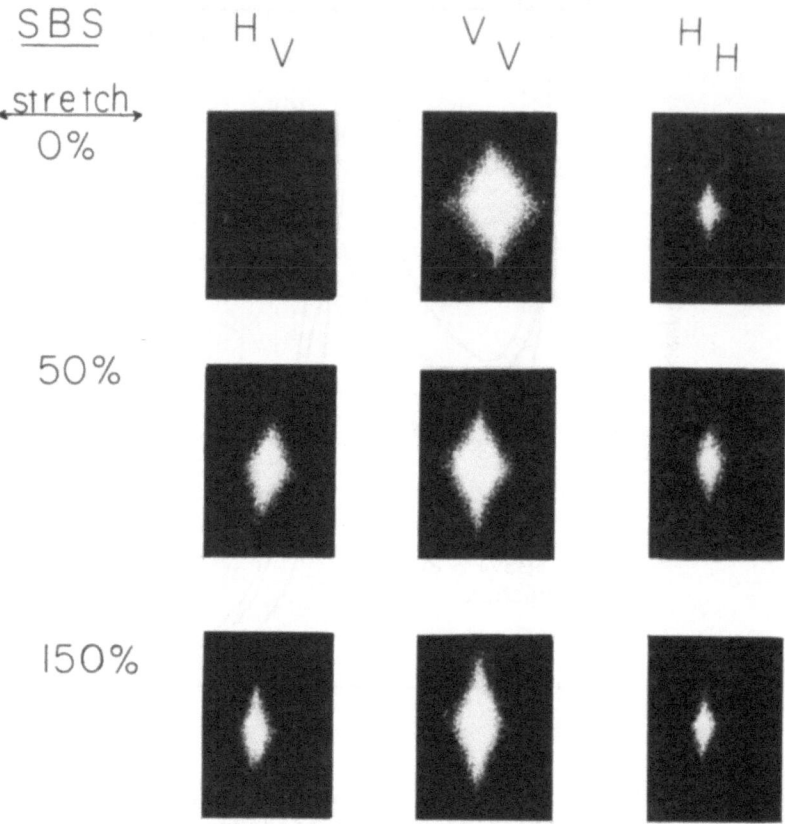

Figure 8. V_V and H_V scattering patterns from film of SBS at
            various elongation.

dominate the scattering (36). For the unstretched samples, the V_V
scattering pattern is almost circular for SBS, indicating a random
correlation of the density fluctuation; whereas for the blends, the
V_V scattering patterns are elliptical, indicating the existence
of a regular domain arrangement in these samples (36,37).   This
is in agreement with the findings from previous work (11) in which
the transmission electron micrographs showed no regularity of
domain arrangement in SBS while in the blends there are regions of
high regularity of domain arrangement.

      Stretching gives characteristic changes of the V_V scattering
patterns. The extent of changes increases with increasing amount
of PS in the blend. The stretched sample of SBS, SBS/PS 10 and
SBS/PS 20 have elliptical scattering patterns with different fine
details. The major axis of the ellipsoid is perpendicular to the
stretching direction. The stretched sample of SBS/PS 30 has a

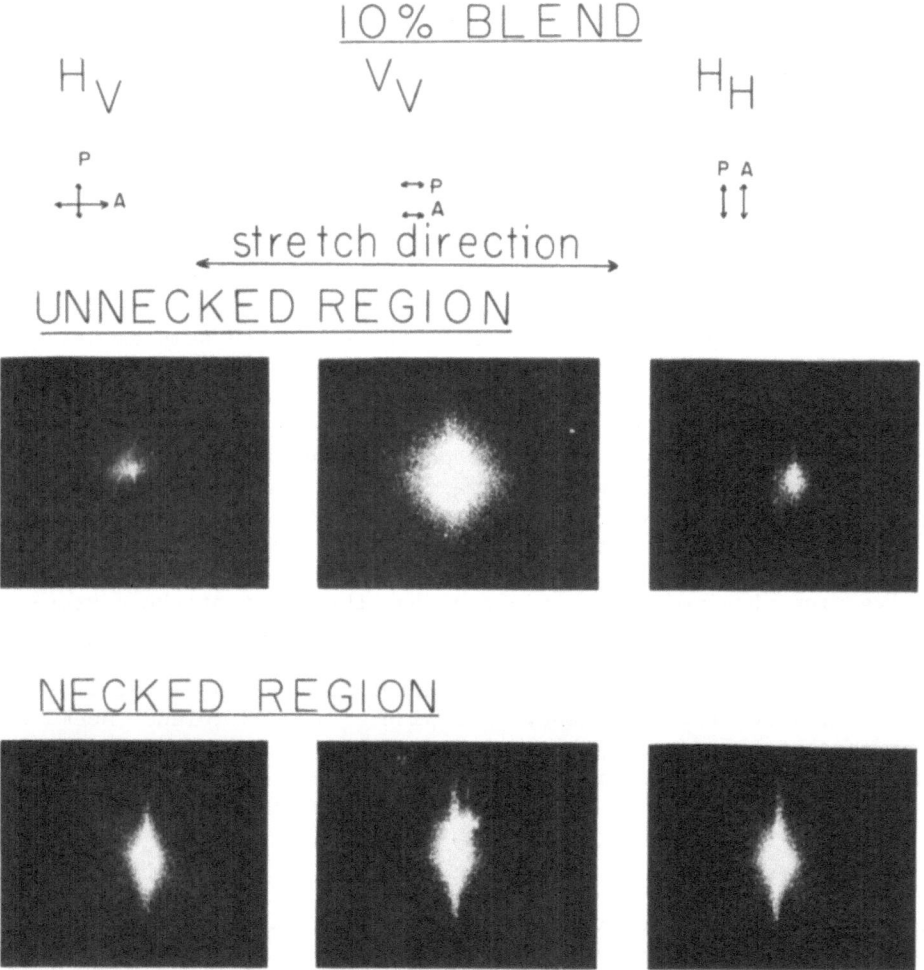

Figure 9. V_V scattering patterns from specimen films of SBS/PS 10, SBS/PS 20 and SBS/PS 30 at unstretched and stretched states.

dumbbell scattering pattern. All the new patterns remain in the same range of scattering angles as those for the unstretched sample, indicating that stretching does not change the range of correlation (36,37) at this elongation.

The characteristic changes of V_V scattering upon stretching may be attributed to changes in the correlation of density fluctuation (36,37) as a result of morphological changes and of another density fluctuation arising from newly formed heterogeneities due

to stretching (16). It has been observed that when an immersion
fluid was applied to the stretched samples, there was a decrease
in scattering intensity and, for SBS/PS 30, a slight change in
scattering pattern. Thus these newly formed heterogeneities must
be correlated to the voids or cracks developed in the styrene
domains upon stretching and are responsible for the plastic-rubber
transition.

Scanning Electron Microscopy (SEM)

Scanning electron micrographs of the fractural surface of the
unstretched SBS and its blends with various amounts of PS are
presented in Fig. 10, and the SEM of the stretched samples are
shown in Fig. 11. The samples were fractured at liquid nitrogen
temperature along the direction normal to the sample surface.

As shown in Fig. 10, the SBS does not show macroscopic phase
separation. On the other hand, macroscopic phase separation
occurs in all the blends forming a surface layer on the sample
surface and inclusions of polystyrene island in the bulk. The
macroscopic phase separation may arise because the added homopolystyrene has molecular weight higher than that of the corresponding blocks in the block copolymer (35). The thickness of the
surface layer is not certain because there appears to be a transition between the surface layer and the bulk of the material,
especially for the 10% blend. The size of the PS islands seems to
increase with increasing homopolystyrene content. The actual
shapes of the islands are not certain. It was suggested from
studies of transmission electron micrographs of these samples (11)
that these islands were rod-like.

Upon deformation, as shown in Fig. 11, both the surface layer
and the inclusions underwent fracture. It is obvious that the
surface layer fractures are attributable to a continuous PS phase.
It is seen that in the polystyrene island fracture occurred across
the islands in the direction perpendicular to the strain rather
than on the boundaries. This seems to suggest that mixing of the
homopolystyrene and the polystyrene component in the block copolymer does occur in these islands. Thus the adhesive force
between the block copolymer matrix and the islands is high enough
to strain the polystyrene islands beyond their fracture point.
It appears that fracture occurs mostly in the larger islands, perhaps because they are more easily strained. On the other hand,
the smaller islands can simply move intact when the sample is
deformed. Another possibility is that the larger islands are more
likely to incorporate imperfections which are easily fractured.
The cracks appearing in the PS islands are not induced during
sample fracture in liquid nitrogen, since no cracks are observable
on the fracture surface of the PS islands in the unstretched samples.

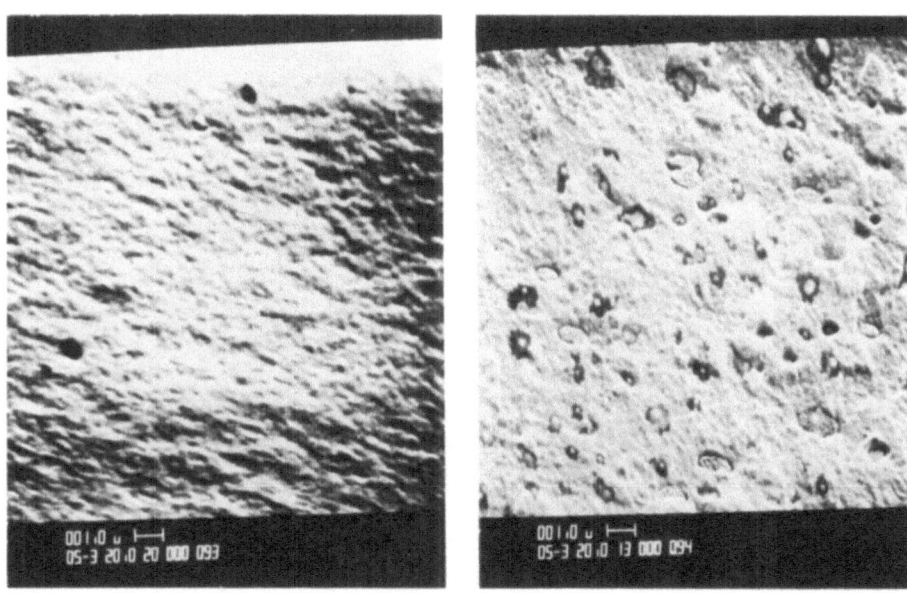

Figure 10.   Scanning electron micrographs of fracture surface of
              SBS and 10% blend with PS. (The samples were
              unstretched.) The fracture surfaces are parallel to
              the normal to the sample surface. (a) The blends all
              have a layer of PS on the sample surface, but the SBS
              does not have it. (b) The blends have inclusion of
              polystyrene islands in the bulk of the sample, the
              SBS shows no inclusion. The sizes of the polystyrene
              islands increase with increasing homopolystyrene
              content.

Fig. 12 illustrates this point with the 30% blend.

    Fig. 11 does not show any cracks formed in the regions outside
the PS islands. It is possible that the cracks formed in these
regions are too small to be resolved by the SEM. Since these
islands are dispersed in the bulk of the material, the fracture of

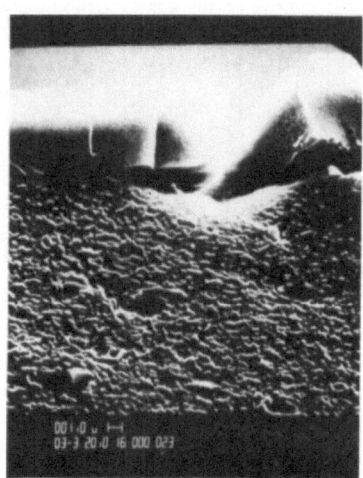 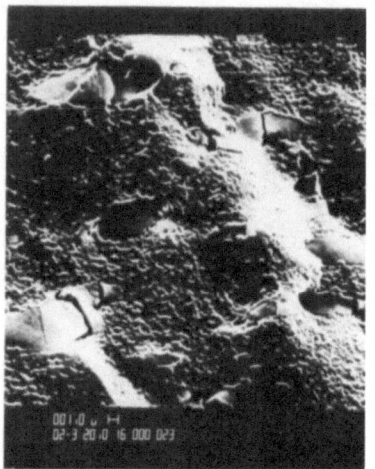

Figure 11.   Scanning electron micrographs of fracture surface of
             the stretched samples of blends of SBS and PS
             and the surface layer of a stretched SBS/30% PS sample.
             Cracks are observed mainly in the larger PS islands.
             (No cracks are seen in the regions outside the PS
             islands.) Fractures occurred in the direction per-
             pendicular to the stretching.

these islands cannot account for the observed plastic-rubber
transformation. The fracture of the surface layer of the blends
may partially account for the yielding of these samples. However,
as shown by the previous transmission electron microscopy studies
(11), these samples have an interconnecting network of the PS as
well as PB domains. Thus the plastic-rubber transformation must
eventually involve the fracture of interconnected polystyrene
domains into the ones which are dispersed in the PB matrix, as
manifested by the results presented previously.

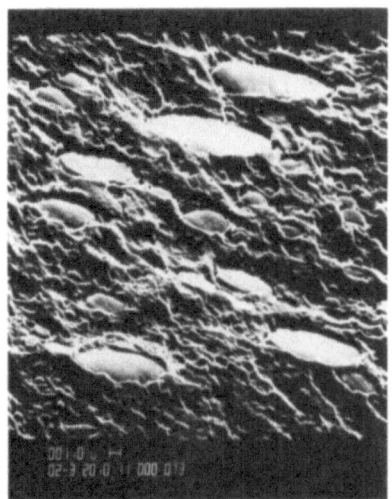 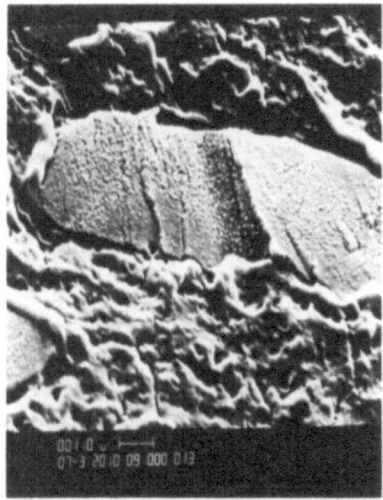

Figure 12. Fracture surface of an unstretched 30% blend indicating
            the absence of cracks in the PS phase.

ACKNOWLEDGEMENT

    This work was supported at the University of California by
the Office of Naval Research, and at the University of Massachusetts by the National Science Foundation, the Army Research Office
(Durham) and the Materials Research Laboratory of the University
of Massachusetts. The assistance of P. Gilmore in SEM experiments
and the helpful discussions with Dr. F. Warner on SAXS analysis
are greatly appreciated.

REFERENCES

1.   L. Bohn, Kolloid-Z. Polym. 213, 55 (1966).

2.   S. Krause, Rev. Macromol. Chem. 8, 251 (1972).

3.   M. Shen and H. Kawai, A.I.Ch.E. J., in press.

4.   J. Mann, R. J. Bird and G. Rooney, Macromol. Chem. 90, 207 (1966).

5.   L. C. Anderson, D. A. Roper and J. K. Rieke, J. Polymer Sci. 43, 423 (1960).

6.   W. W. Barentsen and H. Heikens, Polymer 14, 579 (1973).

7.   W. W. Barentsen, D. Heiken and P. Piet, Polymer 15, 119 (1974).

8.   C. L. Locke and D. R. Paul, J. Appl. Polym. Sci. 17, 2791 (1973).

9.   C. W. Childers, G. Kraus, J. T. Gruver and E. Clark, in G. E. Molau, ed., "Colloidal and Morphological Behavior of Block and Graft Copolymers," Plenum, New York, 1970. p. 193.

10. G. Akovali, J. Diamant and M. Shen, J. Macromol. Sci.-Phys., in press.

11. M. Niinomi, G. Akovali and M. Shen, J. Macromol. Sci.-Phys., in press.

12. L. Toy, M. Niinomi and M. Shen, J. Jacromol. Sci.-Phys. B11, 281 (1975).

13. G. Choi, A. Kaya and M. Shen, Polym. Eng. Sci. 13, 231 (1973).

14. T. Soen, T. Ono, K. Yamashita and H. Kawai, Kolloid-Z.u.Z.f. Polym. 250, 459 (1970).

15. T. Inoue, H. Ishihara, H. Kawai, Y. Oto and K. Kato, Proc. Int'l Conf. Mech. Prop. of Materials, Vol. 3, Soc. Mat. Sci. Japan, 1972, p. 419.

16. G. Wilkes and S. Samuels, in J. Burke and V. Weiss, eds., "Block and Graft Copolymers," Syracuse Univ. Press, Syracuse, N.Y., 1973, p. 225.

17. G. A. Harpell and C. E. Wilkes, in Aggrawal, ed., "Block Copolymer" Plenum, New York, 1969, p. 31.

18. J. F. Beecher, L. Marker, R. D. Bradford and S. L. Aggarwal, J. Polym. Sci. C26, 117 (1969).

19. J. F. Henderson, K. H. Grundy and E. Fisher, J. Polym. Sci. C16, 3121 (1968); E. Fisher and J. F. Henderson, J. Polym. Sci. C26, 149 (1969).

20. G. Akovali, M. Niinomi, J. Diamant and M. Shen, Polymer Preprints 17, 560 (1976).

21. H. Hendus, K. H. Illers and E. Ropte, Kolloid-Z.U.Z.F. Polym. 216, 110 (1967).

22. E. Fisher, J. Macromol. Sci.-Chem. A2, 1285 (1968).

23. L. J. Bellamy, "The Infrared Spectra of Complex Molecules." John Wiley, New York, 1958.

24. C. Y. Liang and S. Krimm, J. Polym. Sci. 27, 241 (1958).

25. M. A. Golub and J. J. Shipman, Spectrochim. Acta 16, 1165 (1960); 20, 701 (1964).

26. J. L. Binder, Anal. Chem. 26, 1877 (1954).

27. R. D. B. Fraser, J. Chem. Phys. 21, 1511 (1953); ibid 24, 89 (1956).

28. P. J. Flory and Y. Abe, Macromolecules 2, 335 (1969).

29. V. Petraccone, I. C. Sanchez and R. S. Stein, J. Polym. Sci. Polym. Phys. Ed., in press.

30. B. E. Read and R. S. Stein, Macromolecules 1, 116 (1968).

31. G. L. Wilkes and R. S. Stein, J. Polym. Sci. A-2, 7, 1525 (1969).

32. M. J. Folkes and A. Keller, Polymer 12, 222 (1971).

33. T. Hashimoto, M. Fujimura, H. Kawai, J. Diamant and M. Shen, to be published.

34. P. R. Lewis and C. Price, Polymer 13, 20 (1972).

35. S. Krause in G. E. Molau. Op. cit. p. 223; in J. J. Burke and V. Weiss, eds., "Block and Graft Copolymers," Syracuse Univ. Press, Syracuse, N.Y., 1973, p. 157.

36. R. S. Stein and P. R. Wilson, J. Appl. Phys. 33, 1914 (1962).

37. R. S. Stein, P. F. Erhardt, S. B. Clough and G. Adams, J. Appl. Phys. 37, 3980 (1966).

BARRIER AND SURFACE PROPERTIES OF POLYURETHANE-EPOXY

INTERPENETRATING POLYMER NETWORKS

H.L. Frisch, J. Cifaratti, R. Palma, R. Schwartz,
and R. Foreman
Department of Chemistry, State University of New York
at Albany
H. Yoon, D. Klempner, and K.C. Frisch
Polymer Institute, University of Detroit

INTRODUCTION

The chemical and physical combination of two or more structurally dissimilar polymers has been of commercial and academic interest for a number of years, since it provides a convenient route for the modification of properties to meet specific needs. It has been used commercially to impart processing, flexibility, tensile and impact strength, chemical resistance, weatherability, flammability resistance, and a variety of other properties (1-4). The physical properties of the combined polymers not only depend on the properties of the constituent polymers but also on the way they are combined.

Interpenetrating polymer networks (IPN's) are a novel type of polymer blend composed of crosslinked polymers. They are more or less intimate mixtures of two or more distinct crosslinked polymer networks with no covalent bonds between the polymers, i.e. polymer A crosslinks only with other molecules of polymer A, as does polymer B. Thus, IPN's may be described as combinations of chemically dissimilar polymers in which the chains of one are completely entangled with those of the others. The entanglements must be of a permanent nature and are made so by this homocrosslinking of the two polymers. IPN's can be of various chemical types and can be synthesized in a variety of ways. There are two basic techniques for producing IPN's. In the first, a sequential technique, a crosslinked polymer (I) is swollen with a second monomer (II), plus crosslinking agents. This is followed by the polymerization and crosslinking of polymer (II) in situ. The second technique consists of combining the linear polymers, pre-

polymers, or monomers of the two polymer types, together with
their respective crosslinking agents, in some liquid form, e.g.
bulk (melt), solution, or dispersion. This is followed by the
simultaneous polymerization and crosslinking of the two polymers.
Care must of course be taken in the selection of the polymers to
prevent reaction occurring between them. It is also preferable
that the polymers be of different chemical types so that the
resulting material will be more than just a copolymer. Interest
centers in these materials for a variety of reasons. Some of the
more important ones are discussed below:

1. IPN's represent a mode of blending two or more polymers
to produce a mixture in which phase separation is not as extensive
as would be otherwise. In fact, it is the only way of combining
crosslinked polymers. Normal blending or mixing of polymers
results in a multi-phase morphology due to the well known thermodynamic incompatibility of polymers. This incompatibility is due
to the relatively small gain in entropy upon mixing the polymers
due to contiguity restrictions imposed by their large chain length
(5). However, if mixing is accomplished on a lower molecular
weight level, and then polymerization accomplished simultaneously
with crosslinking, phase separation may be kinetically controlled
somewhat since the entanglements will have been made permanent by
the crosslinking. In other words, phase separation cannot occur
without breaking covalent bonds. IPN's synthesized to date exhibit
varying degrees of phase separation, dependent principally on the
compatibility of the polymers. With highly incompatible polymers,
the thermodynamics of phase separation is so powerful that it
occurs substantially before the kinetic ramifications (i.e. cross-
linking) can prevent it. In these cases, only small gains in
phase mixing occur. In cases where the polymers are more compatible, phase separation can be almost completely circumvented.
Note that complete compatibility (an almost impossible situation)
is not necessary to achieve complete phase mixing, i.e. interpenetration, since the "permanent" entanglements produced by interpenetration prevent phase separation. With intermediate situations
of compatibility, intermediate and complex phase behavior results.
Thus, IPN's with dispersed phase domains ranging from a few microns
(the largest) (6), to a few hundred angstroms (intermediate) (7),
finally to those with no resolvable domain structure (complete
mixing) (8) have been reported.

2. IPN's represent a special example of topological isomerism (9) in macromolecules (10). Some permanent entanglements
between the different crosslinked networks are inevitable in any
sufficiently intimate mixture of the crosslinked networks. These
represent examples of catenation in polymer systems, i.e. different
ways of imbedding these molecules in three dimensional space.
Permanent entanglements are hindering constraints on the motion of

segments and ought to simulate covalently bound chemical crosslinks (11). Simplified theoretical models of such permanent entanglements (12) exhibit a surprisingly large non-linear elastic
restoring force unlike that expected with chemical crosslinks from
ideal rubber elasticity theory.

    3. The combining of varied chemical types of polymeric
networks in different compositions, often resulting in controlled
different morphologies, has produced IPN's with synergistic
behavior. For example, if one polymer is a glass and the other
is elastomeric, one obtains a reinforced rubber, if the elastomer
phase is the continuous, predominant one, or a high impact plastic
if the glass phase is continuous (13). In the case of more complete phase mixing, enhancement in numerous mechanical properties
is due to the increased physical crosslink density due to this
interpenetration. IPN's have been synthesized with intermediate
maxima versus network composition in bulk properties such as
tensile strength (14-21), impact strength (16, 17), and thermal
resistance (16-18, 20).

    In previous papers, we have investigated more or less system-
atically the mechanical, thermal and electrical properties and
morphologies of simultaneous interpenetrating polymer networks
(SIN's) in which one network component was a polyurethane (6, 8-
10, 14-21). Various, not incompatible views of the microscopic
structure of such interpenetrating polymer networks (IPN's) have
been advanced. Thus, Sperling (22) has advocated a modified
"fringed micelle" picture of such IPN's while we have suggested an
(interpenetrating) sponge model (23). According to the latter
view, we can think of the sponge material per se as one network
and the voids, or holes in the sponge as the other. Since the
voids can be interconnected, neither "sponge" network nor the
"hole" network need be the dispersed or dispersing phase. These
ideas are easily extended to more than two component IPN's. The
apparent domain size of either network depends on the thickness
distribution of the sponge strands or the void channels. Many
limiting cases are possible in such a picture of the IPN. For
example, if the sponge strands and the void channels have a mean
diameter of the order of or smaller than the chain coil diameter
of both networks, we have an optimal mutual dispersion on a
molecular scale of the IPN. On the other hand, if either the
sponge or void network is dispersed with strands whose mean
diameter is of the order of tens of microns we have a relatively
incompatible IPN. Ultimately, one of the networks could be wholly
dispersed in the other, the "holes" in the "sponge" no longer
interconnecting. As the network composition is changed, one can
shift through a large variety of different morphologies each of
which is compatible with this "sponge" model.

Different underlying morphologies are "sensed" differently by various properties of these interpenetrating polymer networks. Thus, mechanical measurements could be more responsive to one aspect of the IPN morphology than thermal measurements, etc. It is, therefore, of interest to bring to bear another class of molecular probes by which an IPN can be studied. The behavior of permeating small molecule penetrants can be as sensitive and useful a probe as thermal or mechanical measurements, providing the penetrant permeabilities of the pure component networks are sufficiently different. This is the case for water vapor permeation in polyurethane (PU) -epoxy (EP) IPN's since one expects the permeability of the PU network to be an order of magnitude larger than the EP network. The variation of water vapor permeability with network composition provides another characteristic diagnostic of molecular and/or phase inhomogeneity in these materials (24). Since some of these materials could potentially be used as coatings, we are further interested in all the barrier properties (permeability, diffusion and sorption) of these substances. We have, therefore, measured the permeability, P, diffusion, D, and sorption, S coefficients of water vapor in our IPN samples.

We have also measured the advancing contact angles of water-methanol mixtures on our IPN's. Wetting properties of surfaces of polymers reflect the composition and structure of surface layers possibly as thin as a few tens of angstroms (23). It is, therefore, of interest to know whether some of the extrema exhibited by some of the bulk properties are also seen in a surface property such as the critical surface tension, γ_c, extrapolated from the advancing contact angle data (25). These contact angles need not be the same as the receding contact angles. We hope shortly to obtain reliable measurements of the contact angle hysteresis, if it is present, for a number of different velocities of the three phase boundary. From the point of view of many practical applications, one is even more interested in the instantaneous, advancing dynamical contact angle (23).

Since these materials may be employed to form adhesive coating formulations, we have also measured the lap shear strength on aluminum plates and peel strengths (180°) of copper and polyethylene sheets on aluminum panels. For comparison, the tensile strengths and elongations at break were measured. Some glass transition temperatures were determined by differential scanning calorimetry.

EXPERIMENTAL

The materials used and their descriptions are listed in Table I. The polyols, [poly(caprolactone) glycols, Niax D-510 and D-560, poly(1,4-oxybutylene) glycol, polymeg 1000], 1,4-butanediol (BD) and trimethylolpropane (TMP) were dried at 80°C for five hours under a vacuum of 1 mm Hg. All other materials were used without further purification.

TABLE I.   MATERIALS

Designation	Description	Source
Hylene W (HW)	4,4'-Methylene bis(cyclohexylisocyanate) Eq. wt. = 131.5	E.I. DuPont de Nemours & Co., Inc.
Polymeg 1000	Poly(1,4-oxybutylene) glycol M.W. = 1000	Quaker Oats, Inc.
Niax D-510	Poly(caprolactone) glycol M.W. = 530 Hydroxyl No. = 212	Union Carbide Corp.
Niax D-560	Poly(caprolactone) glycol M.W. = 2000 Hydroxyl No. = 56.1	Union Carbide Corp.
DER 330	Epoxy resin composed of bisphenol A and epichlohydrin	Dow Chemical Co.
DER 332	Epoxy resin composed of bisphenol A and epichlohydrin; pure adduct	Dow Chemical Co.
DMP 30	2,4,6,-Tris(dimethyl aminomethyl)phenol	Rohm & Haas Co.
TMP	Trimethylol propane	Celanese Chem. Co.
BD	1,4-Butanediol	GAF Corporation
T-12	Dibutyltin dilaurate	M&T Chem., Inc.

The IPN's consisted of polyurethanes as the rubbery components and epoxies as the glassy components.

1. Polyurethanes

The polyurethanes were prepared using the prepolymer technique. The isocyanate-terminated prepolymer was prepared by reacting two equivalents of 4,4'-methylene bis(cyclohexyliso-cyanate) (Hylene W-HW) with one equivalent of polyol at 80°C in a resin kettle under dry nitrogen. The reaction was continued until the theoretical isocyanate content (as determined by the di-n-butylamine method) (26) was reached. One equivalent of the prepolymer was homogeneously mixed with one equivalent of a BD-TMP mixture (4:1 equivalent ratio) for 20-30 minutes and cured at 100°C overnight using 0.02% dibutyltin dilaurate (T-12) as catalyst.

Three polyurethanes were prepared:

PU-I: Prepolymer from poly(1,4-oxybutylene)glycol, MW = 1000 and 4,4'-methylene bis(cyclohexylisocyanate)(Hylene W-HW).

PU-II: Prepolymer from poly(caprolactone)glycol, MW = 530 (Niax D-510) and HW.

PU-III: Prepolymer from poly(caprolactone)glycol, MW = 2000 (Niax D-560) and HW.

Two epoxy resins were employed, DER-332 and DER-330. They are both bisphenol A-epichlorohydrin adducts and differ in that the former is the pure adduct and the latter contains some higher molecular weight material. They were cured with 2,4,6-tris (dimethylaminomethyl) phenol (DMP-30).

IPN's were prepared by mixing the urethane prepolymer, together with chain extender, crosslinking agent, and catalyst, in various proportions with the epoxy resins, containing catalyst, and curing at 100°C overnight. Thus, three IPN's were formed:

IPN-1: PU-I + DER-332

IPN-II: PU-II + DER-330

IPN-III: PU-III + DER 330

In addition, in order to further deduce the effects of interpenetration on the adhesive properties, pseudo-IPN's (PDIPN) in which the polyurethane is linear and the epoxy crosslinked, were produced in various ratios by omitting the trimethylolpropane in the above syntheses.

MEASUREMENTS

1. Lap-shear
        The lap-shear tensile strengths of the IPN's, PDIPN's,
and component polymers were determined on an Instron Tensile
Tester by ASTM D-3163-73 using aluminum plates 3" x 0.5" lapped
0.5" from their edges (see Figure 1). The uncured polymer and
IPN's were used as the adhesive in this test and cured as above.

LAP-SHEAR TENSILE TEST (ASTM D-3163-73)

TEST SPECIMEN

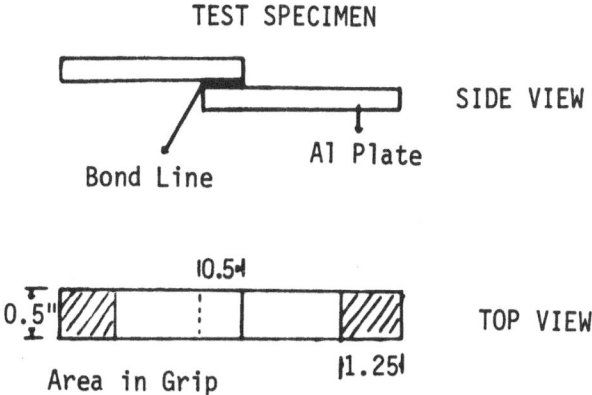

TEST APPARATUS

Instron Model   1130
Load Cell   1000 lb.
Head Speed   10"/min
Chart Speed   5"/min

Figure 1. Lap-shear testing

2. Peel Strength
        The peel-strengths (180°) were determined on an Instron
Tensile Tester according to ASTM D-903-49 using polyethylene and
copper sheets on aluminum panels (see Figure 2).

PEEL STRENGTH TEST (ASTM D-903-49)

TEST SPECIMEN

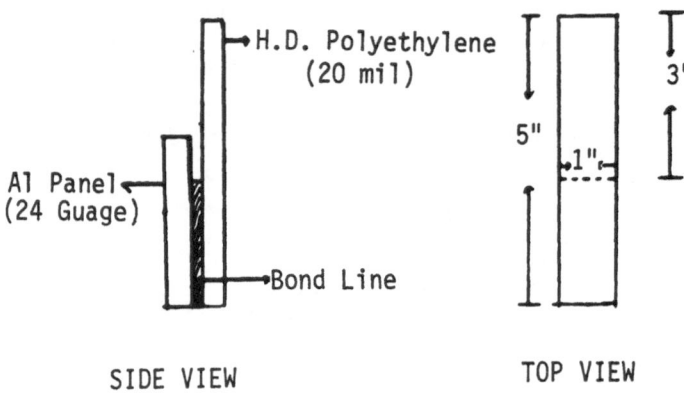

SIDE VIEW   TOP VIEW

TEST APPARATUS

Instron- Model 1130
Load Cell- 10 lb.
Head Speed- 5"/min.
Chart Speed- 5"/min.
Strip Angle- 180

Figure 2. Peel Strength Testing

For the following measurements, specimens were obtained from
films of IPN's cast on polypropylene sheets.

3. Thermal Studies
The glass transitions were determined on a Perkin-Elmer
Differential Scanning Calorimeter, DSC-2. Measurements were
carried out from -120°C to +150°C under helium at a scanning rate
of 10°C per minute. Specimen sizes were on the order of 20 mg.

4. Stress-strain Properties
The tensile strengths were determined on an Instron Tensile
Tester at room temperature at a crosshead speed of 2 in./min. using
dumbbell shaped specimens.

5. Critical Surface Tension
The critical surface tensions of the IPN films were determined by measuring the advancing contact angles, θ, of a series of water-methanol mixtures with a reflected light contact angle goniometer and extrapolating to zero contact angle (where $\cos \theta = 1$).

6. Permeability, Diffusion and Sorption
The water vapor permeability of the IPN-I films was measured by modified cup technique (24,27). A glass cell equipped with a filling arm and a circular orifice of known diameter to which the polymer film to be studied could be affixed with a thick layer of epoxy was suspended on an automatic balance in a dry box over P_2O_5 (anhyd). Once the cell was filled from a syringe it was not further moved. The weight loss was recorded as a function of time. The film thicknesses used were a few mils. The dry chamber and its contents were maintained at 22°C $\pm$ 0.75°C. The equilibrium vapor pressure of water at that temperature is 2.007 cm Hg. From the steady state flow rate the permeability coefficient, P, was determined. The average of at least three measurements was recorded. Permeation time lags, L, were readily obtained for the films up to 40% (by weight) of the EP network. The initial data on the other compositions exhibited too much scatter to reliably deduce L. Diffusion coefficients were calculated from the time lags via $D = \ell^2/6L$ where ℓ is the average thickness of the polymer film. Sorption coefficients, S, were obtained, in turn, by using the relation $S = P/D$.

RESULTS AND DISCUSSION

1. Stress-Strain Results
The variation of the tensile strengths and breaking elongations of IPN-I with composition resembles the behavior of other polyurethane SIN type IPN's which we have previously studied. Again, the tensile stress of IPN-I exhibits a miximum (75% by weight of EP) and minimum (25% EP) at intermediate compositions. More astonishingly as can be seen from Figure 3, this also appears to be the case with the PDIPN-I made without the trimethylolpropane. The elongations at break of IPN-I versus network composition shown in Figure 4 decrease monotonically as the percentage of the harder EP phase increases.

2. Lap Shear Strength
The lap shear strength of IPN-I and PDIPN-I reflect the tensile strength results. As seen in Figure 5, the maximum in the lap shear strength occurs at the same composition as the maximum in tensile strength. In most cases, the adhesive joint failure was cohesive. Thus, from elasticity theory we expect, all other things being equal, a roughly linear relationship between the

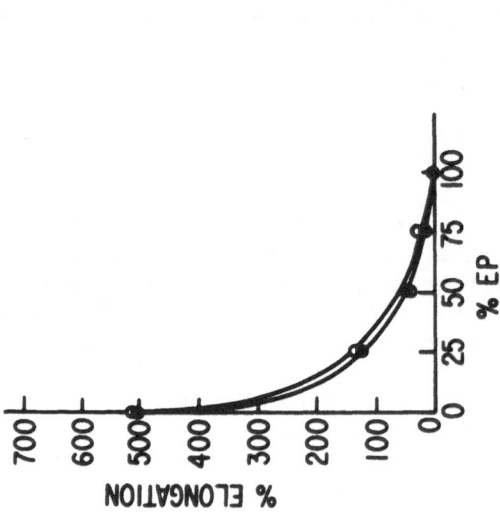

Figure 4. Elongation to break(%) versus network composition. IPN-I(open circles); PDIPN-I (filled circles).

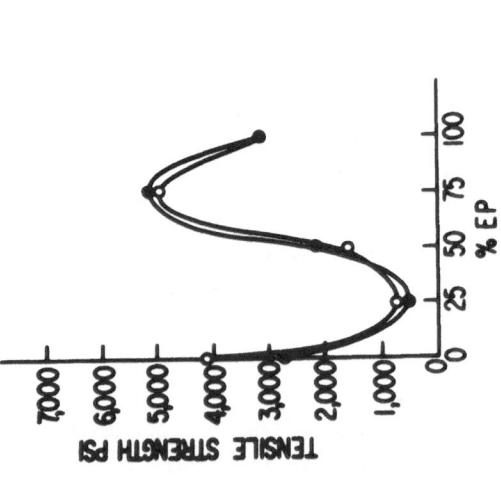

Figure 3. Tensile Stress(psi)to break versus network composition. IPN-I(open circles); PDIPN-I(filled circles).

logarithm of the lap shear strength and the logarithm of the tensile
strength. This would account for the correlation seen between
Figures 3 and 5. In the same figures, we have also plotted the
lap shear strength versus network composition for the caprolactone
type PU IPN's, which exhibit a similar maximum at 50% EP.

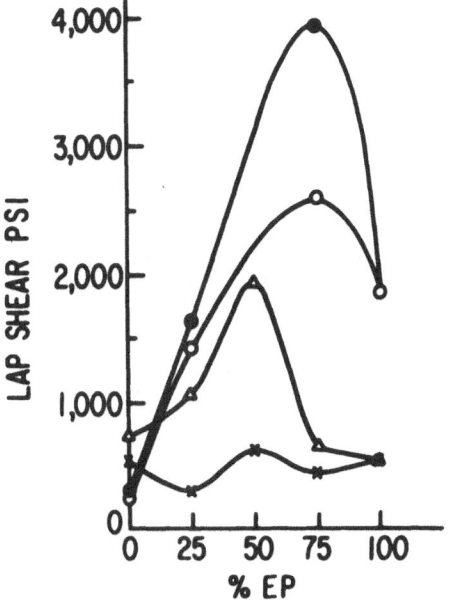

Figure 5. Lap shear (psi) versus network composition.
IPN-I (open circles); IPN-II (triangles); PDIPN-I (filled
circles); IPN-III (crosses).

3. Peel Strength

In contrast, the peel strength versus network composition of IPN-I and PDIPN-I, which we have plotted in Figure 6, decreases with increasing EP network content.

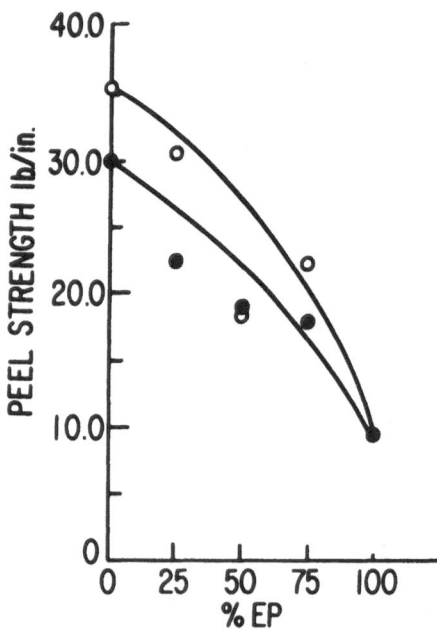

Figure 6. Peel Strength(lb/in) versus network composition IPN-I(open circles); PDIPN-I(filled circles).

4. Glass Transition Behavior

At this point, Tg's have only been measured for a few of the IPN's, so the data are not shown here. However, in the cases measured, one fairly sharp Tg, intermediate in temperature to the Tg's of the components was found indicating a somewhat single phase morphology.

5. Barrier Properties

The permeability of a two phase mixture, $\underline{P}$, of volume fractions ϕ_1, $\phi_2 = 1-\phi_1$, can be strictly bounded in terms of the permeabilities P_1, P_2 of the two phases. Thus $\underline{P}$ must be bounded from above by the permeability of a hypothetical series combination and from below by the permeability of a hypothetical parallel combination of the two phases:

$$\frac{1}{\dfrac{\phi_1}{P_1} + \dfrac{\phi_2}{P_2}} \quad \leq \quad \underline{P} \quad \leq \quad \phi_1 P_1 + \phi_2 P_2.$$

We see immediately from Figure 7 that the water vapor permeability of IPN-I appears to satisfy these inequalities, and is initially convex, passes through an inflection point near 30% EP, and remains concave until 100% EP is reached. This behavior is quite consistent with our "sponge" model (23) or Sperling's "fringed micelle" model (22). It would appear that for small EP% (less than 20% by weight) the epoxide phase may well extend throughout the sample (the "series" permeability model is applicable) rather than be suspended as some convexly dispersed set of domains in the PU network. There is a twenty-fold decrease in P in going from the pure PU network to the pure EP network. The inflection point in the permeability curve is roughly in the vicinity of the minimum in the tensile strength curve (Figure 1).

In the vicinity of the inflection point in Figure 7, D decreases by a factor of 7 while the sorption coefficient, S, increases by a factor of about 2.5.

   6. Critical Surface Tension
        The critical surface tensions obtained from advancing contact angle data at 25°C, of water-methanol mixtures on the IPN-I, IPN-II, and IPN-III samples are shown as functions of network composition in Figure 8. All three IPN's exhibit an absolute minimum at the same composition where the ultimate strength properties (lap shear strength, tensile strength) exhibit a maximum. This result is even more astonishing because of the magnitude of the effect. Amounting as it does to about 4 dynes/cm for two of the IPN's. The minimum in the critical surface tension can also be seen directly in the contact angle data obtained with some of the water-methanol mixtures; e.g. if θ is the advancing contact angle, then for IPN-I and 60% methanol mixture we found for 100% PU θ = 50°, 25% EP θ = 60°, 50% EP θ = 55°, 75% EP θ = 45° and 100% EP θ = 53°. To confirm these results we plan to find the advancing contact angles obtained with another suitable, immiscible liquid series which can be studied at temperatures somewhat higher and lower than room temperature. The contact angles would allow us to extract information about the entropic contribution to the free energy of spreading.

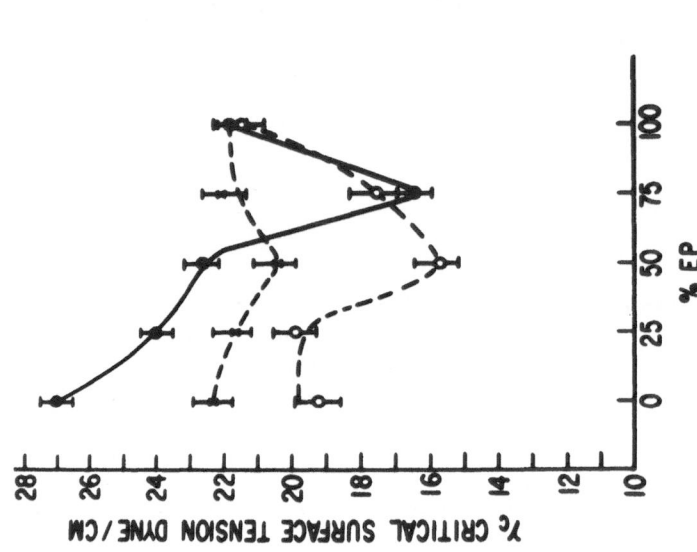
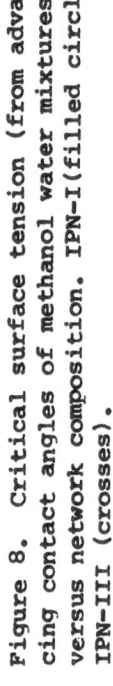

Figure 8. Critical surface tension (from advancing contact angles of methanol water mixtures) versus network composition. IPN-I(filled circles) IPN-III (crosses).

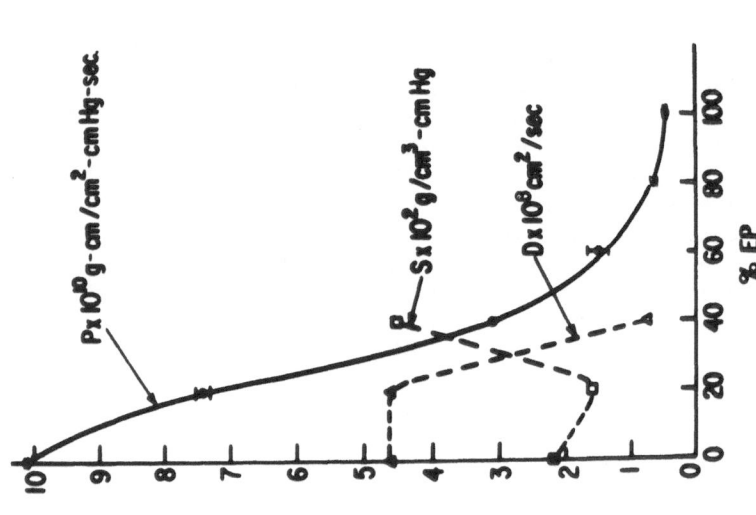

Figure 7. Water vapor permeability, diffusion and sorption coefficients versus network composition for IPN-I.

Acknowledgements

We wish to acknowledge support from National Science Foundation Grants DMR 730 2599 and ENG 740 1954 as well as a grant from the American Can Foundation.

References

1. B. D. Gesner, "Encyclopedia of Polymer Science and Technology", 10, Wiley Interscience, New York, 708 (1969).

2. M. S. Thompson, "Gum Plastics", Reinhold, New York, (1958).

3. H. Keskkula, "Polymer Modification of Rubbers and Plastics", Wiley Interscience, New York, (1967).

4. P. F. Bruins, "Polyblends and Composites", Wiley Interscience, New York (1970).

5. P. J. Flory, "Principles of Polymer Chemistry", Cornell University Press, Ithaca, New York (1953).

6. M. Matsuo, T. K. Kwei, D. Klempner, and H. L. Frisch, Polymer Eng. & Sci., 10, (6), 327 (1970).

7. L. H. Sperling, D. A. Thomas, and V. Huelck, Macromolecules, 5, 340 (1972).

8. K. C. Frisch, D. Klempner, S. Migdal, H. L. Frisch and H. Ghiradella, Polymer Eng. & Sci., 14, (1), 76 (1974).

9. H. L. Frisch and E. Wasserman, J. Amer. Chem. Soc., 83, 3789 (1961).

10. H. L. Frisch and D. Klempner, "Advances in Macromol. Chem.", 2, 149 (1970).

11. P. J. Flory, Chem. Rev., 35, 51 (1944).

12. S. Prager and H. L. Frisch, J. Chem. Phys., 46, 1475 (1967).

13. A. J. Curtius, M. J. Covitch, D. A. Thomas, and L. H. Sperling, Polymer Eng. & Sci., 12, 101 (1972).

14. D. Klempner, H. L. Frisch, and K. C. Frisch, J. Polymer Sci., (A-2), 8, 921 (1970).

15. D. Klempner, H. L. Frisch, and K. C. Frisch, J. Elastoplastics, 3, 2 (1971).

16.   K. C. Frisch, D. Klempner, S. Migdal, and H. L. Frisch, J.
      Polymer Sci. (A-1), 12 (4), 385 (1974).

17.   K. C. Frisch, D. Klempner, S. Migdal, and H. L. Frisch, J.
      Applied Polymer Sci., 19, 1893 (1975).

18.   K. C. Frisch, D. Klempner, S. K. Mukherjee and H. L. Frisch,
      J. Applied Polymer Sci., 18, 689 (1974).

19.   K. C. Frisch, D. Klempner, T. Antczak, and H. L. Frisch,
      J. Applied Polymer Sci., 18, 683 (1974).

20.   S. C. Kim, D. Klempner, K. C. Frisch, H. L. Frisch, and
      H. Ghiradella, Polymer Eng. & Sci., 15, (5), 339 (1975).

21.   S. C. Kim, D. Klempner, K. C. Frisch, and H. L. Frisch,
      J. Applied Polymer Sci. (in press).

22.   L. H. Sperling, "Encyclopedia of Polymer Science and
      Technology, Supplement No. 1," John Wiley & Sons (1976).

23.   H. L. Frisch, G. L. Gaines, Jr., and H. Schonhorn, "Polymer
      Surfaces" in Treatise on Solid State Chemistry," edited by
      N. B. Hannay, Vol VI B, Plenum Press, New York, (1976).

24.   J. Crank and G. S. Park, (editors), "Diffusion in Polymers",
      Academic Press, London, (1968).

25.   W. A. Zisman, "Contact Angle, Wettability, and Adhesion,"
      Advan. Chem. Series No. 43, 1 (1964).

26.   E. J. Malec and D. J. David, "Analytical Chemistry of
      Polyurethanes," D. J. David and H. B. Staley, eds, 87,
      John Wiley, New York, (1969).

27.   A. C. Newns, Shirley Institute Memoirs, No. 24, J. Text Inst.,
      41 T, 269 (1950).

NOVEL PLASTICS AND ELASTOMERS FROM CASTOR OIL BASED IPN'S:

A REVIEW OF AN INTERNATIONAL PROGRAM

L.H. Sperling, J.A. Manson, G.M. Yenwo, N. Devia-Manjarres,
J. Pulido, and A. Conde*

Lehigh University, Bethlehem, Pennsylvania
*Universidad Industrial de Santander
Bucaramanga, Colombia

INTRODUCTION

Among the renewable resources available in the world, plant products rank very high. Examples include cotton, which yields clothing; wood, for construction; and natural rubber, for automotive tires, etc. Many plants yield valuable oils, such as corn oil, linseed oil, and cotton seed oil (1). Besides food uses, these oils provide the basis for paints, adhesives and other industrial uses. The presence of multiple **unsaturated sites allows** for ready polymerization (2). Castor oil, which comes from the castor bean plant, is nearly unique among vegetable oils in containing hydroxyl groups in addition to points of unsaturation. Thus, there are two ways of polymerizing castor oil: through the use of sulfur or oxygen, which attacks the double bonds, or through the hydroxyl groups, to form polyurethanes, or polyesters, etc (3-5). As shown in structure (1), the number of double bonds and hydroxyl groups are identical, at three each per oil molecule (90% pure).

$$
\begin{array}{l}
\overset{\displaystyle O}{\overset{\|}{H_2C-O-C}}-(CH_2)_7-CH=CH-CH_2-\overset{\displaystyle OH}{\overset{|}{CH}}-(CH_2)_5-CH_3 \\
\overset{\displaystyle O}{\overset{\|}{HC-O-C}}-(CH_2)_7-CH=CH-CH_2-\overset{\displaystyle OH}{\overset{|}{CH}}-(CH_2)_5-CH_3 \qquad (1)\\
\overset{\displaystyle O}{\overset{\|}{H_2C-O-C}}-(CH_2)_7-CH=CH-CH_2-\overset{\displaystyle OH}{\overset{|}{CH}}-(CH_2)_5-CH_3
\end{array}
$$

The objective of this paper is to describe the present status of the Colombia-U.S. international research and engineering program. This program, now two and a half years old, has as its objective the synthesis and characterization of toughened polymer materials based on castor oil.

Three prototype engineering materials were investigated (6-9):

(1) The castor oil-urethane elastomer homopolymers,
(2) Impact resistant plastics, where the castor oil elastomers provide the rubber phase, and
(3) Toughened elastomers, composed of 50-70% castor oil-urethane (the remainder being polystyrene), showed high elongation and significant toughness, similar to the thermoplastic elastomers.

EXPERIMENTAL

A.   Synthesis of IPN's Based on Polymerized Castor Oil

1. Polymerization of castor oil using sulfur. (a) By cross-linking castor oil with sulfur, a soft elastomer can be synthesized. The following procedure was used. 100 ml of No. 40 oil (Baker Castor Oil Co.) was heated to 190°C. Two grams of sulfur were added and the mixture stirred vigorously with a magnetic stirrer for 30 minutes. During this period, the sulfur reacted with the oil and the color of the oil gradually changed from deep orange to dark. The viscosity of the mixture increased rapidly during the last 10 minutes of the reaction. The partly formed vulcanizate was then quickly poured between glass plates (sprayed with Price-Driscoll epoxy parfilm mold release agent) heated to 140°C. It was then cured for 10 hours at 140°C. Heating was controlled carefully during the first part of the reaction period so that the temperature of the castor oil-sulfur mixture did not rise above 200°C, at which temperature the oil would degrade. When the castor oil-sulfur elastomer was cured at a temperature higher than 140°C during the second stage of the reaction, bubbling occurred as the oil gelled and this resulted in a porous sample.

(b) IPN Synthesis (10-12). Methyl methacrylate was used as a second monomer in the IPN synthesis, with 0.5% azobisisobutyronitrile as initiator and 0.5% tetraethylene glycol dimethacrylate, TEGDM, as crosslinker. Polymerization of the MMA was carried out at 45°C for 3 days.

2. Synthesis of castor oil - urethane/polystyrene IPN's. (a) Castor oil - urethane synthesis (13,14). Castor oil - urethane elastomers are prepared by reacting 2,4 tolylene diisocyanate (TDI),

80/20:2,4/2,6 TDI, or hexamethylene diisocyanate (HDI) with castor oil. The last reaction was rather slow and thus dibutyltin dilaurate,0.001 gm per gm of HDI,was used as a catalyst. Since TDI hydrolyses significantly in the presence of trace amounts of water, DB grade castor oil from the Baker Castor Oil Company was employed.

The reaction between TDI and castor oil is exothermic and bubbles are produced in the reaction mixture (castor oil contains a few tenths of a percent volatile material that will evaporate as the temperature of the reaction mixture goes up. Some of the bubbles produced are trapped in the mixture as the viscosity increases. Stirring with a teflon coated magnetic spin bar also produces some bubbles). In order to produce elastomers that are bubble free, the reaction is carried out in two stages.

Stage I. A known weight of DB oil is mixed with excess TDI (the excess here refers to the ratio of NCO groups to OH groups being larger than 1.0) at room temperature to produce an isocyanate terminated prepolymer. The mixture is stirred vigorously for at least one hour. The bubbles present are removed by applying a vacuum to the prepolymer for about 15 minutes. This results in a clear bubble-free highly viscous liquid.

Stage II. In this stage the prepolymer is crosslinked with excess castor oil. The degassed prepolymer is mixed with enough DB castor oil to give a final predetermined NCO/OH ratio. The mixture is stirred vigorously for 20 minutes. It is then degassed again as in Stage I and poured into a mold and heated for two hours at 130°C to complete the reaction. The mold is allowed to cool and the product separated. The resulting elastomer is clear and tough, the modulus depending on the NCO/OH ratio. Tables 1-3 present a series of IPN's prepared in this way.

Table 1.   CO/PS* IPN's (Crosslinker 2,4 TDI)

Sample	NCO/OH Ratio	Composition, CO/PS
1	0.95	32/68
2	0.95	40/60
3	0.95	53/47
4	0.85	36/64
5	0.85	40/60
6	0.85	50/50
7	0.75	29/71
8	0.75	36/64
9	0.75	50/50

* Castor oil-urethane/polystyrene.

Table 2.   CO/PS IPN's (Crosslinker 80/20 : 2,4/2,6 TDI)

Sample	NCO/OH Ratio	Composition, CO/PS
1	0.95	38/62
2	0.95	40/60
3	0.95	46/54
4	0.85	37/63
5	0.85	38/62
6	0.85	46/54
7	0.75	31/69
8	0.75	34/66
9	0.75	51/49

Table 3.   CO/PS IPN's (Crosslinker: HDI)

Sample	NCO/OH Ratio	Composition, CO/PS
1	0.95	36/64
2	0.95	38/62
3	0.95	62/38
4	0.85	33/67
5	0.85	35/65
6	0.85	54/46
7	0.75	29/71
8	0.75	31/69
9	0.75	55/45

(b) IPN synthesis.   The urethane elastomer was swelled with styrene containing 0.4% benzoin as initiator and 1% divinyl benzene as crosslinker.   Polymerization of the styrene was carried out by ultraviolet radiation for 24 hours.

3. Synthesis in latex form.   Sodium ricinoleate, the hydrolysis product of castor oil, was used as the soap.   Due to its high water solubility, 15% was used on a monomer basis.

The sodium ricinoleate was synthesized by saponification of castor oil at 80°C, and purified by steam distilling out the glycerine.   The critical micelle concentration (CMC) of aqueous solutions of sodium ricinoleate was found to range between 1.9 and 2.6 grs/liter at 25°C, dependent on the exact degree of hydrolysis and purification.   A standard surface tension method was employed.

The styrene was polymerized at 70°C by emulsion methods using sodium ricinoleate and sodium stearate as unsaturated and saturated emulsifiers, respectively.   This yields a direct comparison: sodium ricinoleate served as first a surface active agent and subsequently as a monomer; sodium stearate served only as a surfactant.   The

recipes of the polymerization mixtures and further steps are shown
in Tables 4 and 5. Two synthetic paths were followed:

Path 1: Latex 1 was based on sodium ricinoleate as the surface
active agent. Afterwards sodium ricinoleate was added to a measured
quantity of latex No. 1 until its surface tension remained unchanged
with further soap additions, as is described in the soap titration
method.

At this point the surface of the particles was assumed to be
completely covered with sodium ricinoleate. Sulfur was then dispersed into the latex by stirring. The mixture was placed into a
stirred autoclave and the vulcanization reaction carried out for 3
hours at 150°C. High levels of coagulation were observed in all
cases. The solids were ground, mixed with their original polystyrene latex in different proportions, coagulated by sodium
chloride addition, washed with distilled water, and dried in an
oven at 90°C for 72 hours before molding. The recipe used for
vulcanization step is also shown in Tables 4 and 5.

Path 2: This latex is based on sodium stearate as the emulsifier. Castor oil and sulfur were added to latex No. 2 to give an
overall composition of 20% of vulcanized oil in the latex. The
mixture was then stirred vigorously and placed into an autoclave
at 150°C for 3 hours. Phase separation was observed and highly
viscous drops of partly vulcanized castor oil separated from the
latex. The vulcanized latex was mixed with the original polystyrene latex at two different levels to give an overall seven and
20% of vulcanized oil. Since some castor oil products were lost
due to the phase separation, the values reported in Table 1 are the
maximum values. The latexes were coagulated with sodium chloride,
washed, and dried. The overall scheme for preparing these materials
is shown in Figure 1.

4. Simultaneous interpenetrating network synthesis. Following
the polymerization of castor oil via the sulfur and urethane routes,
it was decided to prepare castor oil polyesters and simultaneously
synthesize the polystyrene network. When both polymers are synthesized simultaneously to form networks, the products are called
simultaneous interpenetrating networks, or SIN's.

Sebacic acid (itself a castor oil derivative) or sebacyl
chloride was used to crosslink the castor oil. A two step procedure
was employed. First, two prepolymers of castor oil and sebacyl
chloride with different COCl/OH groups ratio (1.65 and 0.6 approx.)
were synthesized the hydrogen chloride gas being removed during synthesis. Then the clear, viscous, bubble-free liquids were mixed
with fresh distilled styrene monomer containing 0.4% of benzoyl
peroxide and 1% of DVB as a crosslinker, poured into a mold and
polymerized at 80°C for 72 hrs.

Table 4. Sample Preparation Techniques for Latex Materials

Path	Styrene Polymerization	Vulcanization — Mixture Composition
1	Emulsifier: Sodium Ricinoleate Emulsifier Concentration: 15.95 grs/lt Volumetric Ratio Water/Styrene: 5.0 Sodium Persulfate (%): 0.27 Yield (%): 96.5° Time of Polymerization (min): 210 Temperature: 70°C	Polystyrene (%wt): 52.2 Soap (%wt): 36.8 Castor Oil (%wt): 0.0 Sulphur (%wt): 11.0 Sulphur Level (%): 30.0 Coagulated Solids After Vulcanization (%wt): 84.1 Temperature: 150°C Time (min): 140
2	Emulsifier: Sodium Stearate Emulsifier Concentration: 16.60 g/l Volumetric Ratio Water/Styrene: 5.0 Sodium Persulfate (%): 0.27 Yield (%): 93.4 Time of Polymerization (min): 210 Temperature: 70°C	Polystyrene (%wt): 80.0 Soap Concentration (grs/100 grs Monomer): 9.17 Castor Oil (%wt): 16.58 Sulphur (%wt): 3.32 Sulphur Level (%): 20.0 Coagulated Solids After Vulcanization (%wt): 18% Temperature: 150°C Time (min): 140

Table 5. Sample Compositions for Latex Materials

Path	No.	Vulcanized and Ground Coagulate %wt	Polystyrene Latex %wt	wt. of Castor Products Max. % Total wt.
	1	4.2	95.8	2.0
	2	10.5	89.5	5.0
	3	16.8	83.2	8.0
1	4	21.0	79.0	10.0
	5	42.0	58.0	20.0
	6	63.0	37.0	30.0
	7	100.0	0	47.7
2	8	55.5	44.5	7.18
	9	100.0	0	20.0

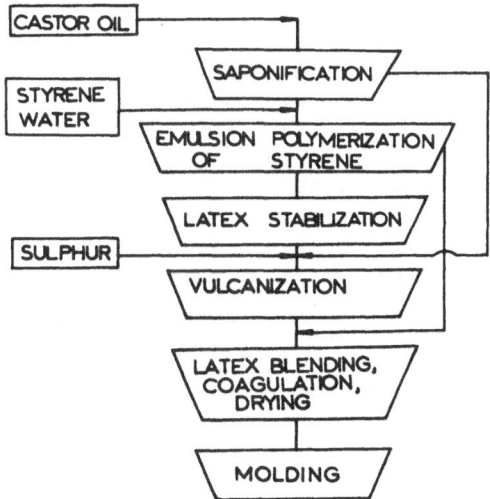

A PROCESS DIAGRAM FOR POLYSTYRENE CASTOR OIL BASED POLYMERS

Figure 1. A flow chart for the preparation of polystyrene/castor
oil based materials, using emulsion polymerization
techniques.

An alternative procedure is the straight mixture of castor oil and sebacyl chloride with the styrene recipe described above, but then the mixture must be kept at room temperature for 2 hours, stirred and degassified before pouring into the mold in order to minimize the bubble formation during the polymerization carried out at 80°C for 72 hrs.

B.   Instrumentation and Measurements

1. Swelling tests.   Several specimens of known dimensions and weight were cut from the elastomers and swollen in several solvents for two weeks at ambient conditions.   The purpose of this experiment was to measure the average molecular weight between crosslinks, and the number of polymer network chains per unit volume.   The data were also used to determine the minimum NCO/OH ratio for the formation of infinite networks.

2. Electron microscopy.   A Philips 300 transmission electron microscope (TEM) was used for all microscopy work.   The technique for preparing samples for electron microscopy has already been described (8).   In brief, specimens about 2 x 2 x 20 mm were cut from the IPN's and one end of each sample was trimmed to a truncated pyramid of sides approximately 0.1mm.   They were exposed to osmium tetroxide vapor for five days and then placed in a vacuum for twenty-four hours.   A Porter-Blum MT-2 ultramicrotome equipped with a diamond knife was used to obtain sections of about 600Å thickness.   They were then observed directly in the electron microscope.

3. Dynamic mechanical spectroscopy (DMS).   All DMS measurements employed a Rheovibron direct reading viscoelastomer, model DDV-11 (Vibron, manufactured by Toyo Measuring Instruments Co., Ltd., Tokyo, Japan).   The temperature range employed was from -100°C to 150°C, with a heating rate of about 1°C per minute.   A frequency of 110 Hz was employed.   As per requirements of the instrument, the sample dimensions were of the order of 10^{-2}cm x 10^{-1}cm x 2cm.   Some modulus-temperature studies utilized a Gehman Torsional Tester.

4. Tensile tests.   The tensile behavior of the materials was studied on an Instron Universal Test Instrument, Model TTDL, at ambient conditions.   The samples were machined to the dimensions of Type IV test specimens and polished to remove all visible flaws. Three specimens of each kind were strained to failure and the average was reported.

5. Impact tests.   Impact resistance measurements were conducted on a Baldwin Impact Tester which is a simple beam (Charpy type) impact machine.   The samples were 6.35cm long by 1.27cm wide.   The

specimens were notched according to ASTM specifications and five samples were tested and the average reported.

6. Stress-strain behavior and crosslink density. Stress-strain measurements were made at room temperature using an Instron tensile tester operated at a strain rate of 0.2 in/min; 4 specimens were used for each test. Results were obtained in terms of Young's modulus, ultimate tensile strength and elongation, and constants of the Mooney-Rivlin equation:

$$\sigma = 2C_1(\lambda-\lambda^{-2}) + 2C_2(1-\lambda^{-3}) \qquad (1)$$

$$\text{or} \qquad \frac{\sigma}{2(\lambda-\lambda^{-2})} = C_1 + C_2\lambda^{-1} \qquad (2)$$

Here σ is the engineering stress, λ the extension ratio L/L_o (the ratio of strained to unstrained gage length). For an ideal elastomer, $C_2=0$, and C_1 varies with crosslink density; $2C_1 \approx \nu kT=G$, where ν is the number of effective network chains per unit volume, k is Boltzmann's constant, T the absolute temperature, and G the shear modulus.

7. Crack propagation behavior. Crack propagation characteristics were determined at room temperature in a static, constant-strain test using rectangular test pieces (2.5cm x 8.9cm) with a 1-mm center notch. The strips were stretched in the Instron Tester at a crosshead speed of 0.05 in/min until the minimum stress for the initiation of crack extension was reached. The crosshead was then stopped, and the load followed as a function of time. The onset of catastrophic tearing was noted from the crack length-time and load-time data.

Tearing energies were calculated from the Rivlin-Thomas equation:

$$-\left(\frac{\partial W}{\partial c}\right)_\ell = Tt \qquad (3)$$

where W is the elastically stored energy, c the crack length, T the tearing energy, t time, and the subscript ℓ implies a fixed initial distance between the grips. Rectangular specimens (2.5cm x 7.6cm) were center-notched to give values of c ranging between 1 and 10 mm, clamped (grips 2.5cm apart), strained at 0.02 in/min, and the point of catastrophic crack propagation determined. Values of W were calculated from areas under the stress-strain curves, and plotted against c; T was then calculated from the slope of the W-c curves at values of c corresponding to the initiation of slow and catastrophic crack growth.

RESULTS

A. Castor Oil-Urethane Elastomers

1. Swelling tests. The solubility of a polymer in a given
solvent depends on the polymer-solvent interaction forces. Solvents
with like solubility parameters are likely to dissolve the same
solutes. This provides an indirect method for determining the
solubility parameter, δ, of a polymer. If the polymer-solvent
interaction forces are stronger than the polymer-polymer forces,
the polymer will dissolve and as the polymer chain expands with
solvent "goodness", the viscosity of the solution will increase.
When the solubility parameters of the polymer and solvent are substantially equal, maximum expansion will occur and the highest
viscosity will be attained. Thus by measuring the viscosities of
the polymer in various solvents, it is possible to determine the
solubility parameter of the polymer. Blanks and Shah have used
this method to determine the solubility of styrene-acrylonitrile
copolymers.

For a crosslinked network, however, the polymer cannot dissolve but individual segments will solvate to give a swollen gel.
The maximum swelling will occur when the solubility parameters of
the solvent and polymer are substantially equal, for the same
reasons as for the linear polymer.

In order to determine X, the polymer-solvent interaction parameter, Bristow and Watson derived the following semi-empirical
equation:

$$X = \beta_1 + \frac{V_1}{RT} [\delta_1 - \delta_2] \qquad (4)$$

where δ_1 and δ_2 relate to the solvent and polymer respectively. β_1
is the lattice constant, usually 0.35 ± 0.1.

Figure 2 shows the swelling curve of an elastomer prepared from
2,4-tolylene diisocyanate, in various solvents of known δ_1 values.
The curve peaks when δ equals 9.2 $(cal/cm^3)^{1/2}$. At this point δ_1
equals δ_2 and thus X equals β_1. Detailed data are given in G. M.
Yenwo's thesis. The average molecular weight between crosslinks,
M_c, and the number of polymer chains per unit volume, N, were calculated from the Flory-Rehner equation given below:

$$\ln(1 - v_2) + v_2 + X_1 v_2^2 = -NV_1 (v_2^{1/3} - \frac{2v_2}{F}) \qquad (5)$$

where v_2 is the volume fraction of polymer in the swollen gel at equilibrium, V_1 is the molar volume of solvent, and F is the functionality of the system which in this case is 3. N and M_C are related by the density, d:

$$N = d/M_C \qquad\qquad (6)$$

Table 6 summarizes the results. The relatively small M_C values indicate rather dense crosslinking of the elastomers. Many elastomers have M_C values near 7500 gm/mole, as opposed to the present values of 1000 - 2000 gm/mole.

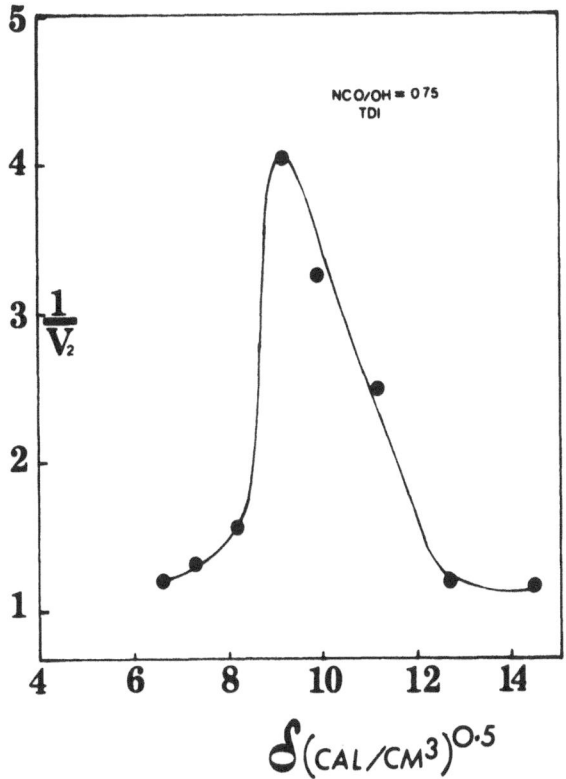

Figure 2.   Swelling data for castor oil-urethane elastomer, NCO/OH = 0.75, in various solvent of differing δ levels.

Table 6.   Values of M_c for Castor Oil-Urethanes

NCO/OH	$N \times 10^4$ ($\dfrac{\text{Polymer Chains}}{\text{Unit Vol.}}$)	M_c, g/mole
0.75	3.7	2947
0.85	7.0	1328
0.95	9.5	964

Figure 3 shows a plot of N versus NCO/OH ratio. The curve crosses the NCO/OH ratio axis at 0.64, i.e. below this point, a branched, finite molecular weight polymer should be obtained and above it, an infinite network should exist. The structure of one of the two possible isomers of the equivalent linear polymer is illustrated in Figure 3. The other isomer arises when the -OH of the middle acid residue is reacted.

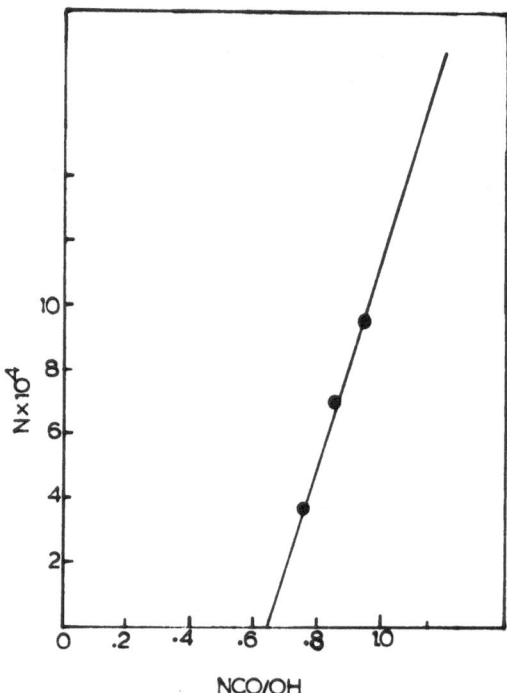

Figure 3.   Number of polymer network chains per unit volume versus crosslink ratio.

2. Tensile behavior. Figure 4 shows that, at least for TDI/CO elastomers, the higher the crosslink density (NCO/OH ratio), the higher the tensile strength, but the lower the elongation. For the HDI/CO elastomers, on the other hand, maximum strength and elongation were noted at an intermediate value of crosslink density (NCO/OH=0.85). HDI/CO elastomers also exhibited Young's moduli 10% to 20% higher than for TDI/CO elastomers and, for NCO/OH ratios of 0.75 and 0.95, greater values of tensile strength and ultimate elongation. It was also noted that the HDI/CO elastomers recovered from the strained state much faster than the corresponding TDI/CO elastomers.

Mooney-Rivlin plots, derived from Figure 4, are given in Figure 5. For values of $1/\lambda$ between 0.8 and 0.65, all specimens exhibited a linear relationship with $\sigma/2(\lambda-1/\lambda^2)$, as predicted by equations 1 and 2; at higher values of $1/\lambda$, deviation was observed. The deviation, probably due to increased errors at low extensions, is greater the higher the crosslink density. As reported earlier by Landel for different castor-oil elastomers (15), and as predicted for an ideal elastomer, the plots are nearly horizontal ($C_2 \to 0$). In agreement with results of Blockland for another polyurethane (16), the intercept, C_1, increases with crosslink density. Thus, the elastomers appear to behave as expected for an ideal elastomer, at least at moderate strains.

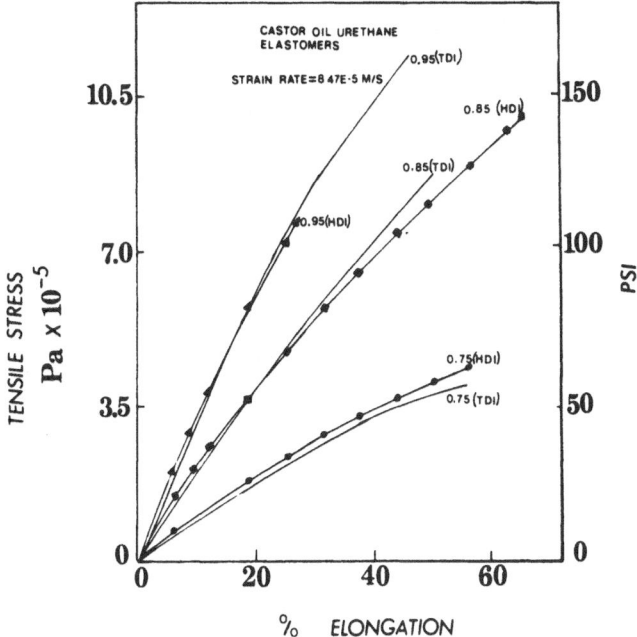

Figure 4. Stress-strain behavior of castor-oil-urethane elastomers.

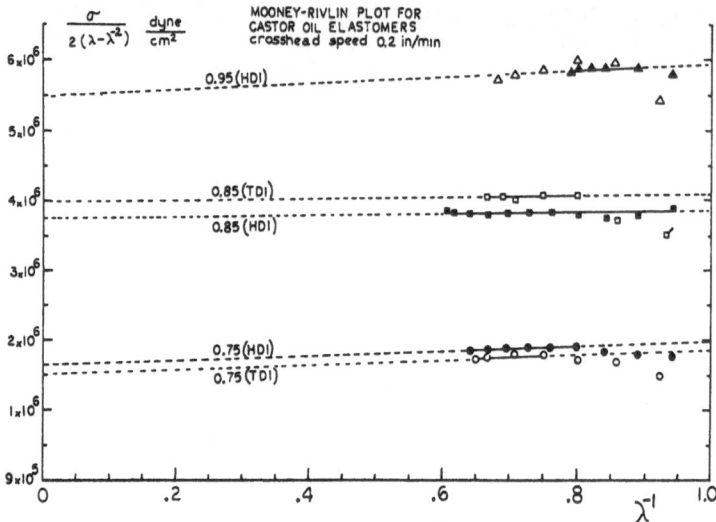

Figure 5.   Mooney-Rivlin plots for castor-oil-urethane elastomers.

3. Network densities.   Table 7 gives network densities calculated both from C_1 (see above) and from the approximation $E \approx 3\nu kT$
(17); clearly excellent agreement was obtained.   It may be noted
that the HDI/CO elastomers are crosslinked slightly but consistently more densely than the HDI ones.   This is as expected due to the
difference in size between HDI and TDI.

Table 7.   Crosslink Density of Castor-Oil Elastomers (65°F)

NCO/OH	$\nu \times 10^{-20}$ from C_1 (chain/cm^3)	$\nu \times 10^{-20}$ from E (chain/cm^3)
0.75 (TDI)	0.73	0.74
0.85 (TDI)	2.00	1.69
0.95 (TDI)	2.54	2.56[a]
0.75 (HDI)	0.83	0.82
0.85 (HDI)	1.89	1.90
0.95 (HDI) .	2.78	2.76

[a] Compare with the value of 2.6×10^{20} obtained by
measurements of G vs. T (6) using the approximation
$E \approx 3G$.

4. Tearing energies and fracture criteria. As shown in Table 8, values of T (incipient) for TDI-elastomers are in the range 2.6×10^4 to 5.8×10^4 ergs/cm^2. For incipient tearing, values may be compared with the values 2×10^4, 2.5×10^4, and 7×10^4 ergs/cm^2 for natural rubber (plain or isomerized) crosslinked by various methods. On the other hand, values of T (catastrophic) are lower than those reported for natural rubber -- 5.3×10^4 to 1.3×10^5 ergs/cm^2 for our specimens, compared to 10^4 to 10^8 ergs/cm^2 for natural rubber. Such a difference is expected, for natural rubber crystallizes on extension, thus providing inherent reinforcement. In any case, the higher the crosslink density, the greater the energy required for the initiation of both slow and catastrophic tearing.

The effect of crosslink density on tearing may also be seen by examining Table 8 and other data. The critical stress and strain for (1) crack initiation and (2) catastrophic crack growth are seen to decrease regularly as the crosslink density (1) decreases and (2) increases, respectively. It may also be noted that these stress and strain criteria for crack propagation in HDI/CO elastomers were invariably less stringent than for TDI/CO elastomers; indeed, crack propagations per se were so high in the former system that measurements by the present technique were impossible. It is interesting that the T_g's for HDI/CO elastomers are all in the range -42°C to -26°C -- much lower than for the TDI/CO elastomers. Evidently, values of T for the former elastomers must be significantly lower than for the latter. Since crosslink densities are only slightly lower with HDI, the reason must lie in decreased inelastic contributions to T with the low-T_g HDI/CO elastomers.

B.   Sulfur-Castor Oil IPN's

Shear modulus (10-sec.) vs. temperature data are presented in Figure 6. The following observations can be made:

(1) Both the PMMA and the castor oil-sulfur have clear, well defined glass transitions near 100°C and near -80°C, respectively.

(2) The very low modulus of the factice in the rubbery state, and the fact that its glass temperature changes to higher temperatures with increasing PMMA both suggest somewhat incomplete network formation. This conclusion is consistent with the fact that a rather small amount of sulfur was used in the factice synthesis, about 2 grams of sulfur per 100 ml of castor oil. Apparently much low-molecular-weight material remains unincorporated in the castor oil-sulfur network.

(3) The flattening and slight drift of the PMMA T_g with increasing factice level (note especially the 78% PMMA curve) suggest

Table 8. Tearing Energies and Critical Stress and Strain
of Castor Oil-Urethane Elastomers

NCO/OH	T (incipient) (ergs/cm^2)	T (catastrophic) (ergs/cm^2)	Av. critical stress[b] (psi)	Av. critical strain[b] (%)
0.75 (TDI)	2.64×10^4	5.30×10^4	23.4	25.4
0.85 (TDI)	5.71×10^4	7.26×10^4	40.4	22.7
0.95 (TDI)	5.77×10^4	1.29×10^5	62.1	22.6
0.75 (HDI)	—[a]	—[a]	23.7	15.6
0.85 (HDI)	—[a]	—[a]	31.3	15.6
0.95 (HDI)	—[a]	—[a]	37.9	12.5

[a] Could not be measured under these conditions; see text.
[b] For 1-mm cut.

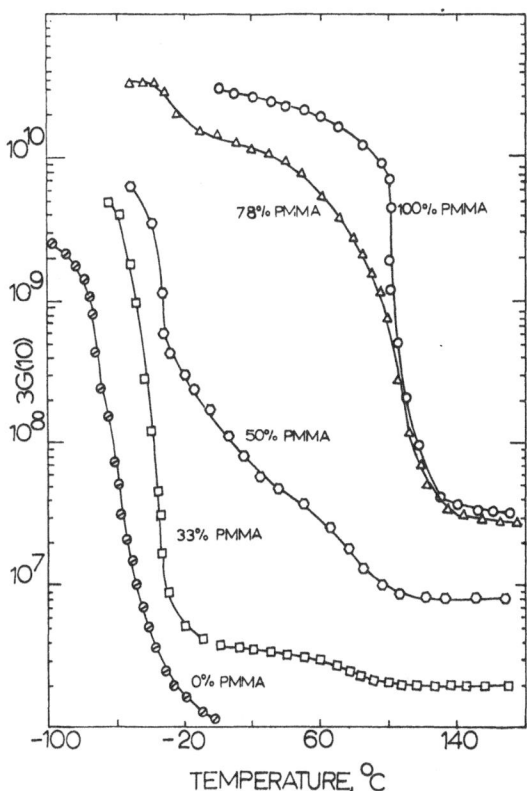

Figure 6.   10-second shear modulus vs. temperature for polymerized
            castor oil based IPN's. The castor oil was crosslinked
            with 6.7% sulfur and the poly(methyl methacrylate), with
            0.5% tetraethyl glycol dimethacrylate.

either extensive grafting, or slight extents of molecular mixing, or
both.

    The castor oil-sulfur elastomer was a soft material with fair
properties as a homopolymer. However, the sulfur in the elastomer
tended to behave as a chain transfer agent towards polystyrene and
poly(methyl methacrylate), making it difficult to produce an IPN
with interesting engineering properties. The castor oil-urethane
IPN (below) on the other hand, yielded extremely handsome specimens
with very promising properties, and the latter synthetic technique
was explored in considerable depth.

C. Castor Oil-Urethane IPN's

     The formation of tough materials via mechanical blending, graft or block copolymer formation or IPN synthesis depends on the morphological detail obtained. Phase size and continuity, degree of phase separation, and the glass transition of the elastomer phase each play a crucial role.

     Figure 7 shows the morphology via electron microscopy of a series of IPN samples. The castor oil component is stained with osmium tetroxide. The phase domains of the polystyrene are between 300Å and 500Å, depending upon composition and crosslinking level. Increased crosslinking level in the castor oil component causes the polystyrene domains to become smaller. These domain sizes are below the optimum required for impact resistant plastics, so it was no surprise that the impact resistance obtained was modest.

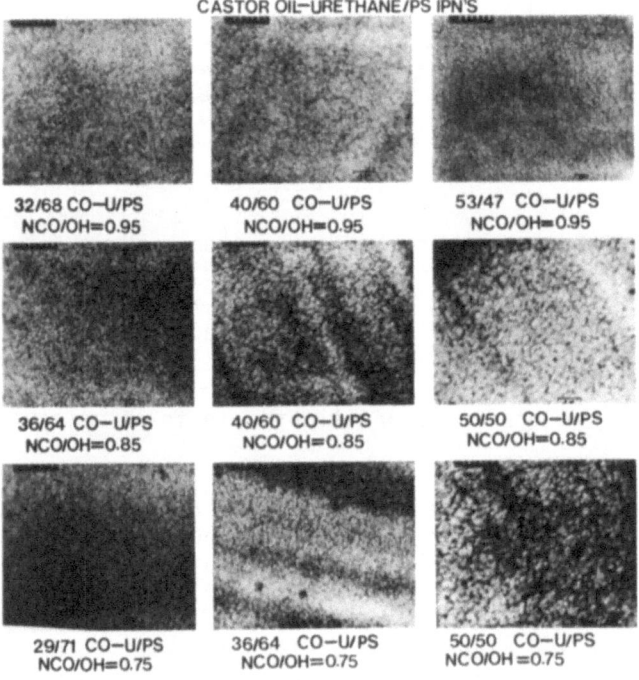

Figure 7. Morphology via electron microscopy of castor oil-urethane/polystyrene IPN's. Low crosslinking levels and low polystyrene levels lead to larger domains.

However, the phase domains are of the size range required for
the formation of tough elastomers, see Figure 8(a). In the cases
where the polystyrene forms the discontinuous phase, the polystyrene
behaves in a manner similar to carbon black or silica reinforcing
agents, or like the thermoplastic elastomers (17-20). (The present
materials are thermoset, though.) Figure 8(a) shows tough plastics
with well defined yield points, compositions 1 and 2, and tough
elastomers, compositions 3-10. In a previous study, IPN's based on
SBR and polystyrene yielded tough elastomers, also (21). Figure
8(b) shows the equivalent data of plastics prepared by the emulsion
route.

Dynamic mechanical spectroscopy (DMS), as a function of temperature, yields information about the glass transitions, and the extent
of mixing of the two component polymers. Figure 9 shows the storage
modulus, E', and the loss modulus, E'', for two IPN compositions as
well as the two homopolymers. The single broad transition observed
for the IPN's indicates extensive but incomplete mixing of the castor
oil-urethane and the polystyrene.

Since many applications of plastics involve cyclic loading,
fatigue behavior is of interest. Figure 10 shows the rate of crack
propagation per cycle, da/dN, as a function of the range of the
stress intensity factor ΔK (ΔK being proportional to the range in
applied load); the maximum value of ΔK attained corresponds to the
fracture toughness (which is a function of the fracture energy).
The toughening effect of the castor-oil-urethane is clearly evident
in the lower values of da/dN at constant ΔK and the higher values
of $\Delta K_{maximum}$, in comparison with polystyrene.

By way of comparison, the works of Frisch, et al. (22,23) have
emphasized IPN's based on polyurethanes, but their exact modes of
synthesis are different, and somewhat different properties emerge.

D. Castor Oil SIN's

Up to this point, this paper has been concerned with a review
of work completed in the first two year Colombia-U.S. international
program. These studies were based on the Ph.D. thesis of Dr. Yenwo,
and the M.S. theses of Mr. Pulido and Mr. Devia-Manjarrés. The
international grant of the National Science Foundation in the United
States and Colciencias in Colombia was extended for another two
years, beginning June, 1976. We now report the initial results from
ongoing laboratory studies by Mr. Devia-Manjarrés, as part of his
Ph.D. thesis work at Lehigh. Mr. Devia-Manjarrés came up from U.I.S.
to Lehigh as part of the international program to continue and complete his castor oil IPN research. Mr. Devia-Manjarrés is investi-
gating the behavior of SIN's (24-27).

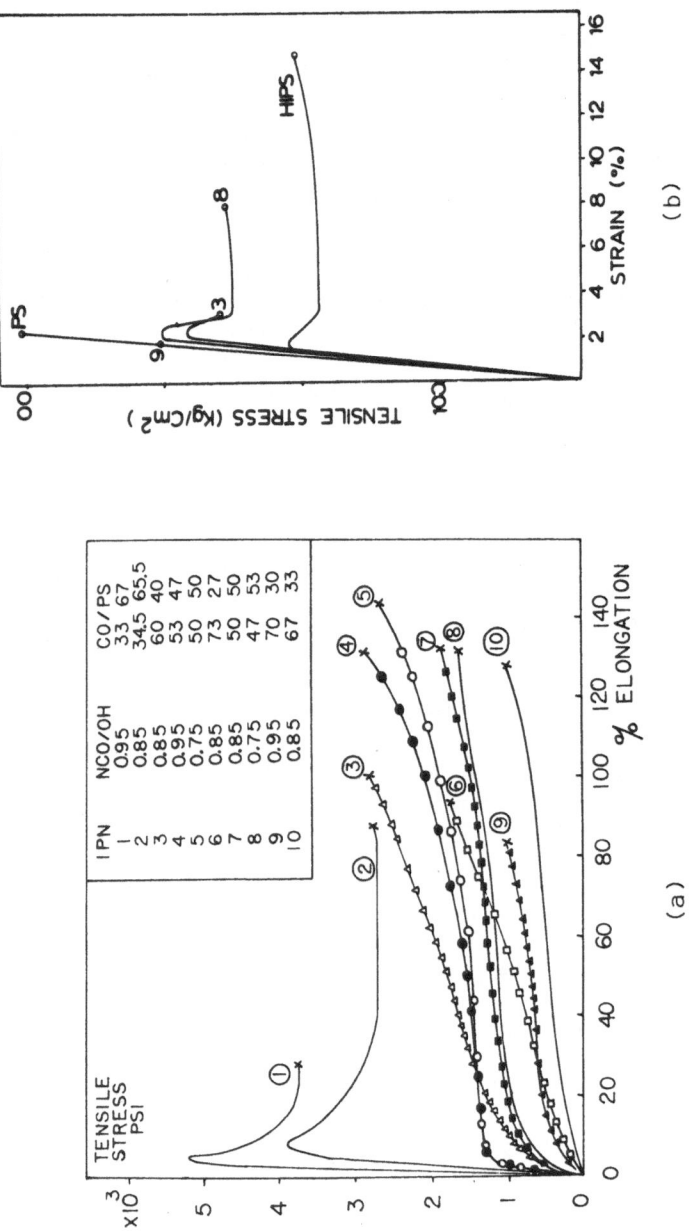

Figure 8.   Stress strain studies on selected materials.   (a) castor oil-urethane/polystyrene IPN's,
(b) materials prepared via emulsion polymerization.   A well defined yield point and high
extensibility indicate tough materials.

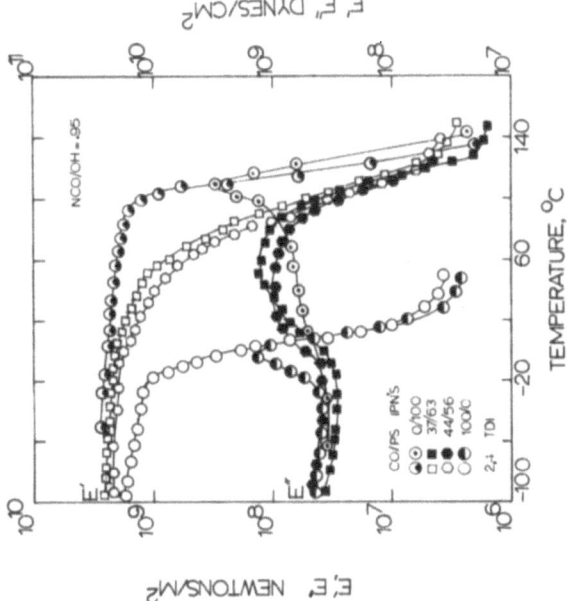

Figure 10. Fracture toughness of selected IPN's. Solid lines indicate linear and crosslinked polystyrene, as labeled.

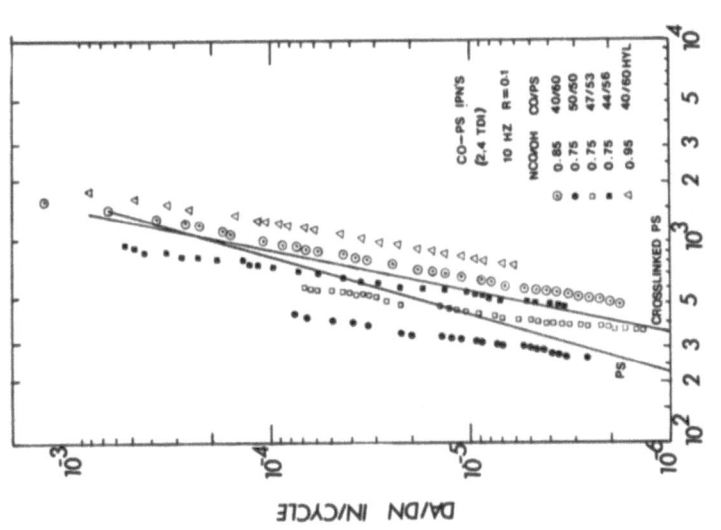

Figure 9. Dynamic mechanical behavior on selected castor oil-urethane/polystyrene IPN's. Compatibility is increased mechanically by high crosslink levels.

The modulus-temperature curves of the castor oil-sebacyl chloride elastomer, polystyrene homopolymer, and three SIN's are shown in Figure 11. Note that the glass temperature of the homopolymer elastomer is near -60°C, considerably lower than the corresponding glass temperature of the castor oil-urethane elastomer. This probably results in part from the use of an aliphatic crosslinking chain, as opposed to the TDI aromatic structure. The lower glass temperature is promising for impact resistance, since recent theories suggest that the glass temperature of the castor oil-urethane elastomer was too high for this property.

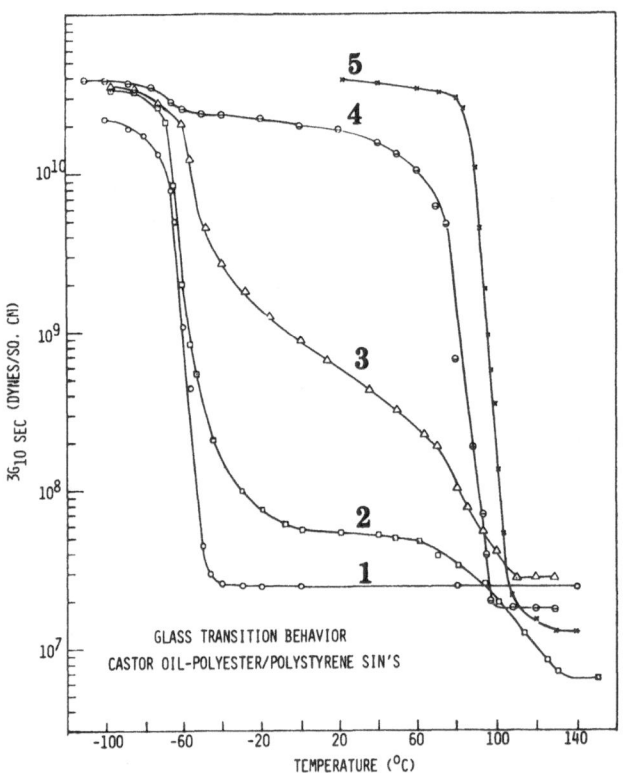

Figure 11. Modulus-temperature behavior of selected SIN's.

      1. Castor oil-polyester elastomer
      2. CO/PS 40/60 SIN (prepolymer route)
      3. CO/PS 40/60 SIN (straight mon. mix)
      4. CO/PS 15/85 SIN (PS prepolymer)
      5. Polystyrene (1% DVB).

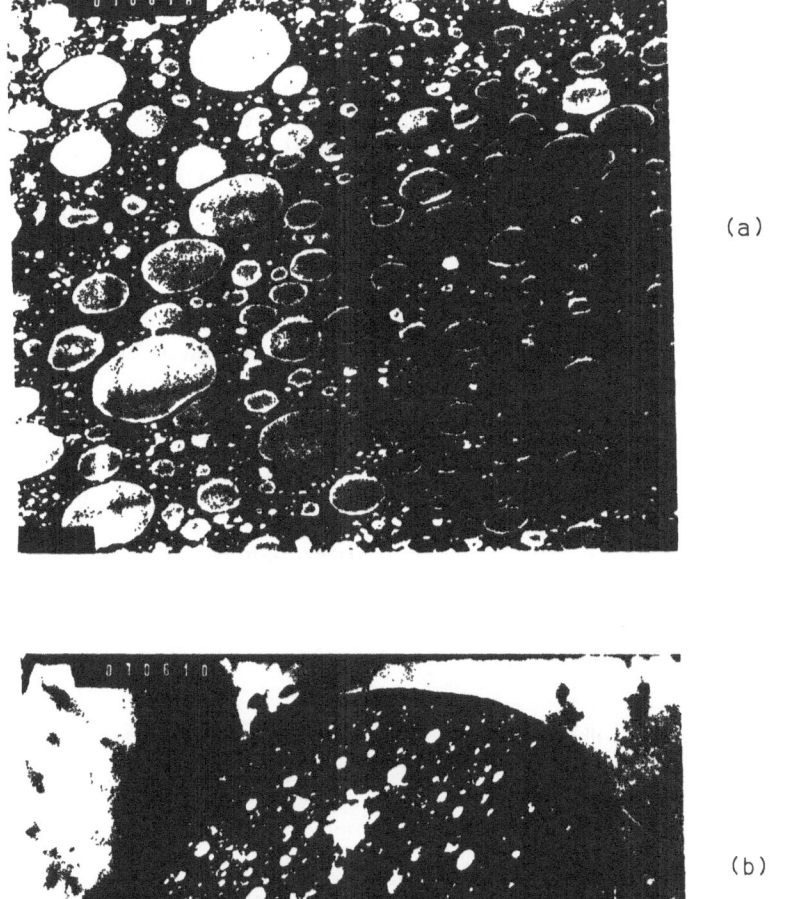

(a)

(b)

Figure 12. Electron microscopy reveals the morphology of castor
oil-polyester SIN's. Castor oil elastomer stained
dark with osmium tetroxide.

Examination of the SIN's reveals that one (the 15/85 castor oil/polystyrene) is plastic in nature at room temperature, and the other (the 60/40 castor oil/polystyrene) is elastomeric. The reason for these differences is revealed by electron microscopy, see Figure 12. Figure 12a shows that the osmium tetroxide stained elastomer forms the continuous phase in the 40/60 composition, while a phase inversion has taken place in the 15/85 composition. The latter morphology has a cellular structure within the elastomer droplet that is reminiscent of the polybutadiene/polystyrene high impact plastics (17).

Stress-strain curves shown in Figure 13 reveal that the SIN plastic material has a well developed yield point, and that both the plastic and the elastomeric SIN's are tougher than the corresponding homopolymers.

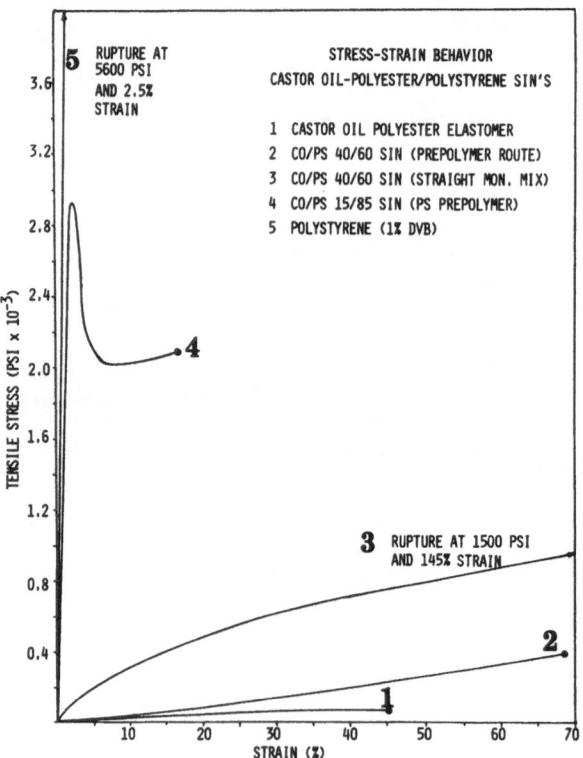

Figure 13.   Stress-strain behavior castor oil-polyester/polystyrene SIN's.
1. Castor oil-polyester elastomer
2. CO/PS 40/60 SIN (prepolymer route)
3. CO/PS 40/60 SIN (straight mon. mix)
4. CO/PS 15/85 SIN (PS prepolymer)
5. Polystyrene (1% DVB)

DISCUSSION

The basis for this international research and engineering program has two points:

(1) Castor oil is a renewable resource, coming from the castor bean plant. Its use will help relieve the shortage of petrochemical, and oil generally.

(2) While castor beans are grown in many tropical countries (including some in the Southern United States), the main world producers are India and Brazil. Climatic conditions are good in Colombia, and as a matter of fact, castor beans now grow wild there, along fences, roadsides, etc. The Colombians wish to develop a domestic castor oil industry, and as such this program is part of a broader effort.

The original idea was to vulcanize the castor oil with sulfur, but it was discovered that the sulfur compounds act as chain transfer agents towards any subsequent free radical polymerization. This led to the urethane derivatives, which attack the hydroxyl group instead of the double bond. The urethane elastomers all had T_g's too high for optimum toughness, however. Perhaps the SIN approach now in progress will yield improved prototype engineering materials.

In this program, as with most others, the bulk of the actual experimental work was done by the students as thesis research. The initial broad assignments were as follows:

(1) G. M. Yenwo, Ph.D. candidate, to "draw a map" of castor oil/polystyrene behavior, with the intention of identifying areas of real interest. As the first student on the program, such exploratory work was important, because castor oil IPN's had never been synthesized before. He emphasized the polyurethane route, and investigated morphology and mechanical behavior.

(2) N. Devia-Manjarrés, M.S. candidate, explored the latex route of synthesizing castor oil IPN's. When the initial route of cross linking castor oil with sulfur as polymer network I proved difficult, he explored using castor oil derivatives (sodium ricinoleate) as the soap, and castor oil-sulfur as polymer II, which works very satisfactorily, as did the polyurethane synthesis.

(3) J. Pulido, M.S. candidate, studied the behavior of the elastomer per se, as well as fracture toughness and fatigue behavior of the IPN's (materials prepared by Mr. Yenwo and Mr. Devia-Manjarrés), and cooperated in important aspects of electron microscopy.

(4) Mr. Devia-Manjarrés is now a Ph.D. candidate, and is working on the castor oil SIN materials. This continuation of the research was the result of the focus made possible by the first, exploratory part of the program.

CONCLUSIONS

As one of the major commercial oils of vegetable origin, castor oil is employed in many industrial uses, such as paints, synthetic fibers, and erasers. Castor oil differs from other oils, however, in that it contains an hydroxyl group on each acid residue. This hydroxyl group allows reactions with diisocyanates to form tough urethane elastomers.

The elastomer so formed was swollen with a solution of styrene, divinyl benzene, and initiator, and polymerized in situ to form the IPN. In the earlier papers in this series, the elastomer networks were characterized for crosslink density. The phase domain size of the polystyrene was ascertained via electron microscopy, and calculated via the Donatelli, et al. equation. In the present paper, the physical and mechanical behavior of the IPN's was evaluated. Electron microscopy and DMS studies are complementary. Comparison of the two experiments show that the domains are small (200-500Å in diameter), and that mixing of the two components is extensive. Mixing increases with increasing crosslink density, probably because of extensive mechanical entrapment of the chains. Many of the products are tough when measured by stress-strain analysis, fracture toughness, and crack propagation analysis, but they lack good impact resistance. There are two probable reasons for the low impact strength: (1) The phase domains of the present materials are below the optimum size range, and (2) The glass transition of the elastomer with the aromatic urethane group is too high. Further studies are in progress to evaluate improved syntheses.

The IPN's with high elastomer contents were tough materials. Their stress-strain behavior was similar to the thermoplastic block copolymers; both materials have a continuous elastomer phase and small, discontinuous plastic phases which served as reinforcement sites. In stress-strain studies, the tensile strength to break of these materials ranged from $6.9 \times 10^6 - 2.8 \times 10^7$ Pa.

ACKNOWLEDGMENT

The authors wish to thank the National Science Foundation for support under Grant No. GF-42712, Grant No. INT74-06791 A01, and Colciencias in Colombia.

REFERENCES

(1)   Source: Agricultural Marketing Service, U.S.D.A.

(2)   Chemical and Engineering News, 52 (37), 8 (1974).

(3)   F. C. Naughton, J. Am. Oil Chemists Soc., 51, 65 (1974).

(4)   A. Conde Cotes and L. A. Wenzel, Revista Latinoamericana de
        Ingeniera Quimica y Quimica Aplicada, 4, 125 (1974).

(5)   D. Swern, Ed., "Bailey's Industrial Oil and Fat Products",
        3rd Ed. Interscience, 1964.

(6)   G. M. Yenwo, J. Pulido, J. A. Manson, L. H. Sperling, A.
        Conde, and N. Devia-Manjarrés, accepted, J. Appl. Polym. Sci.

(7)   N. Devia-Manjarrés, G. Yenwo, J. Pulido, J. A. Manson, and
        L. H. Sperling, accepted, Polym. Sci. Eng.

(8)   G. M. Yenwo, L. H. Sperling, J. A. Manson, and A. Conde, to
        be published in S. S. Labana's edited work, "Chemistry and
        Properties of Crosslinked Polymers", in Press.

(9)   G. M. Yenwo, L. H. Sperling, J. Pulido, J. A. Manson, and A.
        Conde, accepted, Polym. Sci. Eng.

(10)   Volker Huelck, D. A. Thomas, and L. H. Sperling, Macro-
        molecules, 5, 340, 348 (1972).

(11)   L. H. Sperling, Tai-Woo Chiu, R. G. Gramlich, and D. A.
        Thomas, J. Paint Tech., 46, 47 (1974).

(12)   A. A. Donatelli, D. A. Thomas, and L. H. Sperling, in "Recent
        Advances in Polymer Blends, Grafts and Blocks", L. H.
        Sperling, Ed., Plenum, 1974.

(13)   J. H. Saunders and K. C. Frisch, "Polyurethanes: Chemistry
        and Technology", Part I. 1962, Interscience, p. 48-54.

(14)   J. H. Saunders and K. C. Frisch, "Polyurethanes: Chemistry
        and Technology", Part II. 1964, Interscience, p. 340-341.

(15)   R. F. Landel, in "Polymer Networks. Structures and Mechanical
        Properties", A. J. Chompff and S. Newman, Eds., Plenum, 1971.

(16)   R. Blockland, "Elasticity and Structure of Polyurethane
        Networks", Rotterdam University Press, 1968, p. 42.

(17)   J. A. Manson and L. H. Sperling, "Polymer Blends and Com-
        posites", Plenum, 1976.

(18)   S. L. Aggarwal, Polymer, 17, 838 (1976).

(19)   R. Houwink and H. K. deDecker, Eds., "Elasticity, Plasticity,
        and Structure of Matter", Cambridge, 1971.

(20)   N. A. J. Platzer, Ed., "Copolymers, Polyblends, and Com-
        posites", Adv. Chem. Series No. 142, ACS, 1975.

(21)   A. A. Donatelli, L. H. Sperling, and D. A. Thomas, Macro-
        molecules, 9, 671, 676 (1976).

(22)   D. Klempner, H. L. Frisch, and K. C. Frisch, J. Polym. Sci.,
        A-2, 8, 921 (1970).

(23)   M. Matsuo, T. K. Kwei, D. Klempner, and H. L. Frisch, Polym.
        Eng. Sci., 10, 327 (1970).

(24)   S. C. Kim, D. Klempner, K. C. Frisch, W. Radigan, and H. L.
        Frisch, Macromolecules, 9, 258 (1976).

(25)   R. E. Touhsaent, D. A. Thomas, and L. H. Sperling, in
        "Toughness and Brittleness of Plastics", R. D. Deanin and
        A. M. Crugnola, Eds., Adv. Chem. Series 154, ACS, (1976).

(26)   S. C. Kim, D. Klempner, K. C. Frisch, H. L. Frisch, and H.
        Ghiradella, Polym. Eng. Sci., 15, 339 (1975).

(27)   R. E. Touhsaent, D. A. Thomas, and L. H. Sperling, J. Polym.
        Sci., 46C, 175 (1974).

NORMAL AND UNUSUAL MORPHOLOGIES IN BLENDS

OF AB-CROSSLINKED POLYMERS

G.C. Eastmond and D.G. Phillips[†]

Department of Inorganic, Physical and Industrial

Chemistry, Liverpool University, Liverpool L69 3BX, U.K.

INTRODUCTION

The majority of recent studies of multicomponent polymers can be divided into two main categories; those based on well-defined and well-characterised AB and ABA block copolymers and those involving complex mixtures of relatively ill-defined multicomponent species and homopolymers, such as are encountered in impact-resistant materials. The former are the more amenable to scientific investigation in the sense that their morphologies and properties can be related to detailed molecular structures. However, since block copolymers are usually prepared anionically, the range of components which can be incorporated into such well-defined materials is somewhat limited. The AB-crosslinked polymers (ABCPs), with which this paper is concerned, are prepared by free-radical processes and are a class of materials into which a wider variety of components can be incorporated. While ABCPs are structurally related to conventional AB and ABA block copolymers they are more complex but, when prepared under controlled conditions, their structural parameters can be determined. Thus, the formulation of structure-property relations is permitted and extensions of the relations developed for simpler systems to more complex situations can be considered.

Before proceeding with a discussion of the morphologies of ABCPs it will be useful to summarise some pertinent results of previous morphological studies of linear block copolymers.

---

† Present address: CSIRO, Department of Textile Industries, Belmont, Victoria, Australia.

Morphologies of Block Copolymers and their Blends

The morphologies of AB and ABA block copolymers have been studied by many workers. It is now recognised that in bulk polymers the three basic morphologies identified by Matsuo et al. can exist; namely, spherical or cylindrical domains of the minor component within a matrix of the other or alternate lamellae of the two components (1). Under equilibrium conditions the morphologies are determined by the relative sizes of the A and B blocks and in carefully prepared samples very uniform morphologies with a high degree of long-range order have been obtained (2).

From theoretical calculations Meier predicted that characteristic dimensions D of morphological features, i.e. domain radii and lamellar thicknesses, would be given by the equation

$$D = kC\alpha M^{\frac{1}{2}} \qquad (1)$$

in which k is a constant determined by the domain shape and molecular architecture, C is the constant relating unperturbed root-mean-square end-to-end chain dimensions to the molecular weight M of the domain-forming chain and α is an expansion parameter (3,4).

Relatively few morphological studies of block copolymer blends have been undertaken. Hoffmann et al. blended two styrene-butadiene diblock copolymers with different block sizes (5). They reported that in cast films two sizes of styrene domains were observed corresponding to those of the individual polymers, implying that fractionation by molecular weight accompanied microphase separation.

Inoue et al. investigated the effects of simultaneously adding homopolymers of both components to styrene-isoprene diblock copolymers (6). These workers demonstrated that while homopolymers of molecular weights smaller than those of the corresponding blocks in the copolymer were solubilised within the microphase of that component, high-molecular-weight homopolymers were incompatible with the chemically identical blocks in the copolymer and macroscopic phase separation was observed. Inclusion of compatible homopolymers into a styrene-isoprene block copolymer containing 40% styrene caused a disruption of the long-range order of the lamellar structure of the pure block copolymer.

During an investigation into the liquid crystalline structures present in solutions of block copolymers, Douy and Gallot dissolved styrene-butadiene block copolymers in styrene monomer (7). Subsequent polymerization of the solvent produced blends of block copolymer and polystyrene. It was observed that a regular lamellar morphology was retained when up to 45% of polystyrene was included, in the manner described, into a block copolymer which exhibited a lamellar morphology when pure. Inclusion of larger proportions of polystyrene decreased the long range order until in the presence of more than 70% polystyrene the lamellae broke up, ultimately giving rise to a random distribution of very short rods for more than 99% styrene homopolymer.

In an earlier experiment into the colloidal behaviour of block copolymers, Molau and Wittbrodt (8) prepared blends of styrene-butadiene block copolymers with polystyrene in a manner similar to that adopted by Douy and Gallot. For samples containing 5-15% block copolymer (contaminated with a little polybutadiene) supra-molecular structures in the form of elliptical rings ($\sim$ 0.3 μm diameter) of polybutadiene ($\sim$ 0.1 μm thick) were observed if the reaction mixture was stirred during the initial stages of the styrene polymerization. In the absence of stirring irregular lamellae were reported for a material containing 10% block copolymer, while a sample containing only 1% copolymer exhibited isolated regions ($\sim$ 5 μm across) containing polybutadiene ribbons, with little polybutadiene in between.

Bradford (9) examined the morphologies of solvent-cast films of both styrene-butadiene diblock and styrene-butadiene-styrene triblock copolymers in the presence of homopolymers; prior to addition of homopolymer the block copolymers exhibited lamellar morphologies. Some of the results were similar to those of Inoue et al. (6). Addition of 60% low-molecular-weight polystyrene to a 40/60 styrene-butadiene copolymer (containing some polybutadiene contaminent) produced regions ($\sim$ 2-3 μm across) consisting of a polybutadiene matrix and containing small spheroidal polystyrene occlusions. These regions were distributed within a material consisting of a polystyrene matrix containing small (85 nm diameter) polybutadiene spheres. For triblock copolymers containing $\sim$ 50% of each component more complex morphologies were observed in the presence of added homopolymer. Addition of small amounts of homopolymer of either component introduced some disorder into the lamellar morphology. When blended with 99% low-molecular-weight polystyrene the polybutadiene appeared to be contained within ribbons which formed long loops. Addition of 99% high-molecular-weight polystyrene caused the formation of spherical particles ($\sim$ 1 μm across), containing both components and with an ordered internal structure, in a sea of polystyrene.

While some effects of blending with homopolymer, such as the loss of long-range order and even the break-up of lamellae, can be easily understood, the nature and origins of the large supra-molecular structures are less obvious and have apparently not been fully discussed. In this paper we shall consider the morphologies of AB-crosslinked polymers in the presence of one homopolymer.

Features similar to those reported for blends of block copolymers have been observed and the origins of some of these features will be discussed.

SYNTHESIS AND STRUCTURE OF ABCPs

We have developed and previously described a general method of preparing AB-crosslinked polymers which consist of chains of polymer A crosslinked by chains of polymer B (10-12). Synthesis involves a free-radical process and is based on the reaction in which metal carbonyls, under suitable conditions, react with certain organic halides to produce free radicals capable of initiating polymerization (13). To form ABCPs free-radical polymerization of monomer B is initiated from specifically functionalised (halogen-containing) sites on pre-formed A-chains. Combination termination of propagating B-chains generates B-crosslinks between A-chains. Disproportionation termination, irrelevant to materials described in this paper, leads to branch formation. By performing the reactions under conditions such that the occurrence of chain-transfer processes is minimal, little or no B-homopolymer is formed. Careful control of the reaction conditions allows the random introduction of B-crosslinks of known average degree of polymerization between A-chains, avoiding excessive intramolecular crosslinking. The number of B-crosslinks introduced between A-chains is simply governed by the rate of radical formation and the reaction time.

In the earliest stages of reaction the multicomponent species formed have structure (I) in which a single B-chain links two A-

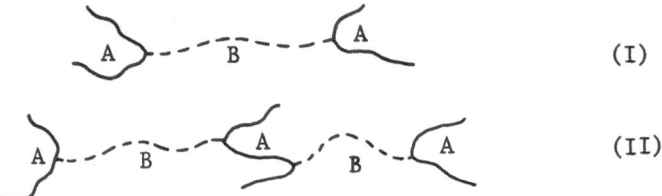

$$(I)$$

$$(II)$$

chains. By analogy with the terminology used for block copolymers, species (I) may be termed an A_2BA_2 block copolymer. Continued random introduction of B-crosslinks into the assembly of A-chains produces not only more species (I) but also more complex structures (II) and even more extended structures formed as a result of initiating from several functional groups on one A-chain. Eventually, with the introduction of one crosslinking point per weight average A-chain gelation occurs as the multicomponent species are linked into infinite network structures. Continued initiation from additional sites on A-chains results in the incorporation of more crosslinks and additional simple species into more highly crosslinked network structures.

Prior to gelation the reaction mixture is a dilute solution of unreacted homopolymer and statistical proportions of multicomponent species of various complexity. If polymer A was initially monodispersed with respect to molecular weight it would be possible to write down the relative populations of species (I),(II) and more complex species and unreacted A chains, as calculated by Flory (14). The situation is not so simple for ABCPs formed by crosslinking A-polymers with various molecular-weight distributions since the relative populations of species (I),(II) etc. are dependent on the initial distribution. We have not calculated the relative populations of species present in ABCPs but preliminary calculations for specific situations indicate that the relative populations of multicomponent species are similar to those which might be anticipated by analogy with Flory's calculations for random crosslinking of monodispersed polymers. For example, dimer species containing two A-chains (i.e. structure (I)) are more numerous than more complex species at low degrees of crosslinking.

Species (I) in ABCPs themselves have distributions of structural parameters. The distribution of sizes of A-chains in species (I) naturally reflects the original molecular-weight distribution of the A-polymer and varies with the degree of crosslinking. The probability of radical formation on an A-chain is proportional to the functionality or molecular weight of that chain. Consequently, long A-chains are preferentially incorporated into multicomponent species, especially in the early stages of reaction. As reaction proceeds shorter A-chains will be incorporated into species (I) while, simultaneously, species (I) having long A-chains will be preferentially converted to species (II) by further initiation from those A-chains. Because crosslinks are introduced randomly between A-chains and each A-chain carries many functional groups, the four sections of A-chain attached to each B-chain in species (I) have different degrees of polymerization. Although the average degree of polymerization of the two A-chains attached to one end of a B-chain will be one-half the degree of polymerization of the original A-chain, the chain lengths will not normally be equal. Assuming equal reactivity of the functional groups on A-chains, the ratio of average degree of polymerization of long to short blocks attached to the same end of a B-chain will be 3:1. B-chains in crosslinked species also have a distribution of molecular weights. Normally the formation of crosslinks, at least prior to gelation, conforms to simple free-radical polymerization kinetics. Consequently, we anticipate that the distribution of molecular weights of B-chains will be that corresponding to free-radical polymerization with combination termination (i.e. $\bar{M}_w/\bar{M}_n = 1.5$).

Soluble multicomponent species present in ABCPs may be considered as linear and non-linear multiblock copolymers with A-branches emanating from junction points such that there are always two A-chains linked to one B-chain at each junction point. Following gelation the polymer contains a mixture of relatively simple species together with a gel fraction; the latter is a three-dimensional multiblock structure. To describe the extent of crosslinking in ABCPs we define a relative crosslinking index, γ_r, as the ratio of the true crosslinking index to that at the gel point. $\gamma_r = 1$ at the gel point and γ_r is proportional to the number of crosslinks in the system and equal to the number of crosslinking units per weight-average A-chain.

When discussing ABCPs with $\gamma_r < 1$ we may treat them as mixtures of relatively complex block copolymers rather than as the conceptually more complex crosslinked materials. Since species(I) is the fundamental structural unit of ABCPs these species may be considered to be a model for the distribution of multicomponent species in ABCPs.

GENERAL MORPHOLOGICAL CONSIDERATIONS

Because the A- and B-components in ABCPs are chemically different they are normally incompatible and multicomponent species will tend to undergo microphase separation. We have previously established that bulk ABCPs exhibit microphase separation at all crosslink densities, at least up to $\gamma_r = 10$ (15,16). If it were possible to isolate the multicomponent species from unreacted homopolymer (for $\gamma_r < 1$) we would anticipate that the materials would exhibit the same simple morphologies as AB and ABA block copolymers; i.e. spheres, rods and lamellae. The particular, or intrinsic, morphology of the polymer would be determined by the relative sizes of the A- and B-blocks. The change from one morphology to another might occur at different relative block sizes than in linear block copolymers as a result of structural differences. Consider an A_2BA_2 polymer constructed from A-chains of molecular weight 25K and B-chains of molecular weight 50K. Taking the molecular weights of the end blocks at the same end of the B-chain together we may compare the polymer to an ABA block copolymer in which the centre block constitutes 50% (w/w) of the total. For such a material we would expect a lamellar morphology. Although this method of estimating intrinsic morphologies is not strictly accurate and over-estimates the sizes of the individual A-blocks, the conclusions are probably reasonable. Considering the polymer as an A_2BA_2 block copolymer and remembering that the A-blocks are of unequal length, as discussed in the preceding section, we would have, completely neglecting the shorter A-blocks, species in which the centre block accounts for 55% w/w of the total and a lamellar morphology would again be expected. In spite of the fact that the effect of a second end-block has not been assessed theoretically we expect, from simple packing considerations, that its influence would be greater than if it was added to the first end block to increase the molecular weight of that block. We therefore expect the morphologies of the pure A_2BA_2 polymers to be shifted slightly towards that expected for larger A-blocks than are actually present in the polymer.

Unreacted A-homopolymer present in the ABCPs must impede the formation of regular intrinsic morphologies. For materials with sufficiently small B-chains the expected intrinsic morphology would be regularly packed B-spheres in an A-matrix, but the presence of A-homopolymer solubilised in the A-matrix would produce an irregular arrangement of B-spheres. Parallel arrangements of rods and lamellae could be disrupted and, in the presence of sufficient A-homopolymer, these extended structures would ultimately be broken down, as found with blends of linear block copolymers. Overall we would expect randomness to be introduced into the packing of domains and lamellae and possibly a shift in the morphology observed from that of the intrinsic morphology towards that expected for a material with relatively shorter crosslinks. A number of these effects have now been observed in practice.

MORPHOLOGIES OF SOME ABCPs

The first indication that microphase separation occurs in bulk ABCPs came from a dilatometric examination of highly-crosslinked networks (γ_r = 1-20) which revealed the existence of multiple glass transitions (17). Although such studies provided no information on the sizes and shapes of the microphases present, the data did provide information on the compositions of those microphases. It was found that microphase separation was essentially complete even in highly-crosslinked networks (γ_r = 1-6) containing no branches of the crosslinking component, the A- and B-components forming pure microphases (17,18). The presence of branches of the B-component hindered microphase separation giving rise to a microphase containing both A- and B-components (12,15,17).

We have also previously published electron micrographs of ultra-thin sections illustrating microphase separation in solvent-cast ABCPs comprised of poly(vinyl trichloroacetate)(PVTCA) and of functionalised polystyrene (PS*) (A-components) crosslinked with polychloroprene (PCp) as B-component (16). All the lightly-crosslinked ABCPs (γ_r $\leqslant$ 1.3) exhibited a random distribution of spherical PCp domains (Fig. 1a). While in most ABCPs examined the PCp chains were the minor component of the multicomponent species, some PS*/PCp ABCPs might have been expected to have a lamellar intrinsic morphology but the overall PCp content was < 16% (w/w) and the excess PS* caused the formation of PCp spheres.

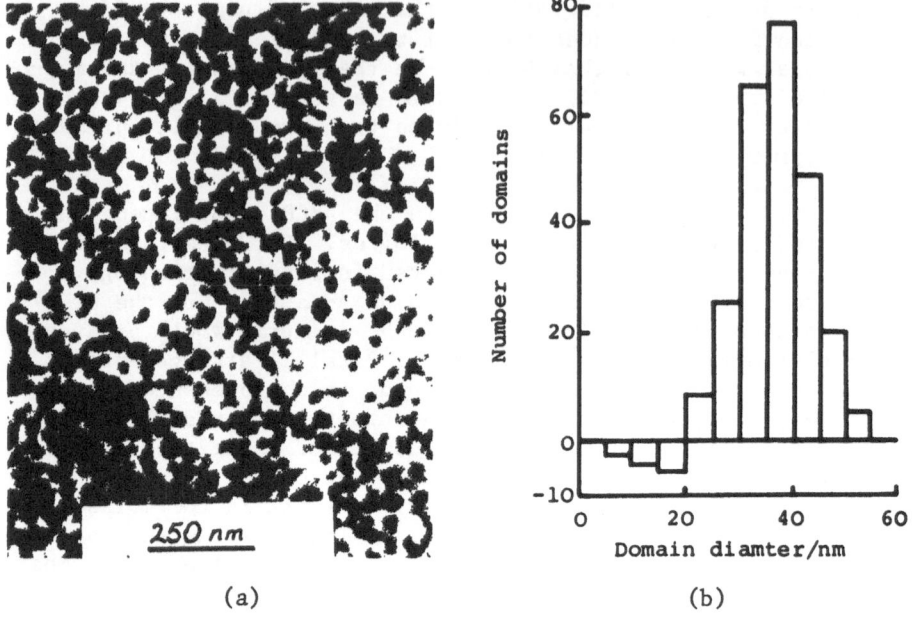

(a)   (b)

Fig. 1 (a) An electron micrograph of a PVTCA/PCp ABCP containing
          4% (w/w) PCp, $\bar{P}_n$ = 900, γ_r = 0.4.

      (b) A histogram depicting the distribution of domain sizes
          in (a) after allowing for the effects of sectioning.

     The PCp domains exhibited a distribution of diameters but af-
ter correction for the effects of sectioning, the distribution of
sizes was found to be quite narrow (Fig. 1b) and much narrower
than the distribution of square roots of degrees of polymerization
of the PCp chains, assuming $\bar{M}_w/\bar{M}_n$ = 1.5 (16). Such results indicated that fractionation of PCp chains with respect to molecular
weight did not accompany microphase separation as previously found
by Hoffmann et al. for mixtures of AB block copolymers (5).

     Domain radii were found to be smaller than expected from equa-
tion (1) for AB block copolymers with the same chain length of the
domain-forming blocks. Assuming values of k, C and α appropriate
for B-spheres containing a diene polymer we obtained $r = 0.21\ M^{\frac{1}{2}}$ nm
compared with, experimentally, $r \sim 0.06\ \bar{M}_n^{\frac{1}{2}}$ (for mean domain radii).
The small domain sizes were attributed to two effects. (i) Both
ends of the domain-forming chain are located at the domain-matrix
interface, an effect recognised by Meier as applying to ABA block
copolymers with B-domains (4). According to Meier, this effect
reduces the domain radius by a factor of 1.33 which alone is insuf-

ficient to account for our observed radii. (ii) The number of A-blocks per unit area of the interface is twice the number of B-blocks. Although this effect has not been assessed quantitatively for a spherical morphology, from calculations appertaining to lamellar morphology, an effect on the value of α is to be expected, probably in the sense of reducing α (5,16). The implications derived from the experimental observations are that individual chains in domains in ABCPs are more tightly coiled than in linear block copolymers in consequence of structural differences.

At high crosslinking indices ($\gamma_r > 2$) microphase separation in a PVTCA/PCp ABCP produced a random arrangement of extremely small irregular B-domains (6). Presumably the restricted mobility of the junction points in highly-crosslinked polymer prevented sufficient reorganisation of chains to allow domains to develop to their equilibrium sizes.

We have now extended our studies to a wider range of components and relative block lengths. An additional A-component which has been used is a polycarbonate (PCarb) ($\bar{M}_n = 16$ K), based on 1,1,1-trichloro-bis-2-(p-hydroxyphenyl)ethane. Structural parameters of such ABCPs having either PCp or polystyrene (PS) B-components are given in Table 1.

Table 1. Structural Characteristics of Multicomponent Species in some ABCPs with a Polycarbonate A-component

Series	Crosslinking Polymer	Relative Molecular Weights ($10^{-3}\bar{M}_n$) in Multicomponent Species (as A:B:A)
A	PCp	16:42.5:16
B	PCp	16:57.5:16
C	PCp	16:97.4:16
D	PS	16:96.0:16

In defining the compositions of multicomponent species (Table 1) and to facilitate prediction of intrinsic morphologies by comparison with ABA block copolymers, we have assumed that all crosslinks are in species with structure (I), that two average A-chains are linked by an average B-chain and we have taken the total molecular weight of A-chains attached to each end of the B-chain. Thus for series A, in which the centre block accounts for 56% (w/w) of the species, we anticipate an intrinsic morphology of lamellae. Figs. 2a,b are electron micrographs of sections of polymers from series A (cast from dilute solution in chlorobenzene) with $\gamma_r = 0.06$ and 0.78, containing 5 and 40% (w/w) PCp, respectively. Fig. 2a shows PCp spheres (stained with OsO_4). Unreacted

PCarb exerts a dominating influence and forces the crosslinks into spheres. Electron micrographs show that as the PCp content
and crosslink density increase the morphology gradually changes
through short rods to lamellae. The lamellar morphology is shown
in Fig. 2b for a polymer in which multicomponent species account
for $\sim$ 70% of the material. At low crosslink densities, excess
PCarb disrupts intrinsic morphologies with extended features
(rods and lamellae) which are usually only observed when multicomponent species account for >40% of the polymer. At higher
crosslink densities (e.g. γ_r= 1, 48% PCp) the morphology of polymers from series A becomes a PCp matrix with irregular PCarb
domains. Although a morphology with a matrix of B-component
would not be expected from the composition of species (I) it
should be remembered that at high degree of crosslinking the longer PCarb chains will be attached to more than one crosslink.
Consequently, the average block lengths of A-chains will be reduced and intrinsic morphologies may change.

   Other examples of lamellar morphologies are seen in Fig.3. Fig.
3a is an electron micrograph of a polymer,from Series B,with γ_r =
0.42 containing 32% PCp with 50% of the polymer in multicomponent
species.The intrinsic morphology expected from data in Table 1 is

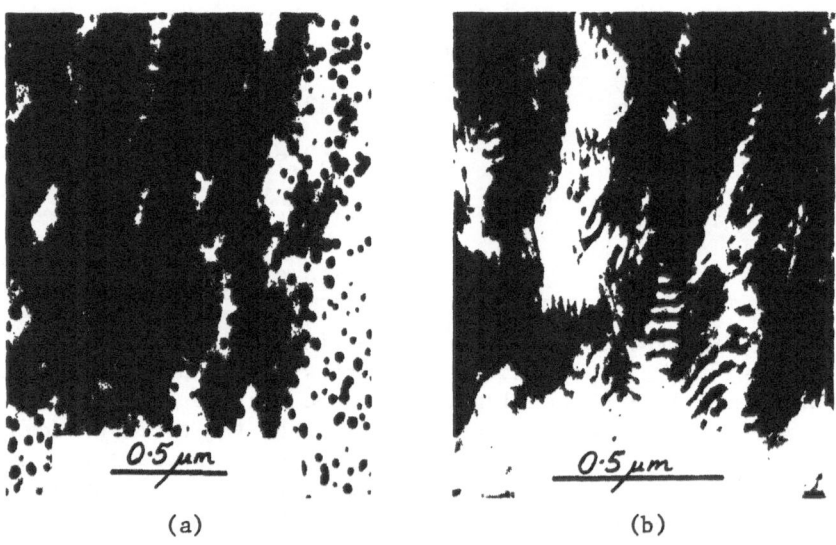

            (a)   (b)

Fig. 2. Electron micrographs of PCarb/PCp ABCPs from series A:
         (a) γ_r = 0.06, 5% (w/w) PCp; (b) γ_r = 0.78, 40% (w/w) PCp.

(a)   (b)

Fig. 3. Electron micrographs of ABCPs: (a) PCarb/PCp from
series B with γ_r = 0.42, 32% (w/w) PCp, the PCp stained
with OsO_4 appears dark; (b) PCarb/PS from series D with
γ_r = 0.35 and 34% (w/w) PS, the sample was unstained and
PCarb appears dark.

lamellae. Fig. 3b refers to a PCarb/PS ABCP from series D with
γ_r = 0.35, 34% (w/w) PS and ∿45% of the material in multicomponent
species; the polymer was cast from dichloromethane and sections
were unstained. The intrinsic morphology for series D is almost
certainly a PS matrix. Fig. 3 illustrates that well-ordered lamellar morphologies can be obtained from comparatively complex
mixtures of multicomponent species and homopolymer. To obtain
such ordered structures it is necessary that either a sufficiently
high proportion of material with the appropriate intrinsic morphology is present, or a material with an intrinsic morphology of
A-domains in a B-matrix may be induced to form regular lamellae
by the presence of unreacted A-homopolymer. The thickness of the
B-lamellae should be determined by the molecular weight of the
B-chains while that of A-lamellae is determined by the overall
composition.

We conclude that we may expect to obtain the same normal basic
morphologies in AB-crosslinked copolymers, in the presence of A-
homopolymer, as are found in pure linear block copolymers. Although
A-homopolymer can produce non-uniformities in the morphologies,
under suitable conditions of structure and overall composition,
materials with a high degree of long-range order can be obtained.

The latter conclusion parallels observations on blends of block
copolymers, as discussed earlier, which show that moderate amounts
of homopolymer may be added to a block copolymer without disrupt-
ing the lamellar morphology or significantly reducing the long-
range order.

MACROSCOPIC PHASE SEPARATION IN ABCP BLENDS

Figs. 4a,b are electron micrographs of PCarb/PCp ABCPs from
series C, Table 1. These and other micrographs from this series
reinforce the comments made in the preceding section. Fig. 4a,
from a sample with γ_r = 0.028 and 5% (w/w) PCp, shows a background
random distribution of PCp spheres together with some large supra-
molecular structures; the latter features are referred to later.
With increasing γ_r and PCp content the morphology changed through
rods to rods + lamellae (Fig. 4b, γ_r = 0.254, 33% (w/w) PCp) to
a PCp matrix (γ_r = 0.52, 49% PCp). PCarb homopolymer was blended
with the ABCP responsible for Fig. 4b to give a material with the
same overall PCp content as that responsible for Fig. 4a. Sections
of this blend revealed the morphology in Fig. 5a consisting of
large supramolecular features in a sea of PCarb containing PCp
spheres. Examination of successive sections has shown that such

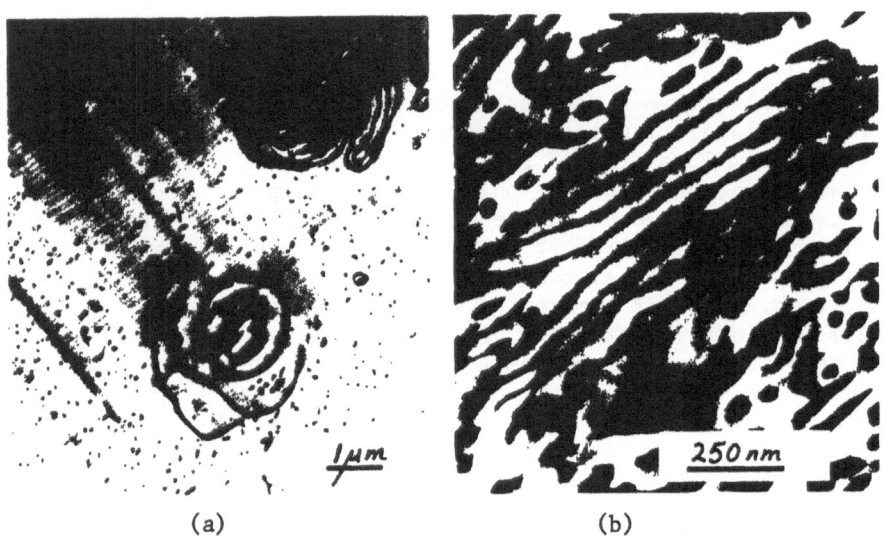

(a)   (b)

Fig. 4. Electron micrographs of PCarb/PCp ABCPs from series C;
         (a) γ_r = 0.028, 5% (w/w) PCp; (b) γ_r = 0.254, 33% (w/w)
         PCp.

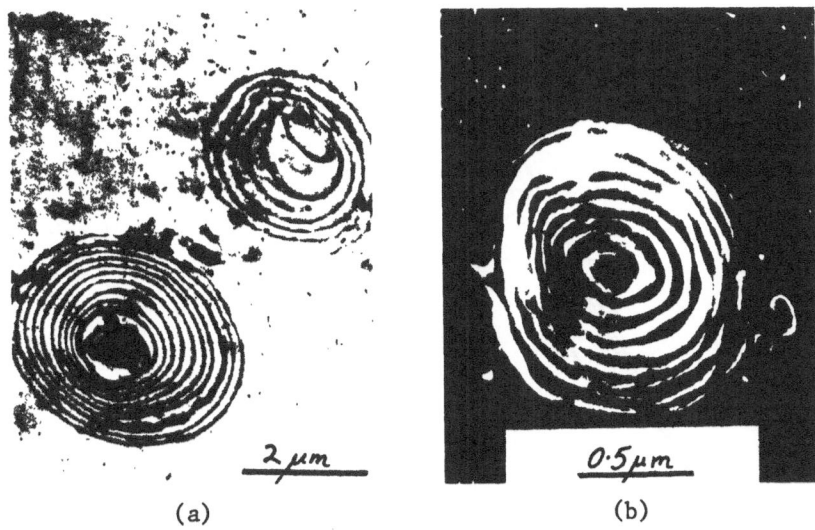

(a) (b)

Fig. 5. Electron micrographs of ABCPs with added PCarb:
(a) Initial PCarb/PCp ABCP as in Fig. 4b, final composition 5% (w/w) PCp; (b) Initial PCarb/PS ABCP as in
Fig. 3b, final composition 10% (w/w) PS.

large structures, which appear as concentric rings of PCp ($\sim$ 10nm
thick) separated by thicker PCarb layers, are sections of large
spherical objects with an 'onion-like' structure consisting of
concentric shells of PCarb and PCp. Similar onion-like structures
were observed when PCarb homopolymer was blended with the sample
responsible for Fig. 3b to give a material containing 10% PS
(Fig. 5b). We have also observed similar features in other ABCPs
blended with A-homopolymer and, in some cases, where no homopolymer has been added but the ABCP contains a very large proportion of unreacted A-chains.

We have established that the onion-like structures arise as a
result of a combination of macroscopic phase separation and microphase separation during the casting process. A possible mechanism
may be understood from the schematic phase diagram, Fig. 6, for
incompatible polymers A, B and a common solvent S, in which it is
impossible to obtain a phase richer in B than that corresponding
to the composition of the multicomponent species, represented by
X. The binodal curve assumes complete phase separation between
A-homopolymer and multicomponent species in bulk polymer at equilibrium. Suppose we start with a solution represented by point K,
a dilute solution of ABCP with a large excess of A-homopolymer.

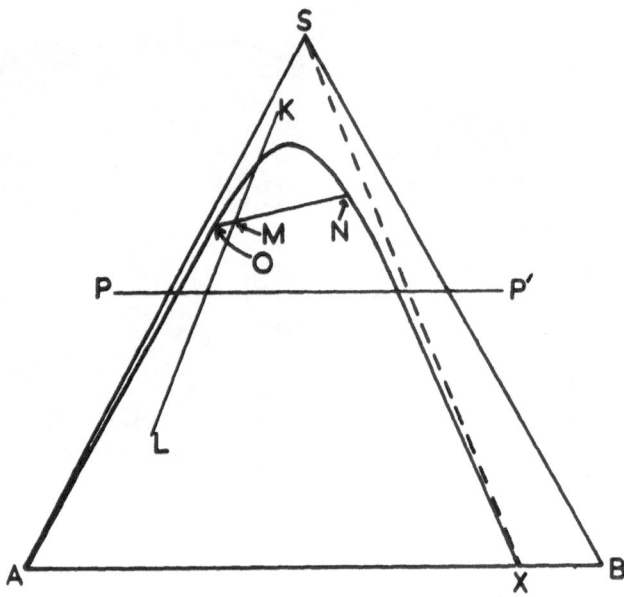

Fig. 6. Schematic phase diagram to show macroscopic phase
         separation and microphase separation in blends of ABCPs.

On evaporating the solvent, the overall composition would follow
the line KL. At point M macroscopic phase separation would have
produced droplets of a phase of composition N in a solution of
composition O. Further drying would cause the compositions of
the two phases to follow the binodal curve until an overall polymer concentration, indicated by the line PP' is reached, at which
point microphase separation is assumed to occur in the individual
phases. The morphologies formed in each phase would reflect the
local compositions of those phases. Thus, 'onions' are considered
to be regions of lamellar morphology formed within spherical droplets of the minor phase, the crosslink rich phase, containing some
A-homopolymer.

   According to the phase, diagram, Fig. 6, the equilibrium morphology of dry polymer would be regions of pure A-homopolymer and
pure multicomponent species and the true equilibrium morphology of
an ABCP blend may approximate to this. The morphologies observed
in the two phases are not necessarily the equilibrium morphologies
but may reflect the kinetics of the casting process. Equilibrium
conditions can probably be readily achieved in the early stages of
casting, prior to microphase separation. Following microphase
separation, e.g. the formation of swollen lamellae in droplets,
equilibration will be slower requiring the transfer of chains of
one component through regions of the other component. Also, as the

polymer concentration increases, the simultaneous increase in
viscosity will retard the approach to equilibrium. Equilibrium
morphologies may be approached only if very slow drying procedures
are used, or if samples are annealed in a solvent atmosphere for
long periods. Fig. 7 shows a large region derived from an 'onion'
by the latter procedure, PCarb having been slowly excluded from a
large droplet to give a region richer in PCp and having a PCp
matrix.

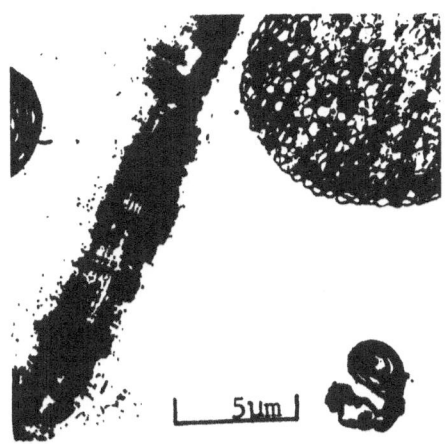

Fig. 7. A PCarb/PCp blend similar to that in Fig. 5a after
annealing in a chlorobenzene atmosphere.

Macroscopic phase separation of the type described above
occurs most readily if the crosslinking chains are very long compared with the A-chains and if large proportions of A-homopolymer
are present. We believe that macroscopic phase separation is also
responsible for the less-regular supramolecular features seen in
Fig. 5a, the sample perhaps having been cast too rapidly to allow
a greater degree of order to develop in the droplets. The structures in Fig. 5a are similar to those reported by Molau and
Wittbrodt (8) and structures similar to those reported by Bradford
(9) have also been observed in the course of these studies. It
is probable that macroscopic phase separation is responsible for
many unusual morphologies observed in blends of ABCPs and linear
block copolymers.

Droplets formed by macroscopic phase separation are very
stable. If solutions dried until macroscopic phase separation has
occurred are allowed to stand, without further drying, for a
period of weeks the droplets may coalesce to form two layers. Subsequent drying produces a two-layer film consisting of a crosslink-
rich phase and a homopolymer rich phase. A study of the interface

in such a film allows the process of coalescence and the approach
to equilibrium morphologies to be observed (19).

REFERENCES

1. M. Matsuo, S. Sagaye and H. Asai, Polymer, 10, 79 (1969).

2. G. Kämpf, M. Hoffmann and H. Krömer, Ber. Bunsenges. physik.
    Chem., 74, 851 (1970).

3. D. Meier, Polym. Prepr. Amer. Chem. Soc., Div. Polym. Chem.,
    11 (2), 400 (1970).

4. D. Meier, in 'Block and Graft Copolymers' (Ed. J. Burke and
    V. Weiss), Syracuse Univ. Press, Syracuse, 1973, p.105.

5. M. Hoffmann, G. Kämpf, H. Kromer and G. Pampus, in 'Multi-
    component Polymer Systems' (Ed. N.A.J. Platzer), Adv. Chem.
    Ser., 99, 351 (1970).

6. T. Inoue, T. Soen, T. Hashimoto and H. Kawai, Macromolecules,
    3, 87 (1970).

7. A. Douy and B.R. Gallot, Mol. Cryst. Liq. Cryst., 14, 191
    (1971).

8. G.E. Molau and W.M. Wittbrodt, Macromolecules, 1, 260 (1968).

9. E.B. Bradford, in 'Colloidal and Morphological Behaviour of
    Block and Graft Copolymers' (Ed. G.E. Molau), Plenum, 1971,
    p. 21.

10. C.H. Bamford, R.W. Dyson and G.C. Eastmond, J. Polym. Sci.(C),
    16, 2425 (1967).

11. C.H. Bamford, R.W. Dyson and G.C. Eastmond, Polymer, 10, 885
    (1969).

12. C.H. Bamford and G.C. Eastmond, in 'Recent Advances in Polymer
    Blends, Grafts and Blocks' (Ed. L.H. Sperling) Plenum,
    1974, p. 165.

13. For a summary of initiation mechanisms see C.H. Bamford in
    'Reactivity, Mechanisms and Structure in Polymer Chemistry'
    (Ed. A.D. Jenkins and A. Ledwith), John Wiley & Sons, 1974
    p. 52.

14. P.J. Flory, J. Phys. Chem., 46, 132 (1942).

15. C.H. Bamford, G.C. Eastmond and D. Whittle, Polymer, 16,
     377 (1975).

16. G.C. Eastmond and E.G. Smith, Polymer, 17, 367 (1976).

17. C.H. Bamford, G.C. Eastmond and D. Whittle, Polymer, 12,
     247 (1971).

18. G.C. Eastmond and E.G. Smith, to be published in Polymer,
     March 1977.

19. G.C. Eastmond and D.G. Phillips, unpublished results.

BLOCK-GRAFT COPOLYMERS BY GRAFTING PIVALOLACTONE TO

POLYISOPRENE AND POLYISOBUTYLENE

W. H. Sharkey, R. P. Foss, and J. F. Harris, Jr.

Central Research and Development Department
Experimental Station
E. I. du Pont de Nemours and Company, Inc.
Wilmington, Delaware 19898

Block and graft copolymers composed of unlike polymeric segments that separate into continuous and discontinuous phases often have unusual and useful properties. Styrene/diene triblock copolymers are familiar examples. These are thermoplastic elastomers in which a network structure is formed by separation of polystyrene blocks into glassy domains that tie together the continuous rubbery phase. Other examples are the $(AB)_n$ or segmented block polymers in which crystalline hard segments are interspersed with soft segments in a polymer chain. Examples are the polyurethane and poly-ether ester elastomers.

Use of crystalline polymers as grafts on thermo-plastic polymers has received little attention. It is probable that synthesis difficulties have discouraged work in this area. Grafting of vinyl monomers often leads to gelation and to formation of homopolymer from the grafting monomer. In addition, low grafting efficiency is often encountered as is backbone degradation.

Further, for efficient domain formation it is necessary to have grafts of relatively high molecular weight. Spacing these large grafts at distances that give typical elastomer crosslink densities would require relatively large amounts of certain vinyl polymers. At high grafting component content, hardening or phase inversion would be expected.

In Du Pont we have been working on ring-opening graft polymerization of pivalolactone on carboxylated polymers, which appears to be free of these problems. A preliminary report of this work has been given in two communications in Macromolecules (1, 2) and two papers at the ACS Centennial Meeting in New York (3,4). A more complete discussion will be given in this and the next two papers. This paper is concerned with the synthesis of thermoplastic elastomers by grafting polypivalolactone to polyisoprene and polyisobutylene.

Pivalolactone is the common name for the lactone of hydroxypivalic acid. Both the monomer (5) and its homopolymer are well known and have been extensively studied.

Caution: Pivalolactone should be handled with great care. Unpublished work at the Haskell Laboratory of Du Pont has shown that pivalolactone caused skin tumors in mice when applied as a 25% solution in acetone for most of the life span of the mice. The time required for tumor formation was greater than that for β-propiolactone, a positive control in the test, and the extent of tumor formation was much less.

FORMATION OF POLYPIVALOLACTONE GRAFTS

Studies on homopolymerization of pivalolactone were reported by Hall (6) in 1969, by Wilson and Beaman (7) in 1970, and by Mayne (5) (1972) and Oosterhof (8) (1974) of Shell International. Kramer, also of Shell International, disclosed grafting of pivalolactone to ethylene/propylene/ diene terpolymers in a Dutch patent application (9), though to our knowledge no follow-up of this work has appeared.

The polymerization of pivalolactone is a remarkably easy, anionic, ring-opening reaction.

Initiation is fast, propagation is fast, and there is no
chain transfer to monomer.

Once formed, polypivalolactone crystallizes very
rapidly to a product of uncommonly high crystallinity.
Heat of polymerization plus heat of crystallization
amounts to about 18.4 kcal./mole (5). Other properties
that are important include

- high melting point (230-235°C.)

- good melt stability

- insolubility in common organic solvents

- resistance to ultraviolet light

- resistance to hydrolytic and
  exchange reactions

The grafting reaction can be applied to almost any
soluble carboxylated polymer. The reaction is best
carried out in aprotic solvents in which reactivity of the
carboxylate is high, e.g., tetrahydrofuran. For fast
initiation and propagation, separation of the propagating
ion pair is essential. For this reason, the carboxyl
group is neutralized with tetrabutylammonium hydroxide.

It is important to note that this hydrogen exchange
between free acid and salt is fast, and therefore all
acid groups act as graft sites regardless of the degree
of neutralization. We customarily neutralized to 80-90%
of the theoretical amount. An important consequence
of this behavior is that grafting frequency is controlled
only by degree of carboxylation. Thus, the number of

carboxyl group sets the distance between grafts and this distance is called the soft segment length, SSL.

Once initiator sites are formed, ring-opening polymerization of pivalolactone (PVL) is carried out by addition of monomer. Initiation and propagation involve nucleophilic attack of the carboxylate anion on the methylene group of the monomer molecule.

$$
CO_2^{\ominus} \; + \; CH_3\!-\!C(CH_3)(CH_2\!-\!O)(C\!=\!O) \quad \longrightarrow \quad CO_2\!-\!CH_2\!-\!C(CH_3)_2\!-\!CO_2^{\ominus}
$$

$$\downarrow \text{PVL}$$

$$
CO_2\!-\![CH_2\!-\!C(CH_3)_2\!-\!CO_2]_n^{\ominus} \quad \xleftarrow{\;H^{\oplus}\;} \quad CO_2\!-\![CH_2\!-\!C(CH_3)_2\!-\!CO_2]_n\!-\!H
$$

Since chain transfer to monomer does not occur in this polymerization and there is no other termination reaction, the growing polymer anion is "living" in the absence of chain terminating impurities. Also, as Hall has reported, initiation and propagation rates are similar As a result, a Poisson molecular weight distribution is obtained in which polydispersity index, M_w/M_n, is very low, approaching 1. Thus, this chemistry allows efficient use of the grafting chains for formation of elastomeric networks.

The effectiveness of polypivalolactone crystalline tie points for converting rubbery backbones to thermoplastic elastomers has been demonstrated for a number of elastomers. This discussion will be concerned with two, namely, poly-cis-1,4-isoprene and polyisobutylene. The goal in the polyisoprene case was two-fold; to compare crystalline crosslinks to sulfur crosslinks and to develop a product that could be melt-spun into an elastic

fiber. The incentive in the isobutylene case was to
ascertain the effect on properties of a non-auto-
oxidizable, rubbery backbone.

POLY(PVL-b-ISOPRENE-b-PVL)-g-PVL

For the polyisoprene case, a difunctional initiator
capable of giving α,ω-dilithio-polyisoprene of very
high cis-1,4 content was needed. The reason is to enable
most efficient use of polypivalolactone blocks by having
them at either end of the polymer molecule as well as
along the chain. Earlier difunctional initiators, which
include 1,4-dilithio-1,1,4,4-tetraphenylbutane (10) and
2,5-dilithio-2,5-diphenylhexane (11), were made using
small amounts of ethers and ethers are known to lead to
decreased cis-1,4 content. After the work described here
was completed, Morton and Fetters (12) reported a new difunctional initiator, which they obtained by adding
lithium to a hexadiene, that appears to give polyisoprenes of high cis-1,4 content.

For our work, we developed a dilithio initiator that
has been shown to give polyisoprene having the same high
1,4 content as obtained with butyllithium. It has the
formula

$$(Et_3N)_{0.1} \quad Li \text{---}(\text{isoprene})_5 \text{---} \underset{\underset{RCH_2}{|}}{\overset{\overset{CH_3}{|}}{C}} \text{---} \bigcirc \text{---} \underset{\underset{CH_2R}{|}}{\overset{\overset{CH_3}{|}}{C}} \text{---}(\text{isoprene})_5 \text{---} Li \quad (Et_3N)_{0.1}$$

R = s-butyl

It is made by addition of s-butyllithium to m-diiso-
propenylbenzene in the presence of sufficient triethylamine, 0.1 equivalent, for promotion of the addition at
a reasonable rate. Then five equivalents of isoprene
are added to insure solubility in hydrocarbon solvents.
This procedure is based on earlier work by Fetters and
Morton and Karoly on dilithio diene initiators. We
named this initiator Diplit 5.1 for convenience (13). It
is a contraction of diisopropenyllithium triethylamine

followed by a number for equivalents of isoprene and a
second number for equivalents of triethylamine. This
initiator has been shown to be exactly difunctional and
it leads to polyisoprenes of very narrow molecular
weight distributions as is shown in Figure 1.

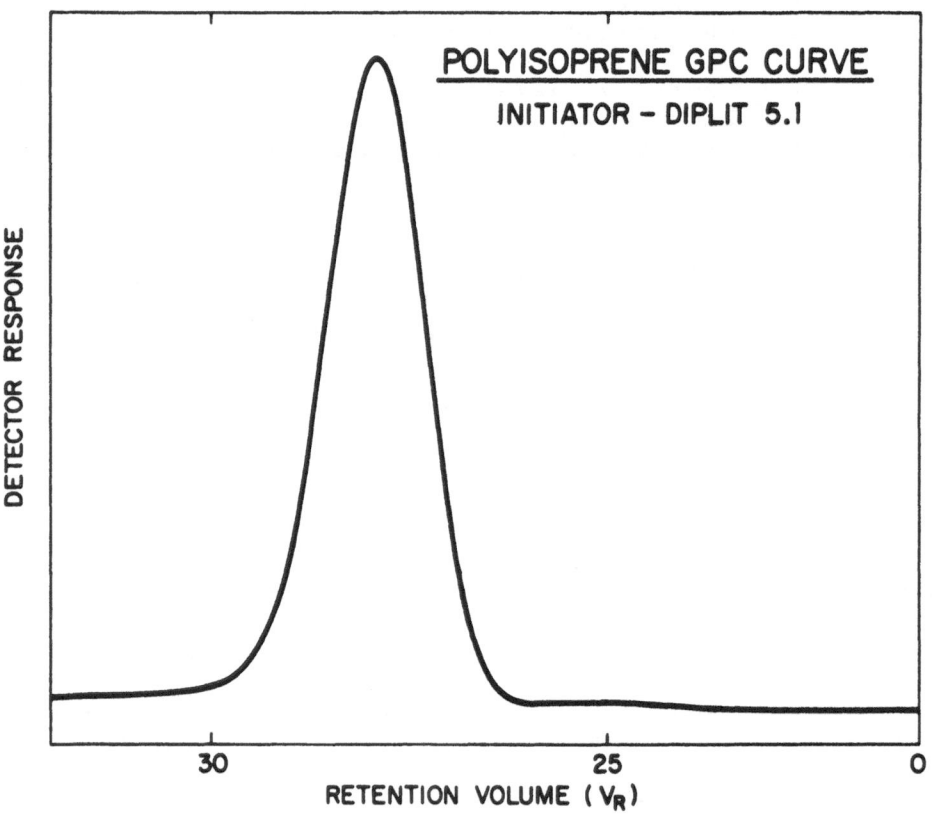

Figure 1

Synthesis of block-graft copolymers using Diplit 5.1 was accomplished by use of the following steps.

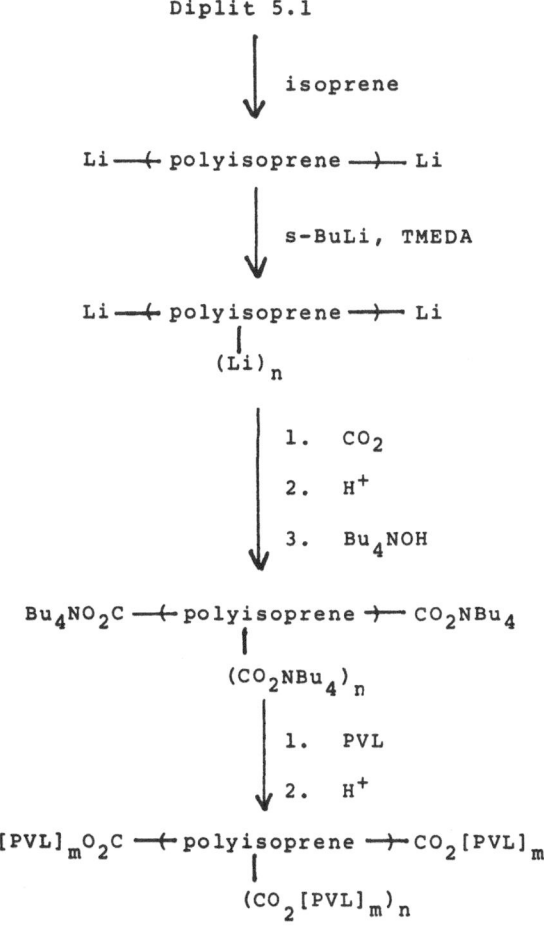

In the first step, isoprene is polymerized to α,ω-dilithiopoly-cis-1,4-isoprene. This product has been used to make ABA triblocks, but they will not be discussed here. For block-graft polymers, additional lithio sites were introduced randomly along the polyisoprene chain by reaction with s-butyllithium in the presence of tetramethylethylenediamine (TMEDA). The amine-activated butyllithium is an exceptionally strong base and reacts at allylic positions to remove protons and form anions. The other product is butane. This reaction was described by Tate, et al. (15) in 1971 and was the subject of a symposium at the 164th ACS meeting in New York, September, 1972.

Reaction of the lithiated polyisoprene with carbon dioxide converted all of the lithium sites to carboxyl groups at either end and along the chain. The carboxyl groups were then 90% neutralized with tetrabutylammonium hydroxide and the polymeric salt was used to initiate polymerization of pivalolactone.

It is necessary to take great care to protect the reaction systems from air and moisture through the first three steps. This was of particular importance because of our desire to have polypivalolactone blocks at the ends of the backbone. The need to preserve functionality in synthesis of triblock copolymers has been emphasized by others and techniques have been developed for doing so. The method we worked out does not require a vacuum train or breakseals. It is illustrated in Figure 2.

SYNTHESIS OF PIVALOLACTONE—ISOPRENE BLOCK-GRAFT COPOLYMER

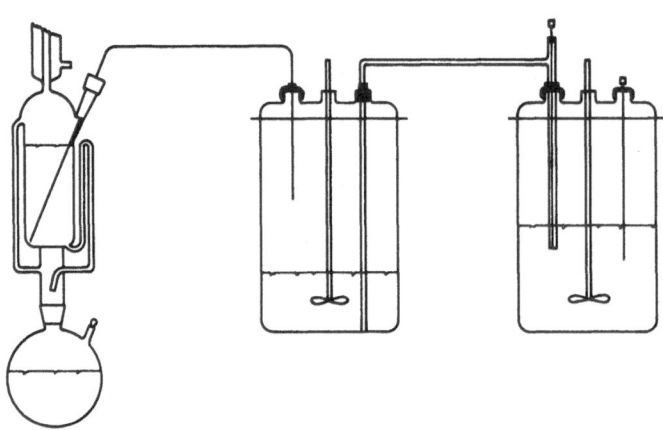

Figure 2

On the left is a solvent purification area, which involves a Soxhlet extractor modified by addition of an exit port and over which was kept a nitrogen or argon blanket. Sodium was placed in the flask. The solvent, cyclohexane, was kept under continuous reflux, which assured that all

impurities were removed by the sodium. To transfer
solvent to the second area, the polymerization reactor,
blanketing gas pressure was increased in the Soxhlet.
After solvent had been transferred, the connection was
removed and the serum capped opening used for addition
of isoprene and initiator.

Isoprene polymerization was done at 45-50°C. for 3
hours, after which tetramethylethylenediamine and s-butyllithium were added. After stirring for one hour at room
temperature, the reaction mixture was pumped into the
third area, the carboxylation reactor, by increasing
blanketing gas pressure in the polymerization vessel.
Prior to this step, the carboxylation reactor had been
charged with a solution of tetrahydrofuran saturated with
carbon dioxide. Additional carbon dioxide was continuously bubbled through the solution. The inlet tube
contained an inner tube through which nitrogen was passed
to protect the polymer solution from carbon dioxide until
it was in the tetrahydrofuran solution where it would be
immediately exposed to a large excess of carbon dioxide.
In this way, formation of gel by chemical crosslinking was
avoided.

After acidification, methanol was added. The upper
layer was separated, washed with water, dried, filtered,
and concentrated by evaporation of solvent. The number
of carboxyl groups was determined by titration and by
comparison of infrared absorption by carboxyls to that by
carbon-carbon double bonds (13).

Conversion of the carboxylated polymer to a tetrabutylammonium salt and use of the salt to initiate
polymerization of pivalolactone was done as discussed
earlier. Graft polymerization at 60°C. is rapid; gelling
usually occurs within a few minutes and the reaction is
probably complete in one hour or so. Conversions in this
step were almost always quantitative.

The products are strong, resilient thermoplastic
elastomers. They can be melt-pressed into films and melt-
spun into fibers at temperatures of 240-280°C. Films
and fibers increase greatly in strength when oriented by
drawing. Orientation requires a number of cycles of
stretching and relaxing. Since strength increases with
orientation, force during stretching can be increased as
the number of cycles is increased up to the point of
maximum orientation.

Properties of oriented fibers from copolymers of backbone Mn 50,000-60,000 containing 25-51% PVL distributed in 4-18 segments per isoprene chain are given in Table 1.

Table 1

Properties of Fibers from PIP-g-P

SSL	% PVL	HSL	T_b g/d	E_b (%)
20,000	25	6,000	0.4	660
14,000	35	6,000	0.6	400
10,200	40	5,700	1.0	440
11,000	50	5,000	0.85	210
3,500	51	4,900	1.38	170

Soft segment lengths (SSL) are statistical averages calculated from degree of carboxylation and Mn. Hard segment lengths (HSL) were determined by distributing PVL evenly among the graft sites. HSL divided by 100 gives the number of PVL units per segment, since molecular weight of PVL is exactly 100.

Tenacities as high as 1.4 grams per denier were obtained. Since one gram per denier represents a strength of about 12,800 psi., this value corresponds to a tensile strength in the 17,000-18,000 psi. range. Strength increased with increase in pivalolactone content up to 50-51%. Products containing still higher amounts of pivalolactone were much less elastic probably because of partial phase inversion.

Apparently the length of the pivalolactone segment, HSL, once beyond an appropriate minimum, has a minor effect on properties. This conclusion is supported by the data in Table 2., which compares fibers from block-graft copolymers having backbones of 50,000 to 90,000 in molecular weight all containing 35% PVL. The fiber with lowest strength had the longest soft segment length. The other samples, which had shorter hard segments spaced closer together, had significantly higher strength.

Table 2

Block-Graft Copolymers
35 Wt. % PVL

$\overline{M}n$ of B Block	SSL	HSL	T_b (g/d)	E_b (%)
53,000	14,000	6,000	0.6	400
52,000	10,000	4,600	0.9	450
75,000	7,500	3,700	0.9	450
92,000	9,200	4,700	0.9	430

The stress-strain curves obtained with these oriented fibers have shapes typical for cured, highly resilient elastomers. Figure 3 shows a hysteresis loop for an oriented fiber extended to 300% elongation.

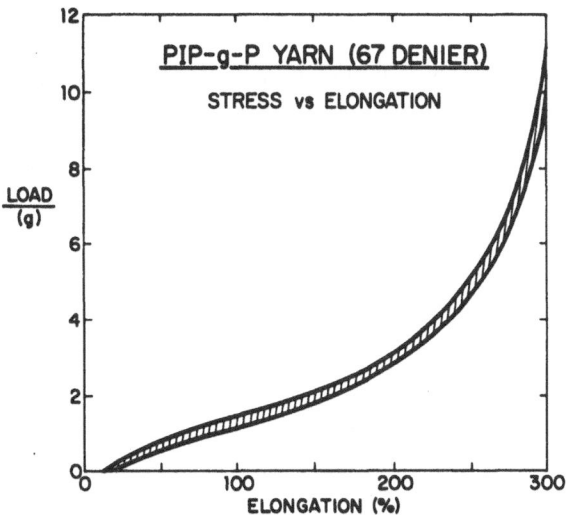

Figure 3

The top curve describes behavior as load is applied and the lower one shows recovery as load is removed. The area between is work lost in the cycle and amounts to about 13% of the area under the load curve. This means

a resilience of 87%, which compares favorably to sulfur-
cured natural rubber. From this result, we conclude that
crystalline polypivalolactone crosslinks are as effective
as chemical crosslinks.

     This conclusion is supported by stress decay data.
When an oriented fiber was protected against autooxidation by incorporation of phenyl-β-naphthylamine,
then extended and held extended for 20 hours, the stress
loss was no greater than that of sulfur-cured gum rubber.
We ascribe this unusual retention of strength to the
effectiveness of the hard segments in providing crystalline tie points for the major rubbery phase.

     POLY(ISOBUTYLENE-co-METHYLSTYRENE)-g-PIVALOLACTONE

     It should be emphasized that the stress decay data
just discussed for copolymers from pivalolactone and
isoprene could only be obtained when the polymer was
protected with large amounts of antioxidant. Apparently,
an important factor in strength loss is rapid destruction
of the backbone by fast air autooxidation. To pursue
this point, we undertook a study of graft copolymers of
isobutylene. In this case, autooxidation of the backbone should be a minor or nonexistent matter.

     The starting products were isobutylene copolymers
containing methylbenzene groups, which were introduced
to provide sites for lithiation and carboxylation (16).
The copolymers were made by cationic copolymerization of
isobutylene with 4-methylstyrene, a mixture of 2,3- and
4-methylstyrenes denoted as vinyltoluene, and a mixture
of 4-methylstyrene and α-methylstyrene.

     Though methylstyrene content was varied over a wide
range, the examples selected for this discussion contained
only 1-2 mol % of comonomer. It was assumed these comonomers were distributed randomly, though it is
recognized this might not have been the case. Typical

preparations, which will be used as examples for the remainder of this discussion, are shown in Table 3.

Table 3

Isobutylene Copolymers

Polymer	Comonomer	Polymer Characterization		
		Mol % Comonomer	Mw/Mn	Mv
I	4-Methyl- styrene	0.9	5	550,000
II	Vinyl- toluene	1.6	21	305,000
III	4-Methyl- styrene α-Methyl- styrene	0.7	3	80,500
IV	4-Methyl- styrene	2.0	3	66,500

Molecular weight distributions were quite broad, especially in the vinyltoluene case. This feature may have contributed to the lower strengths of the final product relative to the isoprene products. Such broad distributions are indicative of large numbers of lower molecular weight chains.

Attachment of polypivalolactone grafts involved the same chemistry discussed earlier.

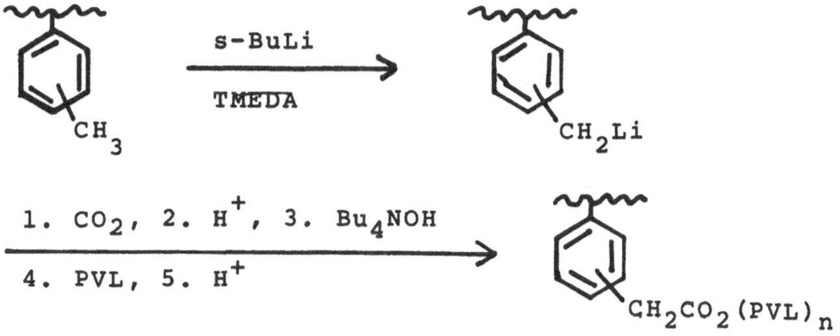

s-BuLi

TMEDA

CH_3 → CH_2Li

1. CO_2, 2. H^+, 3. Bu_4NOH

4. PVL, 5. H^+

$CH_2CO_2(PVL)_n$

Reaction with s-butyllithium in the presence of TMEDA
led to metallation at the methyl group. From this point
on the reactions used were similar to those described
earlier. Carboxylation efficiencies varied from 33 to
68% for the examples discussed earlier (Table 4).

Table 4

Carboxylation Efficiency

Polymer	Eq. wt. per aromatic group	Eq. wt. per carboxyl	Eff. (%)
I	6,250	10,000	63
II	3,650	11,000	33
III	3,540	10,500	68
IV	2,500	6,000	42

     The poly(isobutylene-co-methylstyrene)-g-pivalo-
lactone products ranged from sticky semisolids for those
containing very low amounts of PVL, through a rubbery-
thermoplastic region as PVL content was increased, to
hard, non-elastic thermoplastic solids for PVL contents
over 60%. Melt-pressed films from polymers containing
20-60% PVL were as strong or stronger than conventionally
cured polyisobutylene. Strength increased markedly upon
orientation by drawing. As indicated by the data in
Table 5, high tensile strengths are favored by high
backbone molecular weight and particularly by high PVL
content.

Table 5

Isobutylene Copolymer Properties

Polymer	Mv	SSL	% PVL	HSL	T_b (psi)
I	550,000	10,000	42	7,500	7,500
II	305,000	11,000	25	3,600	4,000
III	80,500	10,500	36	6,000	3,000
IV	66,500	6,000	11	750	1,000

These graft copolymers were spun into fibers at 250-260°C., and the fibers obtained were oriented by alternate stretching and relaxing as described earlier. They were not as strong as the fibers from isoprene/pivalolactone copolymers, as tenacities were in the 0.4-0.5 g/d range. This lower strength may be associated with the broad molecular weight distribution and the inability to place hard segments on the ends of the chain. Oriented isobutylene copolymer fibers did have good hysteresis properties, however. Work lost on 300% extension was only 15% for a fiber from Polymer II. Also, stress decay was outstanding. Fibers elongated 300% dropped only 11% in stress the first hour with a further loss of only 5% in the next 22 hours. This is much better than that for the block-graft isoprene copolymer or sulfur-cured natural rubber and leads to the conclusion that autooxidation is a significant factor in long term stress decay.

REFERENCES

1. S. A. Sundet, R. C. Thamm, J. M. Meyer, W. H. Buck, S. W. Caywood, P. M. Subramanian, and B.C.Anderson. Macromolecules, 9, 371, (1976).

2. R. P. Foss, H. W. Jacobson, H. N. Cripps, and W. H. Sharkey, Macromolecules, 9, 373 (1976).

3. R. C. Thamm and W. H. Buck, Polym. Prepr.,Amer.Chem. Soc., Div. Polym. Chem., 17 (1), 205 (1976).

4. W. H. Sharkey, R. P. Foss, H. W. Jacobson, and H. N. Cripps, Polym. Prepr., Amer. Chem. Soc., Div. Polym. Chem., 17 (1), 211 (1976).

5. N. R. Mayne, Chem. Tech. 1972, 728.

6. H. K. Hall,Jr., Macromolecules, 2, 488 (1969).

7. D. R. Wilson and R. G. Beaman, J. Polymer Science, A-1, 8, 2161 (1970).

8. H. A. Oosterhof, Polymer, 15, 49 (1974).

9. S. K. Kramer, Netherlands application 7,030,074(1971) Shell International Research.

10. L. J. Fetters and M. Morton, Macromolecules, 2 (3), 435 (1969).

11. G. Karoly, Polm. Prepr., Amer. Chem. Soc., Div. Polym. Chem., 10 (2), 837 (1969).

12. M. Morton, L. J. Fetters, J. Inomata, D. C. Rubio, and R.N.Young, Rubber Chem. Tech., 49, 303 (1976).

13. R. P. Foss, H. W. Jacobson, and W. H. Sharkey, Macromolecules, in press.

14. Diplit 5.1 is a contraction of diisopropenyl-lithium triethylamine followed by a number for equivalents of isoprene and a second number for equivalents of triethylamine.

15. D. P. Tate, A. F. Halasa, F. J. Webb, R. W. Koch and A. E. Obster, J. Polm. Sci., A-1, 9, 139 (1971).

16. J. F. Harris, Jr., and W. H. Sharkey, Macromole-cules, in press.

GRAFT COPOLYMERS OF PIVALOLACTONE: SYNTHESIS AND PROPERTIES

S. A. Sundet and R. C. Thamm

Plastic Products and Resins and
Elastomer Chemicals Departments
E. I. duPont de Nemours and Company
Experimental Station
Wilmington, Delaware 19898

INTRODUCTION

The rapid ring-opening polymerization of pivalolactone (PVL), initiated and propagated anionically by the carboxylate anion, occurs under mild conditions relatively free of interfering reactions. The driving force for the reaction is the relief of the strain in the four-membered ring and the heat of polymerization is reported (1) as 18.4 kcal. per mole, including 2.9 kcal. for crystallization. The aliphatic polyester formed is high melting, thermally and hydrolytically stable, and highly crystalline (2-4).

$$(1)$$

These properties of pivalolactone and its polymer make it especially useful and effective for modifying carboxylated polymers through graft polymerization (5-8). The absence of a common termination reaction or chain transfer to monomer in tetrahydrofuran solution leads to grafting efficiencies and pivalolactone conversions approaching 100%.

The purpose of this paper is to discuss the scope and
technique for this facile grafting reaction, the experimental
variables and the property changes observed.

EXPERIMENTAL

Pivalolactone

The pivalolactone (PVL) was 99.9+% pure by gas chromatography,
stored in a freezer at -20 to -30°C as a crystalline solid, and
filtered through neutral alumina just before use to remove traces
of homopolymer sometimes formed during storage.

Unpublished work on its physiologic activity has been
noted (5) and discussed briefly in the preceeding chapter.

Poly(ethylene-*co*-vinyl acetate-*co*-methacrylic acid)

These copolymers were commercial and experimental grades of
Elvax® resins, Du Pont's copolymers of ethylene and vinyl acetate.
Methacrylic acid contents were determined by titration.

Poly(ethylene-*co*-vinyl acetate-*co*-methacrylic acid-*g*-PVL)

In a typical procedure, 200 g of base resin was dissolved in
600 ml of tetrahydrofuran, the solution warmed to a gentle reflux
and 20-50% of the available carboxyl groups in the base resin were
neutralized with tetrabutylammonium hydroxide (1.0 molar in
methanol). The pivalolactone was then added as rapidly as the
exotherm would permit (10-20 min.), and heating and stirring
continued for 30-120 minutes more to complete the reaction. When
the system remained fluid it was acidified with concentrated
hydrochloric acid before precipitating the product in warm water
in a blender. The nontacky solid was washed well with water and
dried in a vacuum oven at 80°C. Conversions of the pivalolactone
to polymer were normally above 95%, often complete.

Poly(ethylene-*co*-propylene-*co*-1,4-hexadiene) (EPDM)

EPDM was a commercial grade of Nordel®, Du Pont's hydrocarbon
elastomer.

EPDM-*g*-thioglycolic acid (EPDM/T) was prepared by the
reaction of EPDM with large excesses of thioglycolic acid and α,α-
azobis-(δ,δ-dimethylvaleronitrile) in tetrahydrofuran (9).

It was isolated by precipitation in acetone, and was further purified by solution and reprecipitation. Carboxyl functionality was determined by a calibrated infrared analysis of unreacted unsaturation.

EPDM-*g*-succinic anhydride (EPDM/S) was prepared by the reaction of EPDM and maleic anhydride in bulk at 260-300°C by reported procedures (10). Unreacted maleic anhydride was removed by solution in THF and precipitation in acetone. Grafted anhydride or derived acid was determined titrimetrically or by infrared analysis.

Poly(EPDM-*g*-PVL)

200 g of EPDM/T containing 0.056 equivalents of carboxyl functionality was dissolved in 4 l of THF; addition of 0.035 mol of TBAH with stirring turned the solution light yellow. The mixture was heated to 60°C, 70 g (0.70 mol) of PVL was added, and heating was continued for 2 h after which the solution was cloudy and amber colored. It could not be cooled without irreversible insolubilization of the product. Conc. HCl (4 ml, ca. 0.05 mol) was added, partially discharging the color. Antioxidant was added and then cold water was slowly and continuously added to the hot, stirred reaction mixture until, at about equal volumes, it abruptly became very viscous. With continued stirring the product precipitated as a fine, white, non-massing powder which could be filtered and dried. Particles were about 50 μm in size and loosely agglomerated. Dried weight was 267 g, corresponding to a PVL conversion of at least 96%.

Poly(ethyl acrylate)

Poly(ethyl acrylate) was prepared by conventional emulsion polymerization using an anionic soap and a sodium sulfite/ potassium persulfate initiator. Dodecylmercaptan was used as a chain transfer agent to control molecular weight, generally so that $\overline{M}_n$ was above 10^6. Rigorous purification of the product by washing with methanol and water was required for best results in the subsequent grafting reaction.

Poly([Ethyl Acrylate]-*g*-Pivalolactone)

40 g of poly(ethyl acrylate) dissolved in 1 l of THF was refluxed (67°C) with 0.0027 mol of TBAH for 2 h. Saponification was 85-90% of theoretical under these conditions.

10 g (0.1 mol) of PVL was added and the mixture was heated at
60-70°C for 2 h. One ml of concentrated HCl was added and the
product was precipitated by adding water slowly to the stirred
mixture. The polymer, which precipitated as a flocculent solid,
was easily removed by filtration, and was purified by washing
with hot methanol to yield 49.3 g of copolymer, corresponding to
a PVL conversion of at least 93%.

DISCUSSION

Scope

Suitable base polymers for this grafting reaction preferably
contain carboxyl functionality, either acid, ester or anhydride.
Such groups may be introduced by appropriate copolymerization
reactions, or by post reactions as in the case of poly(ethylene-
co-propylene-co-1,4-hexadiene)(EPDM). Here, carboxyl groups were
introduced by the peroxide-catalyzed addition of thioglycolic acid
to the unsaturation, or by thermal reaction with maleic anhydride.
Some of the compositions which have been successfully grafted are:

Poly(ethylene-co-methacrylic acid)
Poly(ethylene-co-vinyl acetate-co-methacrylic acid)
Poly(ethylene-co-isobutyl acrylate-co-methacrylic acid)
Poly(ethylene-alt-ethyl acrylate)
Poly(ethylene-co-propylene-co-1,4-hexadiene)
Poly(acrylate esters)
Poly(chloroprene-co-methacrylic acid)
Polydimethyl siloxane with pendant carboxyl groups

The only limitation so far envisioned involves those backbone
polymers whose functional groups may be sensitive to the
nucleophilic attack of the carboxylate anion. In some structures
this attack might lead to gelation, in others to the removal of
hydrogen halide, for example, with the effective termination of
the anionic polymerization. Although unexplored, functional
groups with reactivity that could prove troublesome include epoxy,
nitriles and labile hydrogen functions.

To minimize the side reactions that complicate heterogeneous
polymerization in bulk, the procedures reported here were
conducted in homogeneous solution at temperatures of 60-85°C.
Solvents, while they must be suitable for the backbone polymer,
are preferably polar and must be aprotic. Tetrahydrofuran, in
particular, offers fast initiation, high reaction rates and ease
of product isolation by precipitation in water. Reactions in
toluene were slower but the thermal and mechanical properties of

many of the products isolated from this system by precipitation in ethanol were the same as those obtained from THF reaction.

Graft Site Chemistry and Control of Graft Frequency

Carboxylic acid, ester and anhydride groups are all converted to carboxylate anion initiating sites by reaction with methanolic tetrabutylammonium hydroxide. In these systems the base probably exists as methoxide as well as hydroxide.

$$\text{Polymer-CO}_2\text{H} + \text{Bu}_4\text{N}^{\oplus}\text{OH}^{\ominus} \rightarrow \text{Polymer-CO}_2^{\ominus} + \text{Bu}_4\text{N}^{\oplus} + \text{H}_2\text{O} \quad (2)$$
$$(\text{OCH}_3^{\ominus}) \qquad\qquad\qquad (\text{CH}_3\text{OH})$$

$$\text{Polymer-CO}_2\text{R} + \text{Bu}_4\text{N}^{\oplus}\text{OH}^{\ominus} \rightarrow \text{Polymer-CO}_2^{\ominus} + \text{Bu}_4\text{N}^{\oplus} + \text{ROH} \quad (3)$$

$$\begin{array}{ll}
\text{Polymer-CH-C}{\overset{\nearrow \text{O}}{\underset{\searrow}{}}} & \rightarrow \quad \text{Polymer-CH-CO}_2^{\ominus} + \text{Bu}_4\text{N}^{\oplus} + \text{H}_2\text{O}\\
\quad /\quad \backslash \text{O} + \text{Bu}_4\text{N}^{\oplus}\text{OH}^{\ominus} & \qquad\qquad |\\
\text{CH}_2\text{-C}{\overset{}{\underset{\searrow \text{O}}{}}} \quad +\text{CH}_3\text{OH} & \qquad\quad \text{CH}_2\text{-CO}_2\text{CH}_3 \qquad (4)\\
& \qquad\quad (\text{or isomer})
\end{array}$$

In (2), complete neutralization of the carboxylic acid is unnecessary since all carboxyl groups can participate in propagation through proton exchange.

$$\text{R-CO}_2^{\ominus} + \text{R'CO}_2\text{H} \overset{\leftarrow}{\rightarrow} \text{RCO}_2\text{H} + \text{R'CO}_2^{\ominus} \quad (5)$$

Thus, grafting frequency is determined in this system by the concentration of base polymer carboxylic acid, measurable by analysis.

Complete neutralization is probably undesirable since any residual low concentration of $\text{OH}^{\ominus}$ (or $\text{OCH}_3^{\ominus}$) is further depressed by unneutralized carboxyls

$$-\text{CO}_2\text{H} + \text{OH}^{\ominus} \overset{\rightarrow}{\leftarrow} -\text{CO}_2^{\ominus} + \text{H}_2\text{O} \quad (6)$$

and initiation of homopolymer by $\text{OH}^{\ominus}$ minimized.

The presence of minor amounts of moisture, as a contaminant in the solvents, had little influence on the polymerization or the properties of the graft polymers formed. The unfavorable equilibrium (6), combined with the low concentrations of carboxylate anion and water present, results in ineffective concentrations of hydroxyl ion for nongraft initiation.

In the ester reaction (3), grafting frequency is determined by the amount of base successfully reacted with ester groups. Thus, graft frequency can be varied on a single base resin.

Incompleteness of the saponification, however, can result in free
base initiating formation of homopolypivalolactone. Purification
of the saponified polymer may be required in rigorous fundamental
work.

The reactions of the anhydride (4) involve both the quaternary
base and the solvent methanol. It may be visualized as two
reactions: the first the reaction of methanol (present in excess)
with anhydride groups to form the acid ester; and the second the
partial neutralization of the developed carboxyl function.
Methanolysis of anhydride groups is complete even with less than
stoichiometric amounts of the quaternary base and one initiating
site is developed for each anhydride residue. Evidence for this
reaction included infrared analysis of the products and the
similarity of the physical properties of the graft copolymers
obtained from the anhydride or from methyl hydrogen succinate
sites (8). Succinic acid sites gave products with distinctly
different properties.

In the anhydride system, graft frequency is determined by
the concentration of anhydride residues and the completeness of
reaction with the alcoholic base.

Thus, almost any carboxylate anion is an effective initiator
for the ring-opening polymerization of pivalolactone. The only
restriction for the counterion is that it should not coordinate
strongly with the carboxylate. Tetraalkylammonium ions are
especially effective with little evidence for any advantage from
methyl to butyl. It is conceivable that in less polar solvents the
tetrabutylammonium cation may be advantageous. Tetrahydroxyethyl-
ammonium cations were not effective. Inorganic counterions such as
calcium, lithium and sodium were not useful although some evidence
of grafting was observed when lithium was complexed with agents
like hexamethylphosphoramide and tetramethylethylenediamine. The
products obtained were much inferior to those obtained with
quaternary ammonium cations.

Homopolymer Formation

Homopolymer formation was not observed in systems based on
copolymers containing methacrylic acid. Solvent extraction tests,
gel permeation chromatography and differential scanning calorimetry
all suggest formation of small amounts of homopolymer in grafting
an EPDM functionalized with thioglycolic acid. In one example (8),
10% of the reacted PVL was extracted with hexafluoroisopropanol as
a homopolymer of molecular weight twice that of the grafted side
chains. This was attributed to initiation by a small amount of
$(SCH_2CO_2H)_2$ formed during base polymer modification.

The problems associated with poly(ethyl acrylate) purification and saponification have already been mentioned. Impurities carried through from the emulsion polymerization of ethyl acrylate and unreacted tetrabutylammonium hydroxide contribute to the homopolymerization of up to 20-25% of the added PVL. In these cases, however, the homopolymer molecular weights were usually lower, and never higher than graft polyPVL molecular weight. These homopolymers were probably initiated by non-carboxylic impurities.

Graft Stoichiometry

Since conversion and grafting efficiencies are close to theoretical and chain transfer to monomer does not occur, we have described graft polymer microstructure--average graft degree of polymerization (graft $\overline{DP}$) and the average base polymer segment molecular weight between grafting sites--by calculations based on system stoichiometry. Molecular weight per site is derived from knowledge of the concentration of active sites on the backbone, and graft $\overline{DP}$ can be calculated from mole ratio of pivalolactone monomer to initiator sites.

Effects of Grafting

Under relatively mild conditions a clean ring-opening reaction leads to the growth of polyester sidechains with a strong tendency to crystallize. A two-phase morphology was shown by transmission electron microscopy (11) and the presence of polypivalolactone crystallinity was established by wide angle x-ray studies, to be described in detail by Warren Buck in the chapter to follow. Differential scanning calorimetry also provides evidence for crystallinity. In Figure 1 the strong endotherm marks the melting temperature of the polypivalolactone sidechains and the exotherm on cooling confirms their crystallization.

To the extent that these side chains form microdomains of highly crystalline polypivalolactone they develop a thermoplastic network within the backbone polymer substrate that can exert a profound influence on its mechanical and physical properties.

Flow Behavior. Thermal properties respond sharply to the presence of the higher melting crystalline side chains of poly-pivalolactone. For example, the polymer identified in Figure 1 flows only when temperatures of this last endotherm are reached. For this reason, minimum flow temperatures were used to

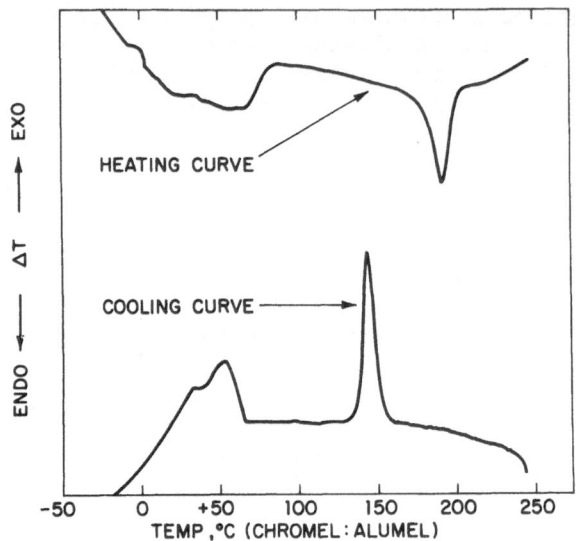

Figure 1 - DSC Analysis of Poly([Ethylene-*co*-Vinyl Acetate-*co*-
    Methacrylic Acid]-*g*-Pivalolactone) (60/21/1/18)
            Mol. Wts.: Base Resin   13,000
                       Per Site   5,700
                       Graft $\overline{DP}$ 12

characterize grafting in those graft copolymers exhibiting
important changes in flow behavior. Automated minimum flow
temperatures (AMFT) were determined in a melt indexer with an
orifice of 0.21 cm at a loading of 3 Kg/cm^3 (43 lbs/in^2) with the
temperature programmed to increase at 1°C per minute. The flow
temperature must correspond to the melting temperature of the
polypivalolactone side chains whose crystallization establishes
the flow-resisting network. With thermoplastic backbones that
exhibit their own softening phenomena, improvements in creep
resistance or minimum flow temperature occur as the melting
temperature of the grafts exceeds the flow temperature of the
backbone resin. Elastomeric backbones, exhibiting no melt
phenomenon of their own but crosslinked by the domain formation
of the polypivalolactone grafts, can be readily processed only
above the melting temperature of such grafts.

    The melting point of these crystalline polymer segments
increases with their length or degree of polymerization. In
Figure 2 the minimum flow temperatures of a series of resins based
on poly(ethylene-*co*-vinylacetate-*co*-methacrylic acid) are
correlated with the degree of polymerization of the

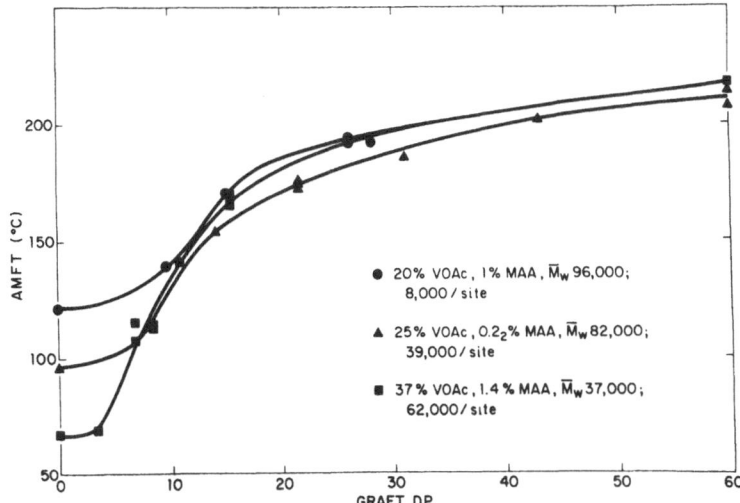

Figure 2 - Effect of Graft $\overline{DP}$ on Minimum Flow Temperatures
of Ethylene Copolymers

polypivalolactone side chains. Similar data were obtained with a
variety of copolymers as backbone resins.

Above their melting points these side chains have an effect
on the melt viscosity which is dependent on the base resin. In
Figure 3 are compared the melt indices at two temperatures of two
graft copolymers and the poly(ethylene-*co*-vinylacetate) from which
they were prepared. Despite modest increases in molecular weight
(2600 and 12,000) these two graft copolymers have slightly higher
flow rates under melt indexing conditions. On the other hand,
bulk viscosity increases were noted when EPDM was similarly grafted.

Solubility. In the initial phases of the grafting reaction
there is an increase in the solution viscosity that may culminate
in gelation of the system. Since polypivalolactone is insoluble
in common organic solvents the solubility of the isolated graft
copolymers is rapidly lost as the degree of polymerization of the
side chains increases. The extractibility of a variety of graft
copolymers based on poly(ethylene-*co*-vinyl acetate-*co*-methacrylic
acid) backbone resins, determined with toluene for five hours in a
Soxhlet extractor, is shown in Figure 4. Solubility in boiling
toluene is complete up to a graft $\overline{DP}$ of about 25, beyond which
percent extracted falls off rapidly. When the extractant is
tetrahydrofuran the limiting degree of polymerization is 7. When
the extractant is the even poorer solvent, cyclohexane, the
limiting graft DP is approximately 2 1/2.

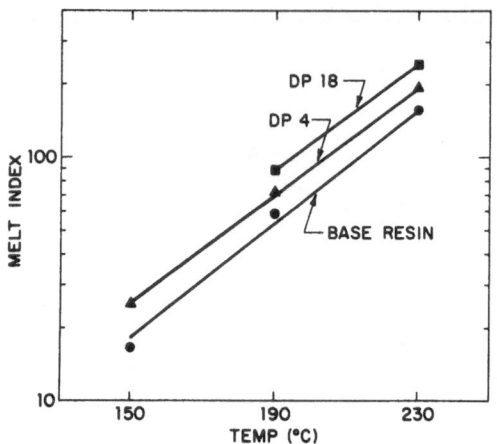

Figure 3 - Effect of Grafting PVL on Flow Behavior of
Poly(Ethylene-*co*-Vinyl Acetate-*co*-Methacrylic Acid)(59/39/1.6)
Mol. Wts.: Base Resin   33,000
Per Site   5,000

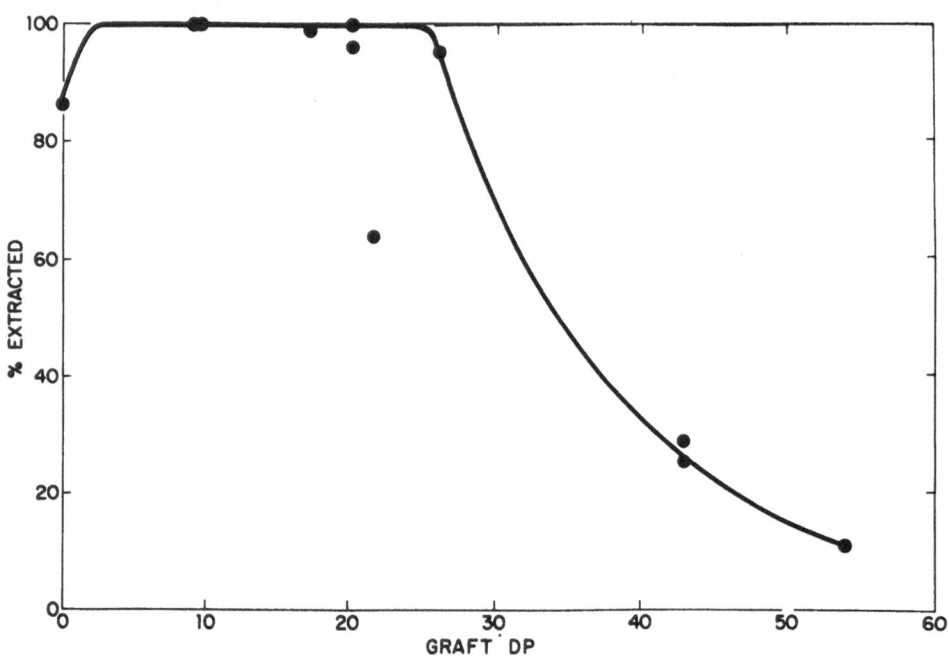

Figure 4 - Toluene Extractions of Poly([Ethylene-*co*-Vinyl
Acetate-*co*-Methacrylic Acid]-*g*-Pivalolactone)

When the backbone is highly nonpolar, as in EPDM, no single common solvent will dissolve the graft system at graft degrees of polymerization above about six. A mixed solvent consisting of nine parts of toluene and one part of hexafluoroisopropanol was effective.

Mechanical Properties. Networks established through the formation of microdomains of grafted polypivalolactone act to reinforce tensile strengths and increase modulus of polymeric substrates which are already basically plastics in form and properties. Figure 5 shows this effect and illustrates the dependence of mechanical properties on the grafted $\overline{DP}$. Apparently, for this polar backbone, effective phase separation does not begin to occur until graft $\overline{DP}$ exceeds about 3.5-4. Potential utility lies in the combination of this reinforcement with higher use temperatures·

In the formation of thermoplastic elastomers, however, these grafts are unique, not only in the use of crystallizable side chains for the reversible formation of crosslinking networks, but also in that they exhibit low Young's Moduli and high tensile strengths typical of conventionally cured and reinforced vulcanizates.

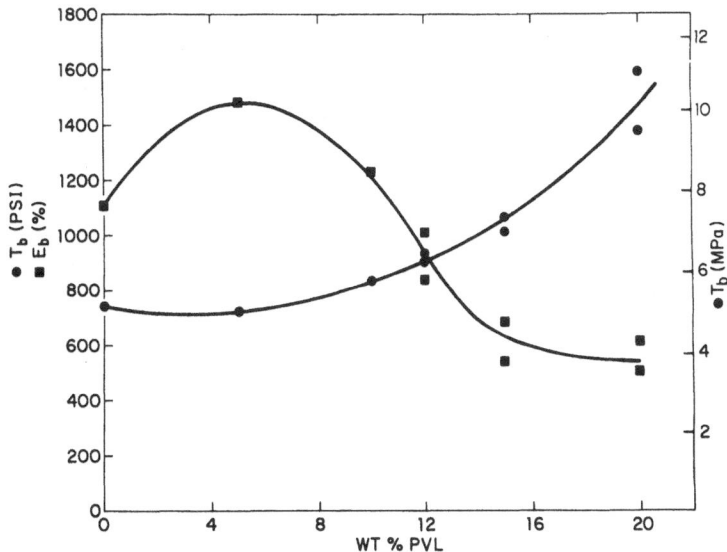

Figure 5 - Effect of Grafted PVL on Break Properties
Base Resin: Poly(Ethylene-co-Vinyl Acetate-co-Methacrylic Acid)
(61/37/1.4)
Mol. Wts.: Base Resin 37,000   Per Site 6,200

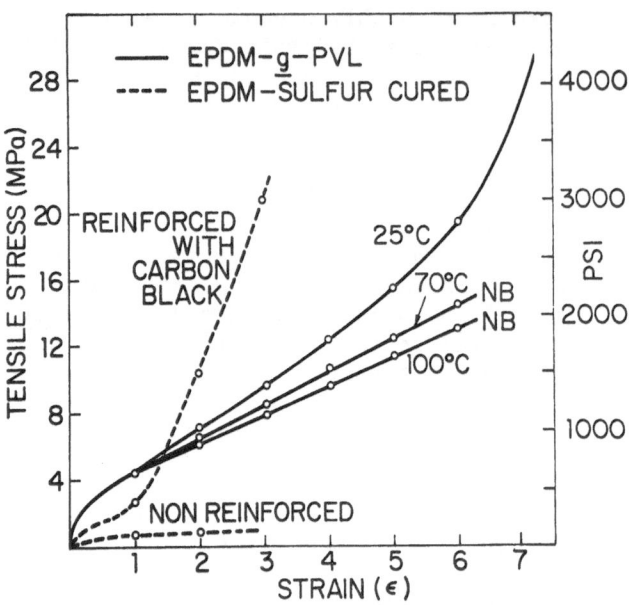

Figure 6 - Tensile Properties of EPDM-*g*-PolyPVL Compared With
Conventionally Cured EPDM. EPDM Segment Mol. Wt. 4300;
Graft $\overline{DP}$ = 13

Of the thermoplastic elastomers examined, that based on the
EPDM backbone was superior in the effective use of very short graft
chains for network formation, probably because of nearly complete
phase separation between backbone and graft domains. Stress strain
response of a typical composition is shown in Figure 6. It has
high strength at much higher elongations than are normally observed
in conventionally cured and reinforced EPDM. Since nonreinforced,
vulcanized EPDM has very low strength, the microcrystalline
polyPVL domains clearly serve as reinforcing particles as well as
interchain tie points. This, of course, is a generally recognized
property of multiphase thermoplastic elastomers. The thermal
integrity of the reinforcing and crosslinking domain is unusually
good, so that high strength and elongation are maintained even at
100°C. Modest improvement in tensile properties is brought about
by annealing.

The effect of graft degree of polymerization is such that at
strains up to 300%, stress increases regularly with graft $\overline{DP}$
(Figure 7), but at $\overline{DP}$ greater than 5, the curve becomes concave
upward with a large increase in tensile strength. Strength
decreases at yet higher HSL values, where an apparent over-
reinforcement or "overcure" by polyPVL domains predominates.

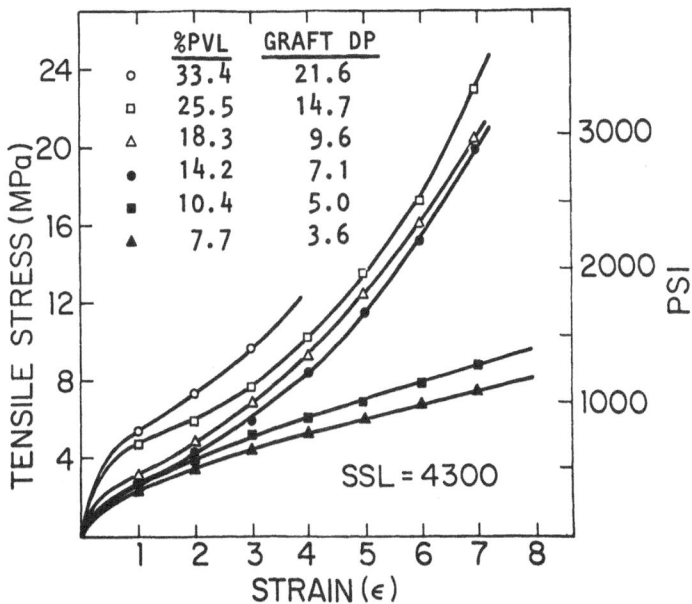

Figure 7 - Effect of Graft $\overline{\text{DP}}$ for EPDM-g-PolyPVL with EPDM Segment
Mol. Wt. = 4300

It was found that frequency of grafting and graft $\overline{\text{DP}}$, when
varied interdependently over a modest range to maintain a constant
composition of 25% polyPVL, had only slight effects upon tensile
properties. This, together with the data of Figure 7, suggests
that the critical factor in strength development is the internal
reinforcement effect of the polyPVL domains, not necessarily the
exact microstructure of the graft copolymer.

The unusual effectiveness of this network is particularly well
illustrated by the creep properties, as measured by compression
set. This test, ASTM D-395, Method B, measures creep under
moderate compressive strain under certain prescribed time-
temperature conditions. By definition, the crosslinks of thermoplastic elastomers are thermally labile, which usually leads to
undesirably high compression set values. In marked contrast, the
materials described in this report have unusually good resistance
to compression set, especially after preliminary annealing
(Figure 8). This we believe to be due to nearly complete
incompatibility of the component polymers and to the stability and
high melting temperatures of the crystalline polyPVL domains.

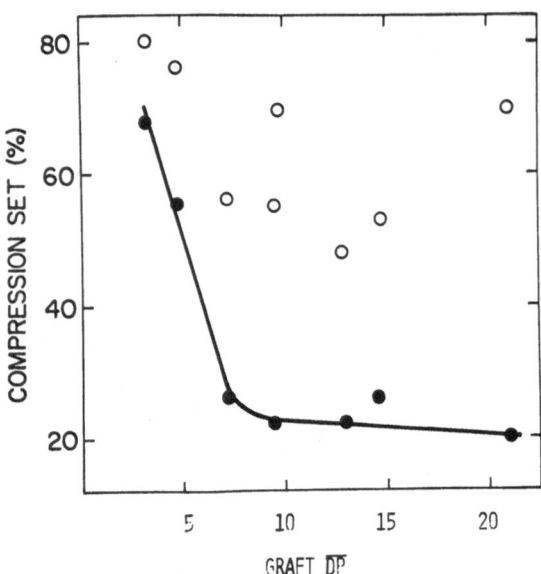

Figure 8 – Compression Set After 70 Hours at 100°C
for EPDM-*g*-PolyPVL with EPDM Segment
Mol. Wt. = 4300
o as molded
● annealed 16 hours at 115°C

REFERENCES

1. N. R. Mayne, *Chem. Technol.*, 728 (1972).

2. G. Peregot, A. Melis, and M. Cesari, *Makromol. Chem.*, 157, 269
    (1972).

3. H. A. Oosterhof, *Polymer*, 15 (1974).

4. R. E. Prud'homme and R. H. Marchessault, *Makromol. Chem.*, 175,
    2706 (1974).

5. S. A. Sundet, R. C. Thamm, J. M. Meyer, W. H. Buck,
    S. W. Caywood, P. M. Subramanian and B. C. Anderson,
    Macromolecules 9, 371 (1976).

6. S. A. Sundet, *Polymer Preprints*, 17, (2), 818 (1976).

7. R. C. Thamm and W. H. Buck, *Polymer Preprints*, 17, (1), 205
    (1976).

8. R. C. Thamm, W. H. Buck, S. W. Caywood, J. M. Meyer and
    B. C. Anderson, *Angew. Makromol. Chemie*, in press.

9. U. S. Patent 3,052,657 (1960), Shell Oil Co.

10. U. S. Patent 3,884,882 (1973), E. I. du Pont de Nemours and
    Company, Inc.

11. W. H. Buck, *Rubber Chem. Technol.*, in press.

PHYSICAL CHARACTERIZATION OF PIVALOLACTONE GRAFTED ELASTOMERS

II. ORIENTATION BEHAVIOR

W. H. Buck

E. I. du Pont de Nemours and Company, Incorporated

Elastomer Chemicals Department, Wilmington, De 19898

INTRODUCTION

Thermoplastic elastomers formed by grafting short, highly crystallizable chains of polypivalolactone (polyPVL) onto various elastomeric backbone polymers have been the subject of several recent publications (1-7). These polymers are synthesized by an anionic solution polymerization of pivalolactone initiated by the tetraalkylammonium salts of carboxyl groups pendent from pre-formed elastomeric backbone polymers. Details of the synthesis and physical properties of the graft copolymers have been des-cribed elsewhere (1-7). The previous paper in this series (6) proposed a morphological model for these polymers based upon the results of characterization by transmission electron microscopy, wide and small angle x-ray diffraction, thermal analysis, and solvent swelling. This paper describes the morphological changes which occur during the uniaxial deformation of the copolymers, citing the specific examples of polyPVL grafts on an ethylene propylene diene elastomer (EPDM) and an alternating ethylene ethyl acrylate (E/EA) copolymer.

Previous work (6) showed that over a fairly wide composition range (10-40% by wt. polyPVL), the general morphology of the polyPVL graft copolymers remained the same; discrete domains of polyPVL in a continuous elastomeric phase. As the molecular weight of the grafted polyPVL side chains and the average spacing between the graft sites exceeded about two thousand molecular weight units, spherulitic structures tended to form. The spheru-lites were composed, however, of discrete polyPVL domains.

191

Various experimental techniques indicated that considerable order exists at the polyPVL crystallite-elastomer interface. Each polyPVL domain is thought to consist of a highly crystalline core of α-form polyPVL surrounded by a layer of ordered elastomer chains. Consideration of polyPVL chain dimensions and domain size has led to the hypothesis that the grafted chains crystallize in a head-to-head fashion, perhaps stabilized by association of the free acid end groups of the polyPVL (6).

The polyPVL domains act as both thermally labile crosslinks and reinforcing filler. The size of these domains depends upon the length of the grafted chain, the average separation of the grafted chains along the elastomer backbone, and, to some extent, upon the structure of the graft site. In general, domain size increases as the length of the grafted chains increases and the average separation between grafted chains decreases.

Annealing of the graft copolymers causes improvement in several important physical properties such as tensile strength and resistance to creep under tensile and compressive stress (3-5). Analysis of annealed samples by transmission electron microscopy and x-ray diffraction has shown that the average domain size decreases and the size distribution is markedly narrowed relative to unannealed samples.

With the aid of wide and small angle x-ray diffraction and infrared dichroism the morphological changes accompanying deformation have been followed. A morphological model of the deformed state has been developed from this data.

EXPERIMENTAL

The graft copolymers of polyPVL of EPDM were supplied by R. C. Thamm of this laboratory. The EPDM base polymer was a commercial grade of "Nordel" hydrocarbon rubber, Du Pont's hydrocarbon elastomer, to which carboxyl functionality had been introduced by thermal reaction with maleic anhydride. Full details of synthesis and physical properties have been given elsewhere (1,3,5).

The polyPVL grafts on a backbone of alternating ethylene-ethyl acrylate copolymer were prepared by J. M. Meyer of this laboratory. The backbone elastomer was synthesized by the free radical catalyzed polymerization of ethylene with the boron trifluoride complex of ethyl acrylate (8). Carboxylate graft sites were generated by partial saponification with tetrabutylammonium hydroxide. The grafting reaction was similar to that described for polyPVL grafts on poly(ethyl acrylate) (1,4,5).

Wide angle x-ray diffraction data were obtained with Ni filtered Cu Kα radiation using a flat plate camera. Small angle data were obtained using a Rigaku-Denki camera with Cu Kα radiation.

Thin films for infrared dichroism were cast from a 1% solution of graft copolymer in 1/9 (v./v.) hexafluoroisopropanol/ toluene onto a mercury surface. Data were obtained with a Perkin-Elmer 221 infrared spectrometer with a stretch holder mounted so that the stretch axis was 45° from vertical. Spectra were obtained at each elongation with the polarizer first at 45°, then at −45° to the vertical. Film thickness was controlled so that the maximum absorbance at all wavelengths was less than 1.7 for the unstretched sample.

RESULTS AND DISCUSSION

NOMENCLATURE

The nomenclature to be used for the graft copolymers in this paper is the same as that described previously (3-6) and is based on the proposals of Ceresa (9). The polyPVL graft on EPDM, properly named poly([ethylene-co-propylene-co-1,4-hexadiene]-g-pivalolactone), is referred to as EPDM-g-PVL. The ethylene/ethyl acrylate graft, properly poly([ethylene-alt-ethyl acrylate]-g-pivalolactone), is named simply E/EA-g-PVL. The substructure of the graft copolymers is described in terms of 1) the average distance, in molecular weight units, between grafted chains on the soft segment (elastomeric backbone), and 2) the average molecular weight of the hard segments (polyPVL chains). These quantities are named, respectively, the soft segment length (SSL) and the hard segment length (HSL). SSL is calculated from analysis of the base polymer and represents a number average value. Since the graft polymerization has been shown to be a "living" polymerization (1,3,5) the molecular weight distribution of the hard segments is narrow, and the HSL should be interpreted accordingly.

WIDE ANGLE X-RAY DIFFRACTION

Figure 1A shows a wide angle x-ray diffraction pattern of a compression molded and annealed sample of EPDM-g-PVL in the relaxed (i.e., unoriented) state. Two sharp diffraction rings characteristic of polyPVL crystallinity are present, together with an amorphous halo due to the EPDM. Transmission electron microscopy has shown that the polyPVL exists in numerous microcrystalline domains in these copolymers (1,3,5,6). The absence of spots or arcs in the diffraction pattern shows that these domains are randomly oriented, as would be expected from the method of preparation. The crystal structure of polyPVL is well known, (10-14, 16) with two crystal forms, the α(helical) and β(planar zig-zag), being commonly found. Since the crystallites are not oriented, and since the (020) reflections necessary to identify the β crystal form are obscured by the amorphous halo, it is impossible to determine the crystal form of the polyPVL in the crystallites. Since the planar zig-zag form is normally found in stressed or highly oriented samples, however, it seems most likely that the crystallites are in the more stable helical (α) form.

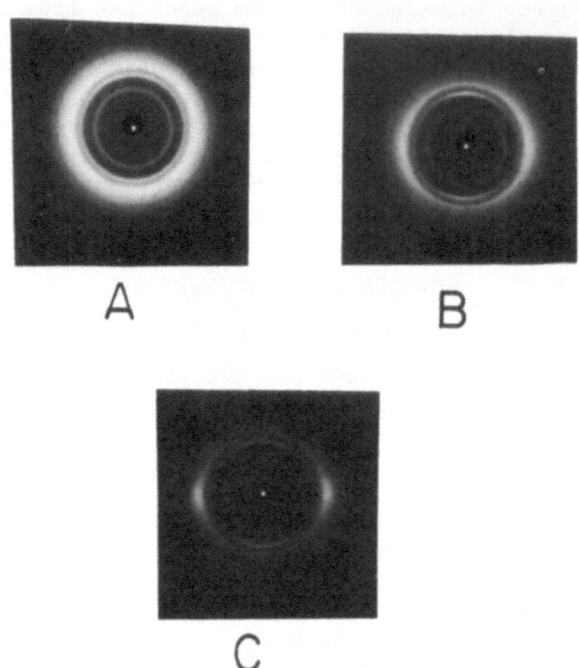

Fig. 1 - Wide angle x-ray diffraction patterns of EPDM-g-PVL, SSL
4800, HSL 1600. A. relaxed, B. 200% elongation, C. 650% elongation. Sample annealed 8 hours at 115°C. Stretch direction
vertical.

As the copolymer is strained, Figure 1B and 1C, the crystalline diffraction ring containing the (1$\bar{2}$0) reflections condenses
to an arc along the meridian, and the amorphous halo thickens on
the equator. The diffraction ring containing the (100) reflections, i.e., the ring closest to the center of the diffraction
pattern, does not show orientation as the sample is elongated.
The stretch direction in Figure 1 is vertical. Since the (1$\bar{2}$0)
planes are tending to orient on the meridian and the amorphous
halo shows orientation, but not crystallization, along the equator, the EPDM chains must be orienting parallel to the stretch
direction and the polyPVL molecular chains perpendicular to the
stretch direction. It should be noted that even at 650% elongation, the elastomer chains show no evidence of stress induced
crystallization. This is due to the random nature of incorporation of ethylene and propylene units in the EPDM backbone.

A quantitative measure of arc formation is given in Figure 2 for both annealed and unannealed samples of EPDM-g̲-PVL. In this figure, the orientation angle is the angular width of the arcs at half maximum intensity measured using a semicircular scan around the ring containing the (1̄20) reflections. If diffraction intensity were constant along the semicircle, as in Figure 1A, then the orientation angle would be 180°, indicative of random orientation. If the crystallites were all perfectly aligned, then the diffraction arc would become a point, and the orientation angle would be 0°. Rather unexpectedly, the annealed sample shows greater orientation at lower strain levels than the annealed sample. Both samples reach equivalent orientation angles, about 90°, at 400% elongation, then show little further change. As mentioned in the introduction, annealing of these samples has been shown to cause a marked narrowing of the polyPVL domain size distribution. These smaller and more uniformly sized domains apparently respond to stress more readily than the unannealed domains. The final orientation angle of 90° is quite high relative to what may be obtained from well spun fibers of polyPVL homopolymers (14). This indicates that although the polyPVL molecular chains tend to orient perpendicular to the EPDM chains, there is still a fairly wide distribution of chain orientation around this perpendicular. This poing will be discussed further in the section concerning small angle x-ray diffraction.

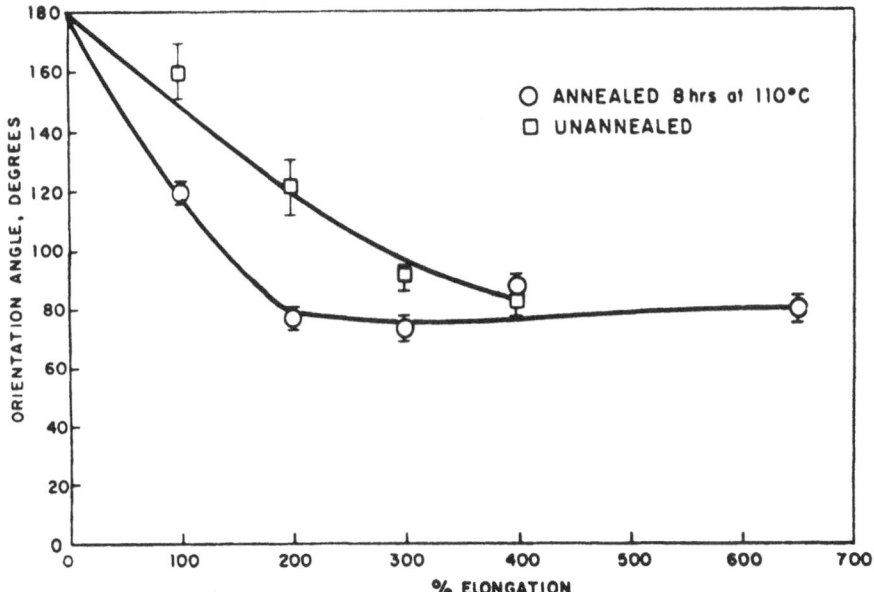

Fig. 2 – Orientation angle vs. percent elongation for EPDM-g̲-PVL, SSL 4800, HSL 1600. Squares, unannealed sample, circles, sample annealed 8 hours at 110°C.

Figure 3 shows diffraction patterns for E/EA-g-PVL. For
this figure, the stretch axis is at -45° from the vertical.
Unlike EPDM-g-PVL the third diffraction ring for polyPVL, containing the (020) reflections, is not obscured by the amorphous scattering. Since the ratio of the intensities of the (020) to the
(100) reflections are approximately 8 to 1 for the β form of poly-
PVL, (12) it can be shown from Figure 3C that the crystallites are
predominantly in the α (i.e., helical) form. Also unlike the
elongated EPDM-g-PVL, new diffraction spots occur at the equator
with spacings different from polyPVL. These new reflections are
attributed to stress induced crystallization of the poly(ethylene-
alt-ethyl acrylate) backbone elastomer. Undoubtedly the regular
alternating structure of this copolymer contributes to the observed phenomenon.

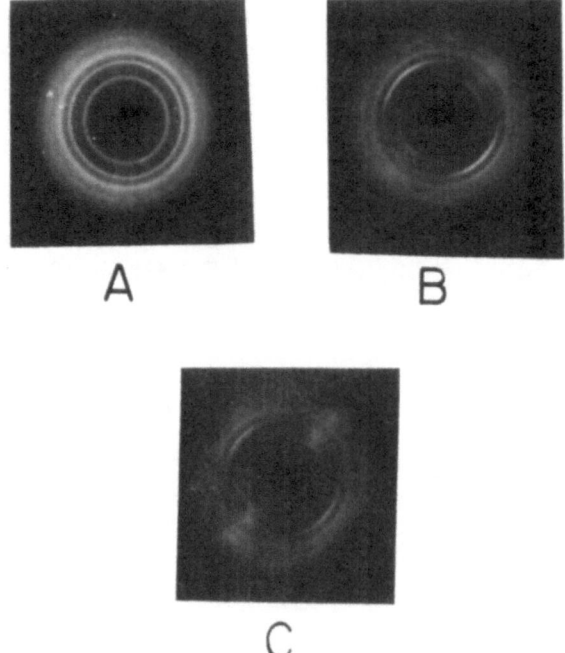

Fig. 3 - Wide angle x-ray diffraction patterns of E/EA-g-PVL,
SSL 3600, HSL 1100. A. relaxed, B. 200% elongation, C. 300%
elongation. Compression molded sample, stretch axis at -45° to
the vertical.

SMALL ANGLE X-RAY DIFFRACTION

Small angle x-ray diffraction patterns of the EPDM-g-PVL
polymer used to obtain Figure 1 are shown in Figure 4. In the
relaxed state, Figure 4A, the pattern is nearly symmetric with a
sharp diffraction maximum. As the sample is stretched, the angle
at which maximum diffraction occurs decreases at the meridian and
remains essentially constant along the equator. Beyond 200%
elongation small angle diffraction shows a four point pattern.

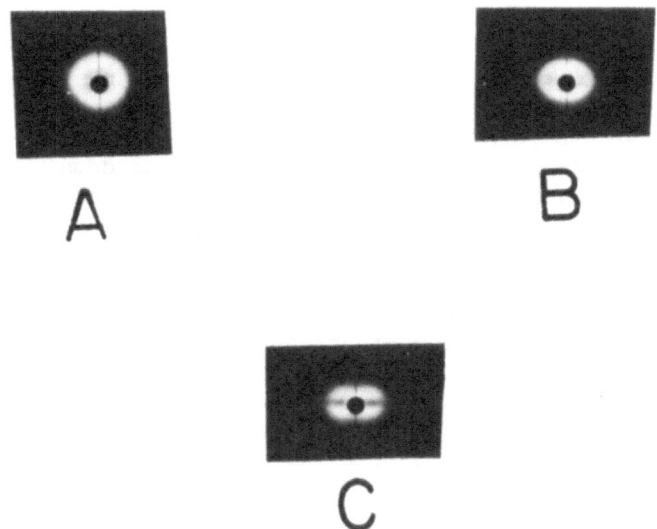

Fig. 4 – Small angle x-ray diffraction patterns of EPDM-g-PVL,
SSL 4800, HSL 1600. A. relaxed, B. 100% elongation, C. 650%
elongation. Compression molded sample annealed 8 hours at 110°C.
Stretch axis vertical.

Previous work (6) has strongly suggested that small angle
x-ray diffraction from these copolymers is due to regularly sized
polyPVL crystalline domains and the associated ordered elastomer
chains at the surface of the crystallites. A schematic of one such
domain is shown in Figure 5. The hypothesized end-to-end crystallization of the polyPVL chains, mentioned in the Introduction,
would require that elastomer chains be chemically bonded to both
the "top" and "bottom" of the polyPVL crystallites. As the bulk
sample is elongated, the elastomer chains will tend to orient in
the stretch direction. This orientation will then exert a torque

which will tend to turn the crystallites such that the polyPVL
chains will be perpendicular to the elastomer chains, and hence
the stretch direction. Because of such factors as elastomer chain
entanglements, the presence of homopolyPVL in the crystallites,
and polyPVL grafted at the end of elastomer chains, the torque on
the "top" and "bottom" of the crystallites will not be equal.
Then on the average, half of the crystallites would be tilted "up"
and half "down" relative to the perpendicular to the draw direction. Such a configuration could give the four point pattern
shown in Figure 4C; the "up" crystallites would give rise to two
of the quadrant spots, and the "down" the other two (15). The
distribution of "up" and "down" crystallites would also explain
the relatively high value of the orientation angle at high strain
levels shown in Figure 2.

     The lateral and longitudinal long period spacings, calcu-
lated from the Bragg equation using the intensity maxima on the
equator and meridian, are shown as a function of elongation in
Figure 6. Previous work (6) has shown a very good correlation
between small angle x-ray long period spacing and domain size as
measured by transmission electron microscopy. Comparison of this
size with crystallite size estimated by wide angle x-ray line
broadening and chain length considerations led to the hypothesis
that the domains consisted of cores of crystalline polyPVL surrounded by partially ordered elastomer chains. Thus the long
period spacings in Figure 6 most probably refer to the dimensions
of the crystalline domains with their associated ordered elastomer
chains in the lateral (horizontal) and longitudinal (vertical)
directions, relative to the stretch direction. The near constancy
of the lateral spacing at modest strain is expected if this interpretation is correct, since the lateral spacing would represent
the thickness in the polyPVL chain direction, which should not
change at the low tensile stress levels in this direction.

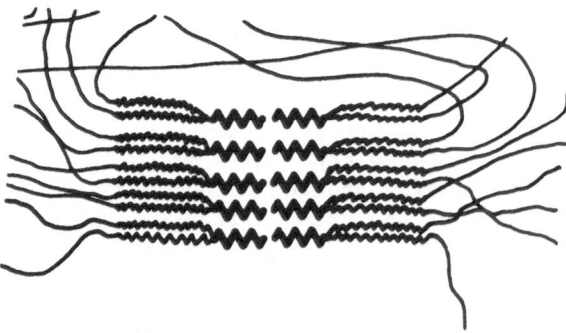

Fig. 5 - Schematic of polyPVL domain. Heavy lines, polyPVL chains,
light lines, elastomer chains.

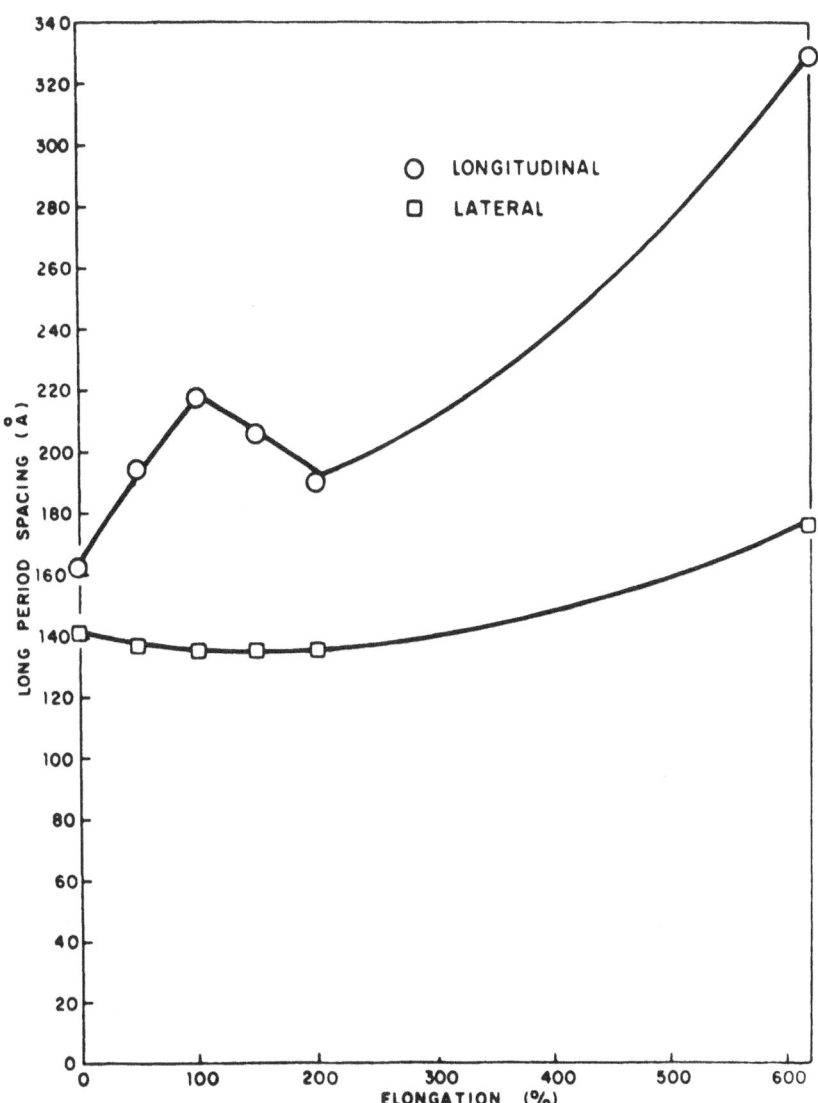

<u>Fig. 6</u> - SAXD lateral and longitudinal long period spacings as a
function of elongation for EPDM-<u>g</u>-PVL, SSL 4800, HSL 1600.
Stretch direction vertical.

The cause of the relative maximum in longitudinal long period
spacing at 100% elongation is not known. At high elongation the
large values of both spacings are probably due to the splaying
apart of EPDM chains prior to ultimate failure.

INFRARED DICHROISM

In order to further elucidate the morphological changes that
occur as EPDM-g-PVL copolymers are elongated, an infrared
dichroism experiment was performed. In this experiment, a thin
film of the graft copolymer was stretched in the beam of an infrared spectrometer. Spectra were recorded with the incident radiation polarized first parallel to and then perpendicular to the
stretch direction. For vibrations in which the transition moment
is parallel to the stretch direction, the absorbance would be
greater with parallel polarized incident radiation. Similalarly,
for molecular vibrations whose transition moment is perpendicular
to the orientation direction, the absorbance will be greater with
perpendicularly polarized radiation. Because the absorbance
of the carbonyl group at ca. 1740 cm^{-1} was too great at practical
film thickness, the asymmetric ester stretch at 1232 cm^{-1} was used
in this experiment. Results were expressed in terms of the dichroic
ratio, defined as the ratio of the absorbance at 1232 cm^{-1} with the
polarizer parallel to the stretch direction, $A_{||}$, to the absorbance
with the polarizer perpendicular to the stretch direction, $A_{\perp}$, i.e.,
to the stretch direction, $A_{\perp}$, i.e.,

$$Dichroic\ Ratio = A_{||}/A_{\perp}$$

It is known from the wide angle x-ray diffraction experiments
that the (1$\bar{2}$0) planes tend to orient on the meridian of the x-ray
pattern as the samples are stretched. If this orientation were
perfect, i.e., the diffraction arcs became points on the meridian,
then the relationship between the base plane of the α-form polyPVL
monoclinic unit cell and the stretch direction would be as shown
in Figure 7. In this projection, the polyPVL helices would be
perpendicular to the plane of the page and the EPDM chains would
be parallel to the stretch direction. The polyPVL helical axis
(c axis) is perpendicular to the a and b axes, and hence is perpendicular to the EPDM chains. The a and b axes make angles of
126° and 247° (measured clockwise), respectively, with the stretch
direction. Reference to the crystal structure of α-polyPVL (16)
shows that the chain makes an angle of about 41° with the a axis
at the ester carbonyl. It has been assumed that the transition
moment of the assymetric ester stretch at 1232 cm^{-1} is parallel
to the polyPVL chain at the ester group. In the case of ideal
orientation, i.e., all (1$\bar{2}$0) planes on the meridian in a highly
stretched sample, the relationship between the stretch direction
and the transition moment of the assymetric ester stretch would

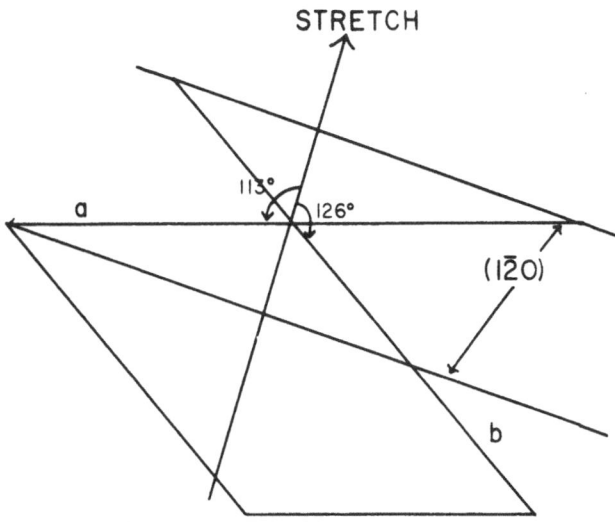

Fig. 7 - Relationship between base plane of monoclinic unit cell of α-form polyPVL and stretch direction in highly oriented EPDM-g-PVL.

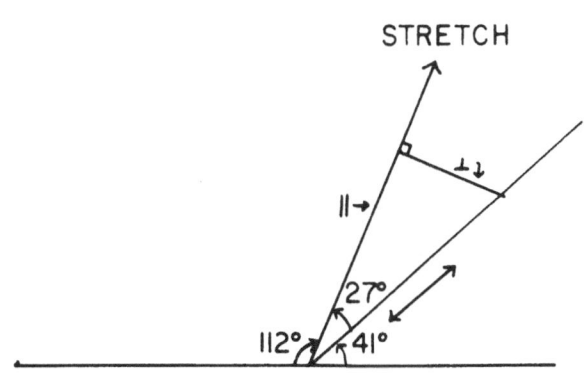

$$\frac{parallel}{perpendicular} = \cot 27° = 1.96$$

Fig. 8 - Relationship between stretch direction and transition moment of assymmetric ester stretch at 1232 cm^{-1}.

be as shown in Figure 8. The angle between the transition moment
and the stretch direction is about 27°. If the transition moment
is resolved into components parallel and perpendicular to the
stretch direction, it can be seen that the parallel component is
larger. It would thus be predicted that the dichroic ratio would
increase with sample elongation. In the limit of perfect orientation, the maximum observed dichroic ration should be 1.96.

Results of dichroic ratio measurements as a function of sample
elongation are given in Figure 9 for an EPDM-g-PVL copolymer. As
predicted the dichroic ratio increases with elongation. The
observed maximum is 1.69, compared to the predicted theoretical
maximum of 1.96. This is consistent with the x-ray data which show
that the polyPVL crystallites are not perfectly oriented even at
very high elongations. As the longitudinal small angle x-ray long
period spacing shows a relative maximum at 100% elongation, so the
dichroic ratio exhibits a relative maximum at the same elongation.
This is true even though the polymers used for the two measurements had different values of HSL and SSL. Such a relative maximum might occur if the a axis of the polyPVL crystallites were to
approach parallelism with the stretch direction at 100% elongation.
The reason for such an orientation is not obvious.

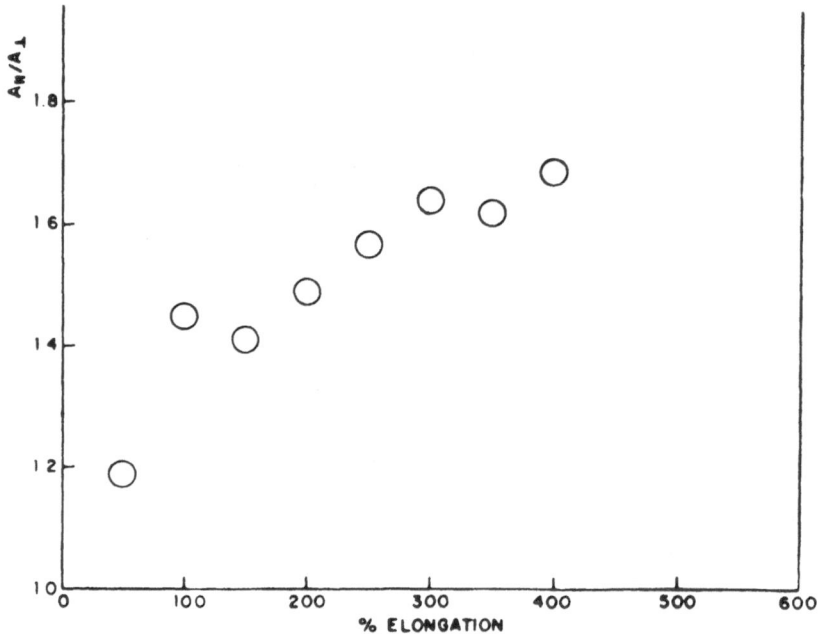

Fig. 9 - Dichroic ratio at 1232 cm^{-1} as a function of elongation
for EPDM-g-PVL, SSL 3600, HSL 1200.

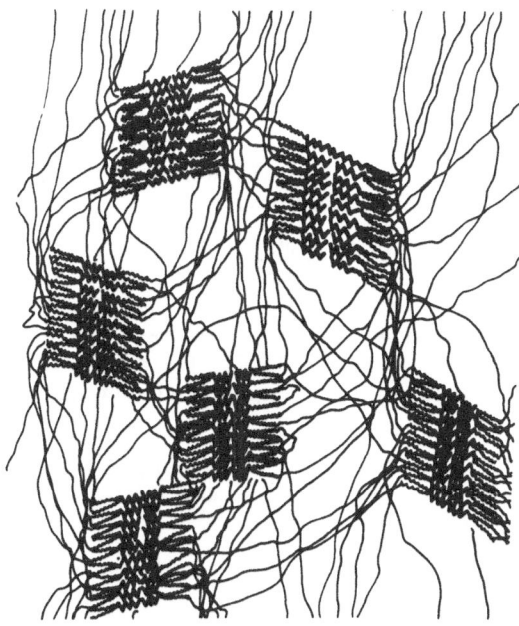

Fig. 10 - Morphological model of oriented EPDM-g-PVL. Heavy lines, polyPVL, light lines, EPDM. Stretch direction vertical.

MORPHOLOGICAL MODEL

Based on the x-ray and infrared dichroism experiments, the morphological model of Figure 10 is proposed for EPDM-g-PVL at high (>500%) elongation. The EPDM chains are aligned preferentially in the stretch direction. The polyPVL chains are aligned preferentially perpendicular to the stretch direction, with about half of the crystallites being tilted "up" and the other half "down" relative to the perpendicular to the stretch direction. This is due to the imbalance in torque applied at the elastomer/crystallite interface. Such an explanation accounts for the relatively high value of the orientation angle (90°) for highly oriented copolymer. The lateral dimensions of the crystallites in oriented samples (i.e., in the polyPVL chain direction) are smaller than the longitudinal dimensions, and the crystalline domains are of relatively uniform size. This latter feature is a necessary consequence of the sharp small angle x-ray diffraction. This model seems to be consistent with all of the currently available experimental data.

SUMMARY

Morphological changes accompanying elongation of EPDM-g-PVL and E/EA-g-PVL were followed with wide and small angle x-ray diffraction and infrared dichroism. During elongation the elastomer backbone chains orient in the stretch direction and the polyPVL

chains perpendicular to the stretch direction. At high elongation
the EPDM chains are ordered, but do not crystallize. The more
regular poly(ethylene-alt-ethyl acrylate) chains crystallize at
high elongation. At high elongation, the polyPVL crystallites do
not orient exactly perpendicular to the elastomer chains, but have
their helical axes tilted either slightly above or below the perpendicular to the stretch direction.

ACKNOWLEDGEMENTS

The author would like to thank R. C. Thamm and J. M. Meyer
for the copolymer samples and H. Thielke for x-ray measurements.
Many helpful technical discussions were held with E. S. Clark,
W. O. Statton and G. M. Estes. J. C. Gallagher, Jr., provided
able experimental assistance throughout the course of this work.

REFERENCES

1. S. A. Sundet, R. C. Thamm, J. M. Meyer, W. H. Buck,
    S. W. Caywood, P. M. Subramanian and B. C. Anderson, Macro-
    molecules, 9, 371 (1976).

2. R. P. Foss, H. W. Jacobson, H. N. Cripps and W. H. Sharkey,
    Macromolecules, 9, 373 (1976).

3. R. C. Thamm, W. H. Buck, Polym. Preprints, 17, 205 (1976).

4. S. W. Caywood, Rubber Chem. Technol., 50, 127 (1977).

5. R. C. Thamm, W. H. Buck, S. W. Caywood, J. M. Meyer and
    B. C. Anderson, Ang. Makromol. Chem., 58/59, 345 (1977).

6. W. H. Buck, Rubber Chem. Technol., 50, 109 (1977).

7. S. A. Sundet, Polym. Preprints, 17, 818 (1976).

8. A. L. Logothetis, J. M. McKenna, J. Polym. Sci., Polym.
    Letters ed., 12, 131 (1974).

9. R. J. Ceresa, "Block and Graft Copolymers", Butterworth,
    Washington, D.C., 1962, pg. 8.

10. G. Carazzolo, Chim. Indust. (Milan), 46, 525 (1964).

11. H. A. Oosterhof, Polymer, 15, 47 (1974).

12. G. Perego, A. Melis, M. Cesari, Makromol. Chemie, 157, 269
    (1972).

13. R. E. Prud'homme, R. H. Marchessault, Macromolecules, 7, 541
    (1974).

14. F. W. Knobloch, W. O. Statton (to E. I. du Pont de Nemours and
    Co.), U.S. 3,299,171 (Jan. 17, 1967).

15. P. Predicki, W. O. Statton, "Small Angle X-Ray Scattering",
    H. Brumberger, ed., Gordon and Breach, London, 1967, pg. 137.

16. J. Cornibert, R. H. Marchessault, Macromolecules, 8, 296
    (1975).

SYNTHESES OF BLOCK COPOLYMERS BY

RING-OPENING POLYMERIZATION

Yuya Yamashita

Department of Synthetic Chemistry
Faculty of Engineering
Nagoya University, Nagoya 464, Japan

INTRODUCTION

The preparation of block copolymers by the living polymer techniques pioneered by Szwarc (1) has been extensively developed in recent years to clarify the fundamental concepts of structure-property relationships such as domain formation and thermoplastic elasticity. The living tendency frequently observed in ring-opening polymerization can provide novel methods for the preparation of tailored polymers, which are not obtained by other methods. Living polymer formation was observed in anionic polymerization of ethylene oxide, propylene sulfide, pivalolactone, α-pyrrolidone, N-carboxy-α-aminoacid anhydride, hexamethylcyclotrisiloxane, chloral and others. In cationic systems tetrahydrofuran is an example of living polymer formation. Basically two methods exist for the formation of block copolymers. The first method consists of the successive addition of monomers to a macromolecular initiator containing effective functional groups for the initiation reaction. The second method consists of the coupling reaction of the independently prepared prepolymers of different kinds which have specified reactive terminal groups. Some of the fundamental problems for the syntheses of the block copolymers by ring-opening polymerization are discussed.

BLOCK COPOLYMERS BY SUCCESSIVE ADDITION

In the successive addition of cyclic monomers to a macro-

molecular initiator, the following factors must be considered:
the purity of the macromolecular initiators with suitable functional groups; the rate of the initiation reaction and the initiator efficiency; the living tendency of the polymerization. If the
purity of the terminal functional groups in the macromolecular
initiator is low, the obtained copolymer becomes a mixture of
block copolymer and homopolymer. Although various methods can be
used to prepare telechelic prepolymers, it is not easy to exclude
the contamination of prepolymers which have lost effective end
groups. When the initiation reaction by the macromolecular initiator is not satisfactory because of slow initiation or because
of non-bonding initiation with transfer, the obtained copolymer
has a wide distribution of compositions. And if the living
tendency of the successive addition is low due to accompanying
chain transfer reactions or depolymerization reactions, considerable amount of homopolymer is formed together with copolymers.

Pivalolactone

The polymerization of pivalolactone by various polystyrene
initiators has been studied (2). There are two modes of cleavage
reaction of β-lactones. A weak nucleophile attacks the CH_2 group
with alkyl-oxygen fission, leading to carboxylate end groups which
can continue the propagation reaction.

$$RCOO^- \; + \; \underset{\underset{O - C=O}{|}}{\overset{\overset{CH_3}{|}}{CH_2-C-CH_3}} \longrightarrow RCOO-CH_2-\underset{\underset{CH_3}{|}}{\overset{\overset{CH_3}{|}}{C}}-COO^-$$

Attack of a strong nucleophile on the carbonyl group causes acyl-
oxygen fission, and although polymerization is known to occur
with these initiators, the exact nature of the propagating species
has not yet been clarified.

$$R^- \; + \; \underset{\underset{O - C=O}{|}}{\overset{\overset{CH_3}{|}}{CH_2-C-CH_3}} \longrightarrow R-\underset{\underset{O}{\|}}{C}-\underset{\underset{CH_3}{|}}{C}-CH_2-O^-$$

Although the polymer precipitated immediately after addition
of the monomer to THF solution of polystyryl sodium or polystyryl
ethoxysodium prepared by the anionic polymerization of styrene in
a high vacuum system followed by end-capping with ethylene oxide
in the latter case, the polymer was completely fractionated by
Soxhlet extraction with cyclohexane into homopolymer of styrene

and pivalolactone and no block copolymer formed. Thus the carbanion or alkoxyanion initiated the polymerization of pivalolactone
with chain transfer.

   To clarify the mechanism of the chain-transfer reaction, the
reaction of pivalolactone with α-methylstyrene tetramer disodium
and its reaction product with ethylene oxide was examined in THF.
The THF soluble part was an oily material and contained small
amounts of pivalolactone segment. The analytical results consist
of the initial reaction of carbanion on the carbonyl group of
pivalolactone by acyl-oxygen fission to form alkoxide anion and
the second reaction of the alkoxide anion on pivalolactone by
acyl-oxygen fission. The alkoxy anion derived from pivalolactone
seems to be sterically unfavorable for the propagation reaction
and a monomer transfer reaction occurs at this stage to form a new
initiating species favorable for the polymerization of pivalolactone. It was also shown that the alkoxy anion derived from
polytetramethylene glycol and sodium naphthalene in THF initiated
the polymerization of pivalolactone by chain transfer and no
block copolymer was obtained.

   On the other hand, macromolecular initiators consisting of
polystyrene, poly-THF and polyacrylonitrile containing carboxylate
end groups were found to be suitable initiators for the block
copolymerization (2)(3). The preparation of macromolecular initiators of molecular weight of several thousands was as follows:
Polystyrene initiator(L) was prepared by anionic living polymerization of styrene in THF with α-methylstyrene tetramer disodium
followed by introduction of carbon dioxide. Initiator for the L'
series was prepared by neutralization of the polystyrene dicarboxy-
lic acid precipitated by acidified methanol with tetrabutylammonium hydroxide. Initiator for the R series was prepared
through radical polymerization of styrene with 4,4'-azobis-4-
cyanovaleric acid after neutralization with sodium hydroxide.
Poly-THF initiator (T) was prepared by reacting poly(tetramethylene)glycol with excess succinic anhydride, and neutralized after
reprecipitation. Polyacrylonitrile initiator (A) was prepared
through radical polymerization of acrylonitrile with bistetra-
butylammonium-4,4'-azobis-4-cyanovalerate in DMSO. The purity of
the end groups was analyzed by titration, and the presence of
2.0±0.2 carboxylate groups in each polymer chain was established.

   Polymerization of pivalolactone in dilute THF solution was
very slow at 50°C using sodium carboxylate. It was necessary

to use higher monomer concentration or to add dipolar aprotic
solvents to accelerate the reaction. If the counter cation was
tetrabutylammonium, the polymerization occurred easily at room
temperature in THF as expected by the dissociation of the
carboxylate anion. The solution becomes turbid by precipitation
of the insoluble block copolymer. The characterization of the
block copolymer was carried out by fractionation. Unreacted
macromolecular initiator was extracted by cyclohexane (polystyrene), methanol (poly THF) or DMF (polyacrylonitrile) and found
to contain a small amount of bonded pivalolactone. Comonomer
content in the block copolymer was determined by nmr in trichloroacetic acid or in mixed solvents containing trichloroacetic acid.
It was difficult to find a good solvent for polypivalolactone and
it was not easy to determine the content of polypivalolactone
homopolymer contaminated in the block copolymer. Increase of
the reduced viscosity in benzyl alcohol of the block copolymer
fraction was observed, supporting the successive addition of
pivalolactone units at the chain ends.

Typical results are shown in Table 1. The presence of considerable amounts of soluble fraction consisting of unreacted
macromolecular initiator is noticeable. This might not be accounted for by the presence of prepolymers which have no carboxylate end group or by chain transfer to contaminated water, because
the purity of the functionality was checked by experiment and the
contamination of water was excluded in the high vacuum experiment.
Thus, explanation by slow initiation in a living polymer formation
seems reasonable. The kinetic scheme of the slow initiated polymerization was formulated as follows:

$$I \quad + \quad M \quad \xrightarrow{\ ki\ } \quad P*$$
$$Pn* \quad + \quad M \quad \xrightarrow{\ kp\ } \quad P_{n+1}$$

The initiator efficiency f is defined as the fraction of consumed
initiator.

$$f = ([I]_0 - [I]) \ / \ [I]_0$$
$$([M]_0 - [M]) \ / \ [I]_0 = f - (kp/ki)[\ln(1-f) + f]$$

Thus kp/ki was calculated from conversion and extracted unreacted
initiator. Some of the results are shown in Table II.

It is noteworthy that there is a difference of nuclephili-
city of the carboxylate group between the initiator and the propagating species. On comparing the pKa values of the model compounds corresponding to conjugate acid of the carboxylate shown
in Table II, it is apparent that the propagating carboxylate
anion is more basic than the initiating carboxylate anion. The
value of kp/ki is larger for less basic carboxylate initiator.

Table I

Polymerization of Pivalolactone with Macromolecular Carboxylates

	L-23	T-29	L'-0-2	A-2
Initiator	St-Na	St-NBu$_4$	THF-Na	AN-NBu$_4$
g	0.707	0.509	0.655	0.500
mmole	0.187	0.124	0.624	0.125
Monomer				
g	0.495	0.989	0.940	0.297
mmole	4.95	9.89	9.40	2.97
Solvent, ml				
THF	7.15	10	30.0	-
DMSO	0.45	-	0.6	5
Temp., °C	20	30	20	30
Time, hrs.	120	168	94	68
Conversion of monomer, %	56.7	100	84.6	81.5
Polymer yield				
g	0.987	1.510	1.450	0.742
soluble part				
g	0.193	0.039	0.808	0.322
insoluble part				
g	0.794	1.471	0.642	0.420
PVL, mole %	28.6	68.1	76.4	59.1

Table II

Initiator Efficiency of Polystyrene Carboxylates

	Unreacted initiator %	Initiator efficiency f	kp/ki	pKa of model compounds	
L-10-5	4.0	0.80	40	CH$_3$CHCOOH 　　Ph	4.37
R- 0-2	2.2	0.85	25	CH$_3$CHCH$_2$CH$_2$COOH 　　CN	4.51
propagating species				CH$_3$COCH$_2$CCOOH 　　O　　CH$_3$ (with CH$_3$ above)	4.76

Polystyrene initiator from living polymer (L) has less basic carboxylate group as kp/ki approaches 40. An increase in the kp by
adding polar solvent did not change kp/ki, but a change in the
countercation from sodium to tetrabutylammonium (L') increased kp
considerably and kp/ki also increased to 120. Polystyrene initiator from radical polymerization (R) has a more basic carboxylate
group and kp/ki approaches 25, thus improving the initiator
efficiency. Poly-THF initiator (T) and polyacrylonitrile initiator (a) showed slow initiation also, and kp/ki was larger than one
hundred.

The block efficiency is affected by kp/ki and by segment
length of the pivalolactone unit. In the case of high conversion
polymer with sufficienty long pivalolactone sequences, block
efficiency is more than ninety percent by using any initiators.
However, the increase of block efficiency for a block copolymer
containing only short segments of pivalolactone unit is made possible only by using well designed high nucleophilic initiators.

Lactams

Anionic polymerization of a lactam is known to propagate by
addition of an activated monomer anion to an N-acyl lactam chain
end, followed by a very fast proton transfer from monomer to polymer anion to give a new monomer anion and an N-acyl lactam of one
unit longer. Therefore, macromolecular initiators containing an
N-acyl lactam group or its precursor group may be used to produce
block copolymers containing nylon segments.

$$
\underset{\underset{O}{\parallel}\quad\underset{O}{\parallel}}{-C-N-C} \quad + \quad \underset{\underset{O}{\parallel}}{{}^{-}N-C} \quad \longrightarrow \quad \underset{\underset{O}{\parallel}}{-C-{}^{-}N} \quad \underset{\underset{O}{\parallel}\quad\underset{O}{\parallel}}{C-N-C}
$$

$$
\xrightarrow{\underset{\underset{H\ O}{}}{\underset{\parallel}{N-C}}} \quad \underset{\underset{O}{\parallel}\;\;\overset{}{H}\;\;\underset{O}{\parallel}\;\;\underset{O}{\parallel}}{-C-N\;\;C-N-C}
$$

The polymerization of $\mathcal{E}$-caprolactam and α-pyrrolidone by poly-
THF and polystyrene initiator was studied (4)(5). Bischloroformate of polytetramethyleneglycol was prepared by the reaction
of excess phosgene in THF and had a molecular weight of 2130.
Bischloroformate of polystyrene glycol was prepared by the reaction of excess phosgene in THF with the addition product of
ethylene oxide on polystyryl sodium obtained in high vacuum system
and had a Mn of 32,200. Polystyrene having N-acyl pyrrolidone
groups was prepared by the free radical polymerization of styrene
with an azo initiator containing N-acyl pyrrolidone groups. The
reaction of 4,4'-azobis (4-cyanovaleric acid chloride) with α-
pyrrolidone yielded 4,4'-azobis (pyrrolidone 4-cyanovalerate),

which was identified by nmr.

$$CH_2-CH_2 \diagdown N-C-CH_2-CH_2-C-N=N-C-CH_2-CH_2-C-N \diagup CH_2-CH_2$$

(structure with CH₃ groups, C=O, and CN substituents)

Polystyrene dipyrrolidone obtained by this initiator had a Mn=
1.38×10^4. These macromolecular initiators were analyzed to have
more than ninety percent functionality. Polymerization of lactams
were carried out in bulk at 80°C for ε-caprolactam and at 30°C for
α-pyrrolidone in a high vacuum system. First, catalyst solution
(sodium naphthalene in THF) was introduced, and second, monomer
was admitted, and last, the macromolecular initiator solution was
admitted. After mixing well, THF was distilled off to initiate
the polymerization. After a desired time, the solid material was
dissolved in formic acid and poured into acetone. The unreacted
macromolecular initiator was separated from the acetone solution
by extracting with cyclohexane and found to be only several
percent of the initiator. Thus, the functionality of the initiator was sufficiently high and the initiation reaction was not slow
compared with propagation reaction. The characterization of the
mixture of block copolymer and homopolyamide is not easy. The
average composition was analyzed by nmr spectra in formic acid or
in trichloroacetic acid-tetrachloroethane. The viscosity was
measured in m-cresol and the increase of the reduced viscosity
with conversion was taken as a measure of the high block efficiency.

Some of the polymerization results with poly-THF initiator
are shown in Figures 1 and 2. Both yield and viscosity of the

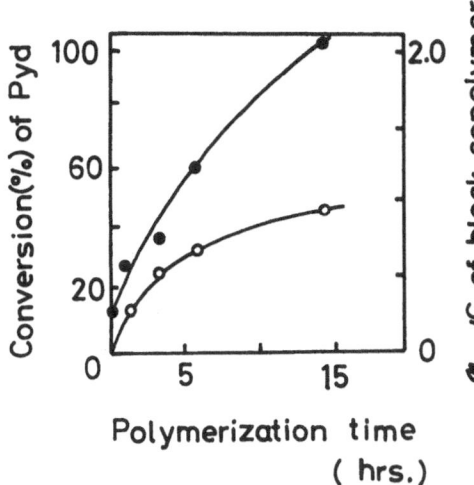

Figure 1. Polymerization of
α-pyrrolidone by PTHF
bischloroformate at 30°C

o conversion of α-pyrrolidone
● η_{sp}/c of block copolymer

PTHF bischloroformate,

2.84×10^{-4} mole
sodium naphthalene,

1.17×10^{-3} mole
α-pyrrolidone,
0.104 mole

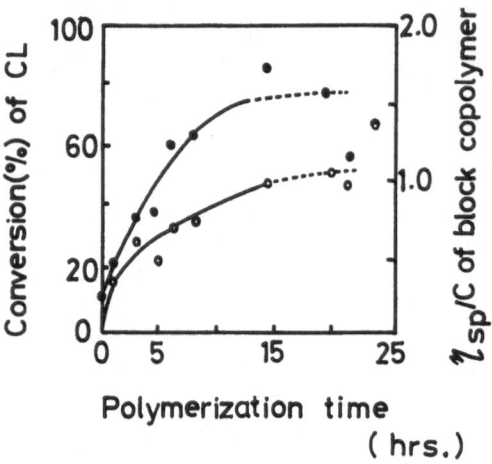

Figure 2.   Polymerization of
ε-caprolactam by PTHF
bischloroformate at 80°C

o conversion of ε-caprolactam
● η_{sp}/c of block copolymer

PTHF bischloroformate,
2.86×10^{-4} mole
sodium naphthalene,
1.17×10^{-4} mole
ε-caprolactam,
0.0728 mole

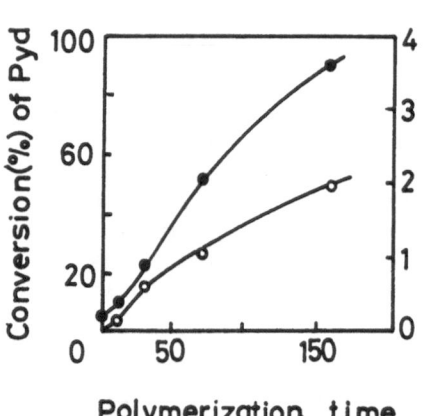

Figure 3.   Polymerization of
α-pyrrolidone by PSt
bischloroformate at 30°C

o conversion of α-pyrrolidone
● η_{sp}/c of block copolymer

PSt bischloroformate,
4.10×10^{-5} mole
sodium naphthalene,
2.48×10^{-4} mole
α-pyrrolidone,
0.0647 mole

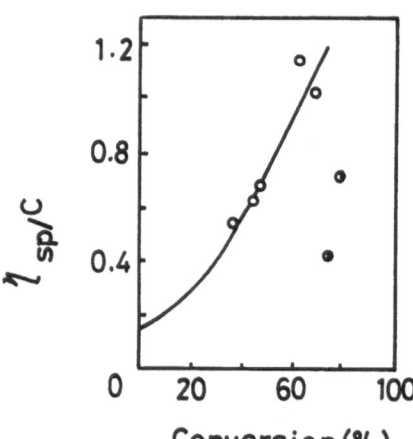

Figure 4.   Polymerization of
α-pyrrolidone by PSt
dipyrrolidone at 30°C

o conversion of α-pyrrolidone
● η_{sp}/c of block copolymer

PSt dipyrrolidone,
6.95×10^{-5} mole
sodium naphthalene,
4.19×10^{-4} mole
α-pyrrolidone,
0.0656 mole

block copolymer steadily increase with polymerization time. In
the case of α-pyrrolidone, continuous increase of the reduced viscosity supports the living nature of the system and also the
purity of the block copolymer. In case of ϵ-caprolactam, the increase of the reduced viscosity is limited to the first 14 hrs.
After that the reduced viscosity decreases showing the
cleavage of polymer chain and thus reducing the block efficiency.
In Figures 3 and 4 are shown the polymerization of α-pyrrolidone
with polystyrene initiator. In this system also, the block efficiency seems sufficiently high if the time of polymerization is
not too long. Preliminary experiments indicated that traces of
water retarded the polymerization and a high vacuum system was
recommended.

BLOCK COPOLYMER BY COUPLING REACTION

In the coupling reaction between prepolymers with specified
reactive groups, the following factors must be considered: the
degree of functionality of prepolymers; the molar ratio of the
functional groups; the rate of the coupling reaction and the
extent of side reactions. Coupling reactions are a type of polycondensation and the extent of reaction strongly depends on these
factors.

Tetrahydrofuran

Since the discovery of the cationic living polymer of tetrahydrofuran, a unique method for forming block copolymer by linking
cationic living polymer with anionic living polymer has been suggested. To prove the formation of living polymers and bond formation by coupling reaction, ion-coupling between poly-St dianion
and poly-THF dication was studied (6). Poly-St dianion is easily
prepared by the polymerization of styrene in THF by using α-
methylstyrene tetramer disodium in a high-vacuum system. Poly-
THF dication was prepared by bulk polymerization of THF at 0°C
in less than 20% conversion by using 2,2'-octamethylene-bis-1,3-
dioxolenium perchlorate as an initiator in a high-vacuum system.

$$BrCH_2CH_2OCO(CH_2)_8COOCH_2CH_2Br \xrightarrow{AgClO_4}$$

$$\begin{matrix} CH_2-O^+ \\ | \\ CH_2-O \end{matrix} \diagdown C(CH_2)_8C \diagup \begin{matrix} O^+-CH_2 \\ | \\ O-CH_2 \end{matrix} \quad 2ClO_4^-$$

The catalyst solution was prepared by addition of bis-2-bromoethyl
sebacate to dry nitromethane solution of silver perchlorate.
After filtration, an aliquot of the solution was transferred in a
flask and the solvent was changed to THF by evaporation.

The reaction of equimolar amounts of poly-THF dication and
α-methylstyrene tetramer dianion in THF was carried out at room
temperature. The molecular weight of the methanol-insoluble fraction did not increase from that of the parent homopolymer. The
ir spectrum of the product showed the presence of ketone carbonyl
group besides ester carbonyl group and nucleopilic attack of the
carbanion on ester carbonyl group with chain fission was suggested.
Less nucleophilic polystyrene diethoxyanion obtained by addition
of ethylene oxide on polystyrene carbanion also failed to increase
the molecular weight by ion-coupling. Preferential attack of
poly-St dianion on the oxonium ion in the presence of ester groups
is expected in the case of carboxylate anion, and the living
carbanion was modified by treating it with carbon dioxide.

$$-(CH_2)_4CO(CH_2)_2(OC_4H_8)_{n-1}-O^+\!\!\begin{matrix} CH_2-CH_2 \\ | \quad | \\ CH_2-CH_2 \end{matrix} ClO_4^- \quad + \quad Na^+ \begin{matrix} CH_3 \\ | \\ {}^-OC-C-CH_2 \\ || \; | \\ O \; Ph \end{matrix}$$

$$\longrightarrow \quad -(CH_2)_4\underset{O}{\underset{||}{C}}O(CH_2)_2(OC_4H_8)_n-O\underset{O}{\underset{||}{C}}\underset{Ph}{\underset{|}{\overset{CH_3}{\overset{|}{C}}}}-CH_2-$$

   30 ml of THF was polymerized at 0°C for 60 hours with 3.12x
10^{-4} mole of the bifunctional cationic initiator to obtain 4.03 g
of p-THF of Mn 12,9000. 3.08×10^{-4} mole of disodium α-methylstyrene
dicarboxylate in 6.2 ml of THF was added and stirred for 1 hour at
room temperature. The occurrence of ion-coupling was indicated by
the precipitation of $NaClO_4$ and also by a gradual increase in the
viscosity of the solution. The molecular weight of the precipitated block copolymer was 53.0×10^4, which is equal to 39 repeating
units of $(THF-MS_4)$. The effect of molar ratio of initiators on
the molecular weight of the block copolymer is shown in Figure 5.
The extent of the reaction at equimolar feed ratios reached 98.7%
based on the initiator concentration. In other words, almost
every initiator molecule formed one living polymer chain and
quantitative coupling reaction between cationic living polymer
and anionic living polymer was observed.

The ion-coupling of the poly-THF dication of Mn 11.1×10^4 and
the poly-St dicarboxylate anion of Mn 6.0×10^4 formed block copolymer of Mn 50.2×10^4, and poly-THF of Mn 0.8×10^4 and poly-St
of Mn 0.5×10^4 formed block copolymer of Mn 40×10^4. These results
show that the ion-coupling reaction is slow, but quantitative
reaction occurs during several days at room temperature. The
formation of multi block copolymer by ion-coupling is a good indication of the high purity of the poly-THF dication and poly-St
carboxylate dianion.

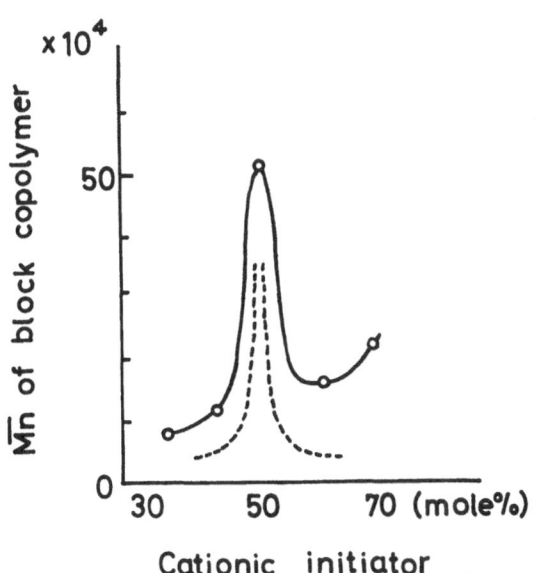

Figure 5. Effect of molar
ratio of initiators on the
Mn of the block copolymer
$(\text{THF-MS}_4)_n$

PTHF Mn=12900
MS_4 Mn= 560

room temperature,
1 hr in THF

REFERENCES

(1) J. F. Henderson, M. Szwarc, Macromolecular Reviews, 3, 317
    (1968) Interscience, New York
(2) Y. Yamashita, T. Hane, J. Polymer Sci., Polymer Chem. Ed.,
    11, 425 (1973)
(3) Y. Yamashita, Y. Nakamura, S. Kojima, ibid, 11, 823 (1973)
(4) Y. Yamashita, H. Matsui, K. Ito, ibid, 10, 3577 (1972)
(5) Y. Yamashita, Y. Murase, K. Ito, ibid, 11, 435 (1973)
(6) Y. Yamashita, K. Nobutoki, Y. Nakamura, M. Hirota,
    Macromolecules, 4, 548 (1971)

THE PREPARATION OF STYRENE BLOCK COPOLYMERS

FROM MACRORADICALS PRODUCED IN VISCOUS MEDIA

G. Allan Stahl [1] and Raymond B. Seymour

Department of Chemistry
University of Alabama
Tuscaloosa, AL 35486
and
Department of Polymer Science
University of Southern Mississippi
Hattiesburg, MS 39401

INTRODUCTION

The control of physical properties one gains by block copoly-merization has sparked much interest in their preparation. Many interesting block copolymers have been reported in,a number of excellent reviews (1, 2, 3, 4).

Homopolymers with reactive end groups such as polystyrene with tribromomethyl end groups have been treated chemically and photolytically to produce a macroradical capable of reacting with a second monomer to produce a block copolymer (5). Bamford (6) produced poly(methyl methacrylate-b-acrylonitrile) by using the chain transfer agent, triethylamine, to cap the methyl methacrylate macroradical, and added this polymer to free radical initiated polystyrene to produce a peroxide, which on decomposition initiated a second monomer producing a block copolymer.

[1]Present address: BF Goodrich Research and Development Center
9921 Brecksville Road
Brecksville, Ohio 44141

Styrene polymers prepared in emulsion systems have been converted into block copolymers by gamma-ray irradiation of the polymer in the presence of a second vinyl monomer (8, 9, 10). These block copolymers show a high degree of monodispersity, and are accompanied by low yields of homopolymers.

Difunctional free radical initiators such as diperoxide and peroxide-azo compounds have also been used to produce block copolymers. Smets (11) produced vinyl acetate block copolymers while Chow and Piirma (12) produced poly(styrene-b-methyl methacrylate) respectively. This technique is not widely accepted as the product is heterodispersed and contaminated by a great deal of homopolymers.

Porter (13) and Fujiwara (14) mechanically degraded poly-(vinyl acetate) in the presence of other vinyl monomers producing block copolymers. O'Driscoll (15) used an extrasonic field to homolytically cleave carbon-carbon bonds, and form block copolymers of styrene. Both techniques produce good yields of monodisperse block copolymers.

Although anionic polymerization was carried out commercially in Germany and Russia as early as 1936, its utility in producing block copolymers was not recognized until the work of Szwarc in 1956, (16,17). Anionically prepared macroions have been called "living polymers" since the termination step of polymerization is absent. When a second vinyl monomer is added to the "living polymers" a block copolymer is produced. High yields of monodispersed polystyrene block copolymers have been produced by this most **versatile technique.**

Block copolymers have also been produced by the addition of vinyl monomers to trapped macroradicals. Long lived macroradicals are produced when vinyl monomers are polymerized in poor solvents (18). These occluded macroradicals have been used to prepare block copolymers of styrene (19, 20), acrylonitrile (21), vinyl acetate (22), and methyl methacrylate (23).

EXPERIMENTAL

All solvents used in the preparation and fractionation of samples were reagent grade and were not purified further. Dow-Corning 200 Fluid® was used as received. The liquid monomers were distilled under reduced pressure so that the temperature of the distillate vapor never exceeded 35°C. Acrylic acid and vinyl acetate were dried over anhydrous sodium sulphate prior to distillation, and vinyl acetate was doubly distilled. All monomers were stored in a refrigerator at 0°C in amber glass bottles.

The polymerization initiator, tert-butyl peroxypivalate (Lypersol 11®, available from Lucidol Division, Pennwalt Corp., Buffalo, N.Y., a 75% solution in mineral spirits with a half-life of 12h at 50°C) was used as received. Degussa Inc., Aerosil 200 fumed silica was used as received. Zero grade argon was used to blanket all reactions.

Most samples were prepared in one fluid ounce amber glass bottles with foil-lined phenolic screw caps. The monomers and solvents were deoxygenated by bubbling zero grade argon for at least 30 minutes. The solutions containing monomers were then prepared volumetrically in a deoxygenated 17" x 17" disposable glove bag. The samples were placed in a water bath at 50.0 ± 0.1°C for 96h. The charge of second monomer was then added volumetrically in the glove bag. At the end of 96h the sample was removed from the bath and quenched in 100 ml of solvent.

Relative polymerization rates were determined by precision dilatometry (29) .

The block copolymers obtained from the reaction of styrene macroradicals were purified by precipitation with selected solvents. The mixture of homopolymers and block copolymer was first dissolved in a good solvent. After dissolution the mixture was fractionally precipitated by the gradual addition of a poor solvent.

Poly(styrene-b-methyl methacrylate) was purified by first dissolving in benzene. Polystyrene was precipitated by the addition of 0.5 volume parts of methanol. The block copolymer was removed by the addition of 0.8 volume parts methanol. The remaining poly(methyl methacrylate) was precipitated by adding excess methanol.

Poly(styrene-b-acrylic acid) was purified by extracting with acetone in a soxhlet extractor for 24h, followed by the addition of 0.7 volume parts petroleum ether to precipitate any poly-(acrylic acid). The block copolymer was precipitated by adding 1.2 volume parts petroleum ether.

Poly(styrene-b-ethyl methacrylate) was also purified by the benzene/methanol technique. Polystyrene was removed by the addition of 0.6 volume parts methanol, while the block copolymer precipitated when 0.8 volume parts petroleum ether was added.

Poly(styrene-b-acrylonitrile) was purified by benzene extraction in a soxhlet extractor for 24h. The supernate was fractionally precipitated by the addition of 0.4 volume parts petroleum ether which precipitated the block copolymer. Polystyrene was precipitated by the addition of more than 2.0 volume parts petroleum ether.

Poly(styrene-b-vinyl acetate) was purified by precipitation from a benzene solution. The addition of 0.6 parts petroleum ether removed the poly(vinyl acetate), while the block copolymer was precipitated by the addition of about 1.0 part of petroleum ether. An excess of petroleum ether precipitated the polystyrene.

Pyrolysis gas chromatography was performed using a Varian Aerograph Pyrolyzer Model A425 fitted with a Gow-Mac, Rh-W 13-002 pyrolyzing coil. The filament coil was connected to a Wilkins Model A100C thermal conductivity gas chromatograph with a 6 ft. column of 60/80 mesh chromosorb W and 20% SE-20 (General Electric Corp.)

A DuPont Model 990 thermal analyzer equipped with a DSC cell was used for differential scanning calorimetry. All runs were made in a nitrogen atmosphere over a temperature range from room temperature (usually 22°C) to 150°C.

Gel permeation chromatograms were obtained by using a GPC/ALC Waters Model 301 gel permeation chromatograph. Molecular weights were determined by comparison with calibrated Waters Associates standard polystyrene samples.

RESULTS AND DISCUSSION

Previous work has shown that when a vinyl monomer is polymerized in a poor solvent, an occluded macroradical is produced (18). Block copolymers can be produced if the solubility parameter of the growing polymer and solvent differ by at least L 8 H (24), and the polymer and second monomer differ by less than 3.1H (25).

The increased reaction rates in viscous media have also been explained by long lived macroradicals (26, 27, 28). In the absence of the usual termination steps, it should be possible to produce block copolymers by the addition of a second vinyl monomer to a macroradical in viscous solvent.

As shown in Figure 1, the rate of polymerization of styrene in a nonviscous good solvent, benzene, was slower than that in a nonviscous poor solvent, hexane, and both were slower than that in a viscous good solvent, benzene with 5% fumed silica, and nonviscous, poor solvent 1-propanol. The fastest polymerization rates were observed in viscous poor solvents, diethylene glycol and nujol.

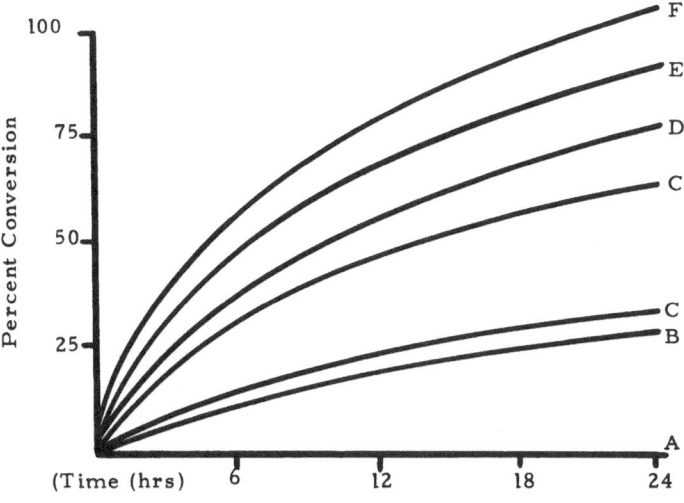

Figure 1. Rate of Polymerization of Styrene in Various Solvents at 50°C. (A) benzene, (B) hexane, (C) benzene with 5% fumed silica, (D) 1-propanol, (E) diethylene glycol, (F) nujol.

The increase in rate of polymerization in viscous poor solvents is also shown in Figure 2. Dow-Corning 200 Fluid®(silicone oil) was selected as the viscous poor solvent because it is a poor solvent for polystyrene, and is commercially available in a wide range of viscosities. The rate of formation of polystyrene is rapid in 50cs silicone oil, yet is faster in 200cs silicone oil. The fastest rate of polymerization occurs in the more viscous system, 500cs silicone oil.

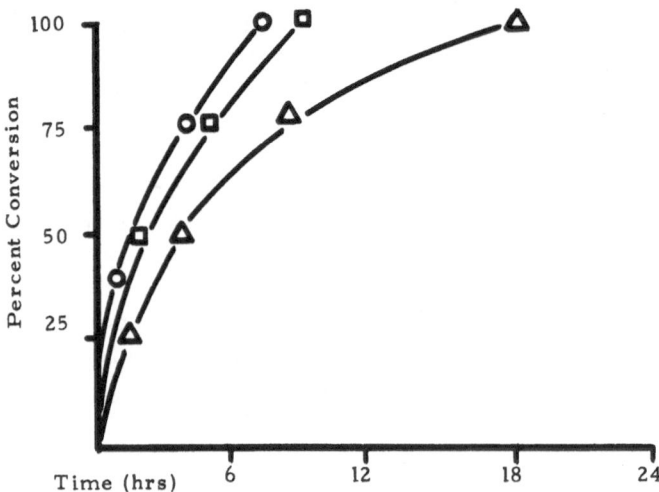

Figure 2. Rate of Polymerization of Styrene in Silicone Oil of Different Viscosities at 50°C. (Δ), (□) 200cs, (O) 500cs.

The existence of long lived styrene macroradicals was best demonstrated when styrene was added to styrene macroradicals produced in silicone oil. The styrene was added four days after initiation to insure that there were no longer any primary radicals present. One would not expect long lived macroradicals to be present when styrene is polymerized in benzene, a nonviscous good solvent. As shown in Figure 3, no increase in weight was found. When styrene was heated with macroradicals produced in 50cs silicone oil, a weight increase of 71 percent was found in 32h . Styrene macroradicals produced in 200cs silicone oil added styrene more rapidly. In 32h an increase in weight of 86 and 96 percent was noted respectively.

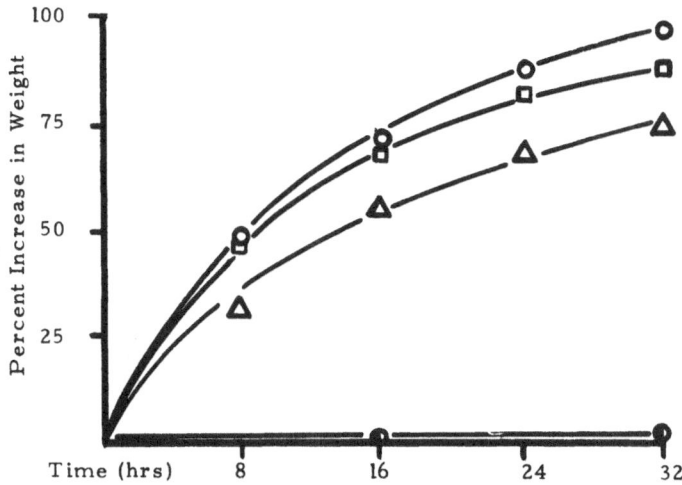

Figure 3. Rate of Addition of Styrene to Styrene Macroradicals in Silicone Oil of Different Viscosities at 50°C. (●) benzene, (Δ) 50cs, (□) 200cs, (O) 500cs.

As shown in Table I the molecular weight of polystyrene produced in 500cs silicone oil is greater than that produced in benzene. The molecular weight distribution(Mw/Mn) is also more narrow. When styrene is heated with styrene macroradicals in 500cs silicone oil, the resultant polymer is of higher molecular weight than the original polystyrene. This is presumably because the styrene adds to the existant styrene macroradical. The resultant polystyrene is also more narrow in molecular weight distribution.

Table I. Molecular Weights of Polystyrene Produced when Styrene Was Heated with Styrene Macroradicals at 50°C in Silicone Oil (500cs) for Varying Lengths of Time.

Solvent	$Mw \times 10^{-3}$	$Mn \times 10^{-3}$	Mw/Mn
benzene	131	24	5.46
silicone oil	148	30	4.93
silicone oil, 8h	158	38	4.15
silicone oil, 16h	166	46	3.60
silicone oil, 32h	187	57	3.28

When methyl methacrylate was added to polystyrene produced
in benzene, no block copolymer was formed. Macroradicals of
styrene are stabilized in a nonviscous poor solvent, 1-propanol,
and 72.1 percent poly(styrene-b-methyl methacrylate) was produced. The largest amounts of block copolymers were produced
in a viscous benzene (89.6 percent), and silicone oil (500cs)
(88.9 percent). The composition of the solvent fractionated polymers produced when methyl methacrylate was added to styrene
macroradicals is shown in Table II.

Table II. Composition of Polymeric Product Obtained When Methyl
          Methacrylate and Styrene Macroradicals Were Heated
          at 50°C.

Solvent	PS[a](%)	PMMA[b] (%)	P(S-b-MMA)[c] (%)	Total Yield (%)
benzene	91.7	8.3	0	51.6
1-propanol	20.8	7.1	72.1	98.4
benzene with 50% fumed silica	9.4	1.1	89.6	96.8
silicone oil, 500cs	8.6	2.7	88.9	100

[a]abbreviation for polystyrene
[b]abbreviation for poly(methyl methacrylate)
[c]abbreviation for poly(styrene-b-methyl methacrylate)

Pyrolysis gas chromatography demonstrated the presence of
both monomers in the block copolymer. Differential Scanning
Calorimetry (DSC) shows a single endotherm at 89°C. A single
endotherm would be expected since the two homopolymers have
the same glass transition temperature, and are mutually soluble.

As shown in Table III, only 2.9 percent poly(styrene-b-acrylic
acid) resulted when acrylic acid was added to polystyrene produced
in benzene. Only 21.4 percent block copolymer, and 90.3 percent
total conversion resulted when the polystyrene was produced in poor
solvent 1-propanol. The low yields are presumably because 1-pro-
panol solvates the growing acrylic acid segment allowing it to

terminate readily. To increase the yield, styrene macroradicals
were prepared in hexane, a poor solvent for poly(acrylic acid).
The yield of poly(styrene-b-acrylic acid) was increased to 37.5
percent, and the total conversion to 95.7 percent. The highest
yield of block copolymer, 69.9 percent, was obtained when acrylic
acid was added to polystyrene prepared in silicone oil (500cs).

Table III. Composition of Polymeric Products Obtained When
Acrylic Acid and Styrene Macroradicals Were Heated
at 50°C

Solvent	PS[a](%)	PAA[b](%)	P(S-b-AA)[c] (%)	Total Yield (%)
benzene	92.2	4.9	2.9	55.4
1-propanol	73.2	5.4	21.4	90.3
hexane	57.3	5.0	37.5	95.7
benzene with 5% fumed silica	67.1	6.4	26.6	93.0
silicone oil, 500cs	25.7	4.5	69.9	92.6

[a]abbreviation for polystyrene
[b]abbreviation for poly(acrylic acid)
[c]abbreviation for poly(styrene-b-acrylic acid)

Pyrolysis gas chromatography showed the presence of both
monomers in the block copolymer. DSC showed the block copolymer to have two glass transitions (80°C and 100°C) indicative of
a molecule with two heterogeneous segments.

Good yields of poly(styrene-b-ethyl methacrylate) are produced
when ethyl methacrylate is heated with styrene macroradicals in
benzene with 5% fumed silica (77.4 percent), and silicone oil (500cs)
(86.8 percent). The composition of the polymers produced is
listed in Table IV.

The characteristic peaks of the two monomers were found in
pyrolysis gas chromatograms. DSC thermograms showed two
glass transitions (54°C and 86°C) in the block copolymer. This is
characteristic of a polymer with two heterogeneous segments.

Table IV. Composition of Polymeric Products Obtained When
Ethyl Methacrylate and Styrene Macroradicals Were
Heated at 50°C

Solvent	PS[a](%)	PEMA[b] (%)	P(S-b-EMA)[c] (%)	Total Yield (%)
benzene	89.8	0	10.2	68.8
1-propanol	30.2	19.0	50.8	96.8
benzene with 5% fumed silica	16.6	5.7	77.4	94.1
silicone oil, 500cs	9.2	3.8	86.8	99.5

[a]abbreviation for polystyrene
[b]abbreviation for poly(ethyl methacrylate)
[c]abbreviation for poly(styrene-b-ethyl methacrylate)

Table V. Composition of Polymeric Products Obtained When
Acrylonitrile and Styrene Macroradicals Were Heated
at 50°C

Solvent	PS[a](%)	PAN[b](%)	P(S-b-AN)[c] (%)	Total Yield (%)
benzene	88.0	7.5	4.5	62.9
1-butanol	35.1	3.9	61.0	93.0
benzene with 5% fumed silica	16.8	4.1	79.1	100
silicone oil, 500cs	12.4	3.3	84.3	97.3

[a]abbreviation for polystyrene
[b]abbreviation for polyacrylonitrile
[c]abbreviation for poly(styrene-b-acrylonitrile)

It has been shown that no block copolymer was produced when styrene was heated with acrylonitrile macroradicals (21). Presumably styrene (ζ = 9.2H) cannot diffuse into polyacrylonitrile (ζ = 12 - 15H) since the difference in solubility parameter is greater than 3.1H. There are no such restrictions when acrylonitrile (ζ = 10.5H) was added to styrene macroradicals (ζ = 9.1H) in a nonviscous poor solvent, and 61.0 percent poly(styrene-b-acrylonitrile) was produced. As shown in Table V, the greatest yields of block copolymer were produced by heating styrene macroradicals and acrylonitrile in a viscous good solvent, benzene with 5% fumed silica (79.1 percent), and viscous poor solvent, silicone oil (500cs) (84.3 percent).

The DSC thermogram of the block copolymer had one endotherm (96°C) characteristic of a glass transition. The homopolymers, styrene and acrylonitrile, have the same glass transition temperature. Pyrolysis gas chromatography indicated the presence of both monomers.

As shown in Table VI, low yields of block copolymer were produced when vinyl acetate and styrene macroradicals were heated in a nonviscous poor solvent, 1-propanol (19.8 percent). The greatest yield of poly(styrene-b-vinyl acetate) (44.4 percent) was produced from styrene macroradicals produced in silicone oil (500cs).

Table VI. Composition of Polymeric Products Obtained When Vinyl Acetate and Styrene Macroradicals Were Heated at 50°C

Solvent	PS[a](%)	PVAC[b] (%)	P(S-b-VAC)[c] (%)	Total Yield (%)
benzene	84.3	15.7	0	58.1
1-propanol	73.5	6.7	19.8	71.5
benzene with 5% fumed silica	78.5	9.3	12.2	75.3
silicone oil, 500cs	43.9	11.7	44.4	89.2

[a]abbreviation for polystyrene
[b]abbreviation for poly(vinyl acetate)
[c]abbreviation for poly(styrene-b-vinyl acetate)

Pyrolysis gas chomatograms of the block copolymer show the characteristic peaks of both monomers. DSC indicated two glass transitions (22°C and 87°C). This would be expected from a polymer with two heterogeneous segments.

CONCLUSIONS

As evidence of the stabilization of macroradicals, the rate of styrene polymerization was increased in viscous good and poor solvent, and in nonviscous poor solvents. Block copolymers were prepared by adding a second vinyl monomer to these longer lived macroradicals.

Styrene block copolymers with methyl methacrylate, acrylic acid, ethyl methacrylate, acrylonitrile, and vinyl acetate were produced in good yields in viscous diluents. Polystyrene produced in silicone oil was characterized by gel permeation chromatography and found to increase in molecular weight when styrene was heated with styrene macroradicals.

The block copolymers of styrene were characterized by solvent fractionation, pyrolysis gas chromatography, and differential scanning calorimetry.

ACKNOWLEDGMENTS

The authors wish to thank Dr. David P. Garner, Case Western University, for his open discussion of this work and initial laboratory screening; and Professor Charles U. Pittman, Jr. for use of his gel permeation chromatograph. This investigation was supported by a grant from the ACS/Petroleum Research Fund.

BIBLIOGRAPHY

1. Ceresa, R. J. (Ed.), "Block and Graft Copolymerization", McGraw-Hill Co., New York, 1973.

2. Burke, J. J. and Weiss, V. (Eds.), "Block and Graft Copolymers", Syracuse Press, Inc., Syracuse, N.Y., 1973.

3. Allport, D. C. and Janes, W. H. (Eds.), "Block Copolymers", Applied Science Publishers, London, 1973.

4. Szwarc, M. "Carbanions, Living Polymers, and Electron Transfer Processes", Interscience Publishers, New York, 1968.

5. Dunn, A. S., Stead, B. D., and Melville, H. W., Trans. Faraday Soc., 50, 279 (1954).

6. Bamford, C. H. and White, E. F. T., Trans. Faraday Soc., 54, 268 (1958).

7. Urwin, J. R., J. Polym. Sci., 27, 580 (1958).

8. Allen, P. E. M., Burnett, G. M., Downer, J. M., Hardy, R. and Melville, H. W., Nature, 182, 245 (1958).

9. Nozaki, K., U.S. Patent 3,069,380, December 18, 1962.

10. Nozaki, K., U.S. Patent 3,069,381, December 18, 1962.

11. Smets, G., Convent, L., and Vander Bought, X., Makromol. Chem., 23, 162 (1953).

12. Chow, L. P. H. and Piirma, I., paper presented at American Chemical Society Meeting 1975; Polym. Preprints, 16 (2), 506 (1975).

13. Casale, A. and Porter, R. S., Rubber Chem. and Technol., 44 (2), 543 (1971).

14. Gato, K. and Fujiwara, H., J. Polym. Sci., B 1, 505 (1963).

15. O'Driscoll, K. F. and Sridharan, A. U., J. Polym. Sci., Polym. Chem. Ed., 11, 1111 (1973).

16. Szwarc, M., Levy, M., and Milkovich, R., J. Am. Chem. Soc., 78, 2656 (1956).

17. Szwarc, M., Nature, 178, 1168 (1956).

18. Hiemeleers, J. and Smets, G., Macromol. Chem., 47, 7 (1961).

19. Minoura, Y. and Ogata, Y., J. Polym. Sci., A-1, 7, 2547 (1969).

20. Seymour, R. B., Stahl, G. A., and Wood, H., Appl. Polym. Symp., 26, 249 (1975).

21. Seymour, R. B., Owen, D. R., Stahl, G.A., Wood, H., and Tinnerman, W. N., Appl. Polym. Symp., 25, 69 (1974).

22. Seymour, R. B. and Stahl, G. A., J. Polym. Sci. Polym. Chem. Ed., 14, 2545 (1976).

23. Seymour, R. B., Stahl, G. A., Owen, D. R., and Wood, H., Advan. Chem. Ser., 142, 309 (1975).

24. Seymour, R. B., Tsang, H. S., Jones, E. E., Kincaid, P. D., and Patel, A. K., Advan. Chem. Ser., 99, 418 (1971).

25. Seymour, R. B., Owen, D. R., and Stahl, G. A., Polymer, 14, 324 (1973).

26. Norrish, R. G. W. and Smith, R. R., Nature, 150, 336 (1942).

27. Trommsdorff, E., Kohle, H., and Legally, P., Makromol. Chem., 1, 169 (1948).

28. Benson, S. W. and North, A. M., J. Am. Chem. Soc., 81, 1339 (1959).

29. Seymour, R. B. and Stahl, G. A., Rev. Sci. Instrum., 46, 1467 (1975).

PHASE SEPARATION IN STYRENE-α-METHYL STYRENE BLOCK COPOLYMERS

Sonja Krause and Magdy Iskandar

Department of Chemistry, Rensselaer Polytechnic

Institute, Troy, N.Y.   12181

INTRODUCTION

In previous work from this laboratory, it has been shown that
diblock and triblock copolymers of styrene, S, and α-methyl
styrene, MS, are homogeneous up to considerably higher molecular
weights than are mixtures of the corresponding homopolymers (1,2).
Homogeneity was inferred from the presence of a single T_g, intermediate between those of polystyrene, PS, and of poly (α-methyl
styrene), PMS, in the samples. The glass transition temperatures
were measured using DTA, DSC, and dilatometry. The same
techniques showed that, if low molecular weight PS or PMS was
mixed with one of the block copolymers, phase separation was
always enhanced (3). The molecular weight and composition at
which microphase separation occurred in the block copolymers could
be predicted quite well using a theoretical treatment developed by
one of us (4). The data indicated that the interaction parameter
between PS and PMS was between 0.0030 and 0.0036 for samples with
$\overline{M}_w/\overline{M}_n < 1.3$. Block copolymer samples with broader molecular
weight distributions showed enhanced microphase separation, i.e.,
they exhibited microphase separation even when their $\overline{M}_w$ was so
low that homogenity was predicted theoretically (4). Furthermore,
the addition of homopolymer to the block copolymers always
enhanced microphase separation, even though it was predicted (5)
that the presence of very low molecular weight homopolymer should
serve to compatibilize the system.

It seemed of interest to observe the presence or absence of microphases in our block copolymer samples and in our mixtures of homopolymers with block copolymers by a method more direct than the measurement of glass transition temperatures. Electron microscopy seemed appropriate, even though staining of one of the phases with osmium tetroxide, as is usually done when observing styrene-butadiene or styrene-isoprene block copolymers, did not appear possible.

Before OsO_4 became a popular staining agent in polymer electron microscopy, several workers (6,7) found that unstained rubber mixtures could exhibit recognizable phase patterns in electron micrographs; these patterns were identical to those observed using phase contrast microscopy (7). It was postulated (7) that electron micrographs could be obtained on these unstained samples because of thickness differences between the phases.

We have succeeded in obtaining electron micrographs on un-stained films of styrene-α-methyl styrene block copolymers and their mixtures with homopolymers. Different morphologies have been observed, although the lamellar morphology seems to be the easiest to see. This may have something to do with the fact that nonlamellar morphologies are expected only in cases in which the lengths of the blocks in the block copolymer are very different (8), so that one component is present in excess, making the other one very difficult to see in an unstained sample. Sometimes it was necessary to prepare several films before micro-phases could be observed in a particular block copolymer.

In general, one expects a spherical morphology, i.e., spherical inclusions of A in a matrix of B when the percentage of A in the AB block copolymer is very low, usually below 25%; one expects a cylindrical morphology, i.e., cylindrical inclusions of A in a matrix of B when the percentage of A in the AB block co-polymer is more or less in the range 25-40%; and one expects a lamellar morphology of alternate lamellae of A and B when the percentage of A in the AB block copolymer is more or less in the range 40-60%. Meier (8,9) has devised a theoretical treatment by which one may calculate not only the morphology to be expected for different diblock and triblock copolymers, both in solution (8) and in bulk (9), but also the expected microphase dimensions in the bulk samples (8,9). We used the following equations adapted from Meier's work (8,9):

For A-B diblock copolymers:

lamellae (i = A or B): $T_i = 1.40\ \alpha_i\ \bar{K}_i\ M_i^{1/2}$ (1a)

spheres (i = A or B): $R_i = 1.33\ \alpha_i\ \bar{K}_i\ M_i^{1/2}$ (1b)

cylinders (i = A or B): $R_i = 1.00\ \alpha_i\ \bar{K}_i\ M_i^{1/2}$ (1c)

For A-B-A triblock copolymers:

lamellae: $T_A = 1.40\ \alpha_A\ \bar{K}_A\ M_A^{1/2}$ (2a)

              $T_B = 1.20\ \alpha_B\ \bar{K}_B\ M_B^{1/2}$ (2b)

spheres: $R_A = 1.33\ \alpha_A\ \bar{K}_A\ M_A^{1/2}$ (2c)

              $R_B = 1.00\ \alpha_B\ \bar{K}_B\ M_B^{1/2}$ (2d)

where T_i is the thickness of a lamella composed of monomer i

R_i is the radius of a spherical or cylindrical inclusion composed of monomer i

M_i is the weight average molecular weight of the block(s) composed of monomer i

$\bar{K}_i$ is an experimentally determined constant which relates the unperturbed end-to-end chain dimensions of the homopolymer composed of monomer i to its molecular weight. In our calculations, we used $\bar{K}_i = 7.5 \times 10^{-9}$, an average value for polystyrene, the same value used previously by Meier (10).

α_i is a chain perturbation parameter for the blocks composed of monomer i

The chain perturbation parameter, α_i, has a complex dependence on the type of block copolymer and on the particular morphology (9). In the case of diblock copolymers, Meier calculated the following relationships (8,9):

lamellae:

$$\frac{\alpha_B}{\alpha_A} = \left(\frac{M_B}{M_A}\right)^{1/2} \tag{3a}$$

spheres:

$$\frac{\alpha_B}{\alpha_A} \simeq 1 \tag{3b}$$

cylinders:

$$\frac{\alpha_B}{\alpha_A} = \left(\frac{M_B}{M_A}\right)^{1/6} \tag{3c}$$

where α_A can be calculated from:

$$\alpha_A^3 \left(1 + \frac{\alpha_B^2}{\alpha_A^2}\right) - 2\alpha_A = \frac{CM_A^{1/2}\gamma}{\rho_A RT\overline{K}_A} \tag{4}$$

where C equals 1.43 for lamellae, 2.25 for spheres, and 2.00 for cylinders (8), γ is the interfacial tension between microphases, ρ_A is the density of the microphase, R is the gas constant, and T is the absolute temperature. Since Meier indicated that the calculation is not sensitive to the exact value of γ chosen, and since the concept of interfacial tension between microphases is somewhat ambiguous, we chose the value $\gamma = 0$ for our calculations.

We used equations 3 and 4 for calculations on triblock co-polymers as well as diblock copolymers, using the idea (11) that the results of diblock copolymer calculations by Meier may be taken as applicable to triblock A-B-A copolymers if these are taken, for purposes of these calculations, as if they were A - 0.5 B diblock copolymers. Meier stated this assumption specifically for calculations of the interfacial energy between microphases, but it seems fairly reasonable for our calculations as well.

EXPERIMENTAL

Table I gives information, most of it taken from a previous publication (2), on the polymers used in this work. Sample numbers shown in Table I are the same as those used previously (2). All polymers were prepared by anionic polymerization in tetra-hydrofuran at -78° using 2-butyl lithium as initiator for the PMS homopolymer and for the S-MS and S-MS-S block copolymers, and sodium naphthalene as initiator for the MS-S-MS block copolymers.

The compositions of the copolymers were obtained using proton
magnetic resonance and infrared spectroscospy. The composition
shown in Table I is the average of the values obtained by the two
methods. Molecular weight averages were obtained using Gel
Permeation Chromatography, GPC, of the samples dissolved in
toluene. The GPC was calibrated using standard polystyrenes and
molecular weight averages were calculated assuming that the
calibration for PMS and for the block copolymers were unchanged.
Glass transition temperatures shown in Table I were obtained by
differential scanning calorimetry using a Perkin-Elmer DSC-2 at
heating rates of 20°/minute. The first run on any film was
usually discarded because it depended on sample history and was
not reproducible; the results of the first two reproducible runs
were usually recorded. Glass transition temperatures on samples
DB4 and TB2 were also obtained using other methods and
instruments (2); only the data obtained using the Perkin-Elmer
DSC-2 are recorded in Table I.

Specimens were prepared for transmission electron microscopy
(TEM) by room temperature evaporation of toluene from a 1%
solution of polymer. In some cases, the solution was cast onto a
microscope slide so that a polymer film about 5 μm thick was
formed after evaporation of solvent. This film was cut, floated
onto water, and picked up on an electron microscope specimen grid.
In other cases, the solution was evaporated directly on a specimen
grid, forming a film estimated to be thinner than 5 μm. The TEM
used for these observations was a Hitachi HU-125 with an
accelerating voltage of 125 kV. The magnification of the TEM was
calibrated using a replica of a standard diffraction grating with
2160 lines/mm leading to an estimated average error in the
calibration of approximately 10%.

TABLE I. Polymers Used in These Studies

Sample	Type of Sample	$\overline{M}_w \times 10^{-5}$	%MS	$T_g(°C)$
PMS	Homopolymer	1.49	100	not measured
DB4	Diblock (S-MS)	10.6	58	112, 180
TB2	Triblock (S-MS-S)	16.6	50	113, 180
TB4	Triblock (MS-S-MS)	4.0	45	very broad transition
TB8	Triblock (MS-S-MS)	9.7	25	111
TB9	Triblock (MS-S-MS)	11.6	36	112, 188

RESULTS AND DISCUSSIONS

The T_g data in Table I indicate that samples DB4, TB2, TB8, and TB9 definitely contained two phases and that sample TB4 probably had a distribution of composition within the sample (2). In samples such as TB8, containing only 25% MS, it is not unusual to detect only the T_g of PS.

Table II shows the results of the TEM observations on the samples. Dark and light areas were observed on the electron micrographs and are so indicated in the Table. Figures 1 through 7 are electron micrographs of the different samples and mixtures. Samples whose morphology is labeled "spherical" may actually have cylindrical morphology seen "end on"; for purposes of discussion, this morphology will be considered spherical.

TABLE II.   Results of TEM Observations

Sample	Figure	Morphology	Dimensions of Phases (Å)	
DB4	1	lamellar	Light:	800 - 1600
			Dark:	400 - 800
TB2	2	lamellar	Light:	1000 - 2000
			Dark:	1000 - 2500
TB2	3	spherical	Dark:	2500 - 3000
TB4	4	lamellar	Light:	300 - 600
			Dark:	600 - 900
TB8		none observed		
PMS/TB8:	5	lamellar	Light:	600 - 1800
3/7 by wt. (48% MS)			Dark:	600 - 1200
TB9	6	spherical	Light:	1340
PMS/TB9: 2/8 by wt. (52% MS)	7	spherical	Light:	2000

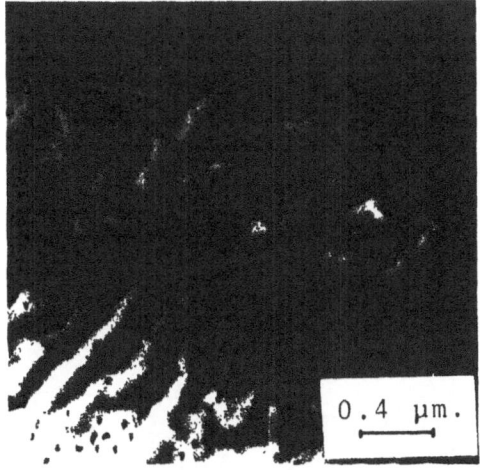

Figure 1.   Transmission electron micrograph
of an unstained film of Sample DB4.

Figure 2.   Transmission electron micrograph
of an unstained film of sample TB2 showing
lamellar morphology.

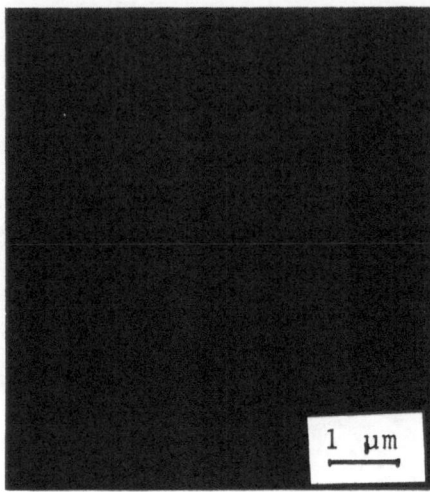

Figure 3.   Transmission electron micrograph of
an unstained film of sample TB2 showing spherical
or cylindrical morphology.

Figure 4.   Transmission electron micrograph of an
unstained film of sample TB4.

Figure 5. Transmission electron micrograph of an
unstained film of a 3/7 by weight mixture of PMS
with sample TB8.

Figure 6. Transmission electron micrograph of sample TB9.

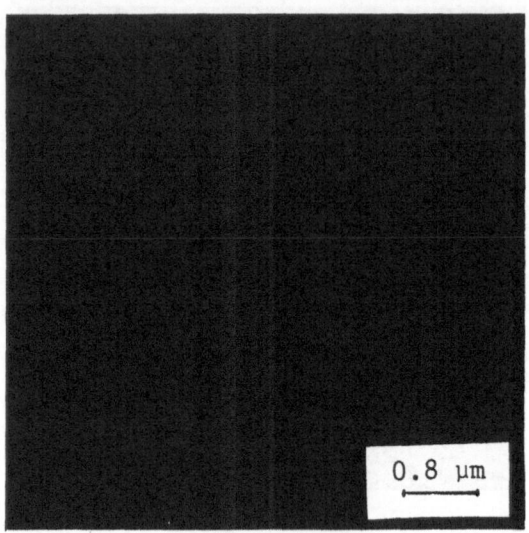

Figure 7. Transmission electron micrograph of an unstained
film of a 2/8 by weight mixture of PMS with sample TB9.

TABLE III   Calculations of Microphase Sizes

Sample	Type of Phase	Dimensions (Å)
DB4	PS Lamella	660
	PMS Lamella	1130
TB2	PS Lamella	680
	PMS Lamella	580
	PS Sphere	1820
	PMS Sphere	1820
TB4	PS Lamella	305
	PMS Lamella	310
TB9	PS Sphere	1290
	PMS Sphere	965

Table II and Figures 2 and 3 indicate that sample TB2 showed a lamellar morphology in one preparation and a spherical morphology (spherical or possibly cylindrical inclusions in a continuous matrix)in another preparation. Since sample TB2 is a 50/50 copolymer, the lamellar morphology is the expected equilibrium morphology. However, the electron microscopy results indicate that it seems possible to obtain a nonequilibrium morphology rather easily!

The dimensions of the microphases given in Table II are widths of lamellae in lamellar morphology and diameters of spheres in spherical (or cylindrical) morphology. We cannot be sure of the identity of the light and dark areas in the electron micrographs. We leaned originally toward a belief that the dark areas are poly(α-methyl styrene) because of its slightly higher density, but this belief may be incorrect.

Table III shows the dimensions of the microphases as calculated from equation 1-4. Only dimensions of polymers which contained no admixed homopolymer were calculated. As can be seen from the Table, the calculated dimensions are in reasonable agreement with the measured dimensions, no matter which micro-phase comprises the light or dark areas. A comparison of the experimental data in Table II and the theoretical calculations in Table III leaves the identification of the dark and light areas ambiguous. The results on sample DB4 indicate that the light areas are PMS but it is difficult to confirm this for the other samples and mixtures. It may be, if the observability of the microphases in unstained specimens is related to different thickness of the microphases in these films, that the light areas may sometimes be PS and sometimes PMS.

In sample TB9, one would expect that the spherical inclusions are MS, since this is the minority constituent of the sample. Adding PMS to TB7 increased the size of the spherical inclusions, as would be expected if these were MS in the unmixed sample. The fact that the inclusions appear light on the electron micrographs also indicates that they are probably MS. However, the calculated size of the microphase in sample TB9 agrees better with the observed size if it is assumed to be composed of styrene.

Sample TB4 exhibited lamellar morphology (Table III and Figure 4) even though DSC data did not indicate microphase separation (Table I). The DSC data indicated a broad composition distribution within the sample, but no actual phase separation (2). Nevertheless, phases appeared in the TEM. This probably indicates again the ease with which metastable phase states can be observed in block copolymers. It is likely that the phase separated state observed in the TEM after evaporation of solvent from a solution of the block copolymer with no further treatment of the polymer

film is the metastable state; the DSC data which indicate that
microphase separation did <u>not</u> occur in the sample were recorded
after at least one temperature cycling of the polymer film during
which there was some opportunity to attain an equilibrium state.

No phases were observed in the TEM when sample TB8 was
observed by itself. This sample contained only 25% PMS; it may
not have been possible to observe existing phases without
staining. At any rate, lamellar phases were observed when
sufficient PMS was added to the sample to increase the MS content
to 48%.

CONCLUSIONS

1. It is possible to observe different morphologies in unstained
    films of block copolymers of styrene and α-methyl styrene
    and their mixtures with poly(α-methyl styrene).

2. The same sample may exhibit different morphologies, depending
    on details of preparation of the polymer film. One of these
    morphologies is probably metastable with respect to the other.

3. The dimensions of the microphases observed were in reasonable
    agreement with those predicted using a theoretical treatment
    devised by Meier (8-11).

ACKNOWLEDGEMENTS

We would like to acknowledge support of this work from the
National Science Foundation under Grant No. DMR 72-02986 A01. One
of us (SK) would like to thank the National Institutes of Health
for a Research Career Award.

REFERENCES

1. D. J. Dunn and S. Krause, J. Polym. Sci. Polym. Letters
    Edition, <u>12</u>, 591 (1974).
2. S. Krause, D. J. Dunn, A. Seyed-Mozaffari, and A. M. Biswas,
    submitted to <u>Macromolecules</u>.
3. S. Krause and A. M. Biswas, submitted to <u>J. Polym. Sci, Polym.
    Phys. Ed.</u>
4. S. Krause, Macromolecules, <u>3</u>, 84 (1970).
5. S. Krause, in "Colloidal and Morphological Behavior of Block
    and Graft Copolymers," G. E. Molau, Ed., Plenum Press, 1971,
    p. 223.

6. M. H. Walters and D. N. Keyte, Rubber Chem. Technol., $\underline{38}$, 62 (1965).

7. P. A. Marsh, A. Voet, and L. D. Price, Rubber Chem. Technol., $\underline{40}$, 359 (1967).

8. D. J. Meier, Polym. Preprints, Amer. Chem. Soc., Div. Polym. Chem. $\underline{11}$, 400 (1970).

9. D. J. Meier, in "Block and Graft Copolymers," J.J. Burke and V. Weiss, Eds., Syracuse University Press, 1973, p. 105.

10. D. J. Meier, J. Polym. Sci, Part C, $\underline{26}$, 81 (1969).

11. D. J. Meier, Polym. Preprints, Amer. Chem. Soc., Div. Polym. Chem. $\underline{15}$, 171 (1974).

THE DEPENDENCE OF THE DYNAMIC MECHANICAL PROPERTIES OF POLY[(1,4-CYCLOHEXANE)BISMETHYLENE ISOPHTHALATE]-b-POLYCAPROLACTONE COPOLYMERS ON SEGMENT SIZE AND CRYSTALLINITY

J. E. Herweh, J. L. Work, and W. Y. Whitmore

Armstrong Cork Company, Research and Development Center

Lancaster, Pennsylvania 17604

INTRODUCTION

The influence of segment size on the crystallizability of block copolymers has been studied (1-5). Merrill and Petrie (1) found that the bisphenol A polycarbonate segment of a compatible poly(oxyethylene)-b-bisphenol A polycarbonate copolymer was crystallizable while the high-molecular-weight bisphenol A polycarbonate homopolymer was not crystallizable in the absence of solvent. A compatible block copolymer is defined as one having mutually soluble segments. Hall, et al (2) found that if the molecular weight of poly(oxytetramethylene) (POTM) segments was not greater than 2000 and the POTM content was less than 60 w/w% in ether-b-ester copolymers containing crystallizable ester segments, only the ester would crystallize. If the POTM-segment molecular weights were greater than 2000, both segments would crystallize. However, if the ester segment was noncrystalline, POTM segments would crystallize even at 1000 molecular weight. Cella (3) likewise found that at a POTM-segment molecular weight of 1000, poly(tetramethylene terephthalate) segments readily crystallized from POTM-b-poly(tetramethylene terephthalate) copolymers. He further found that the Tg of the block copolymer increased as the ester content increased, suggesting that segments of the block copolymer were compatible.

In this study, poly[(1,4-cyclohexane)bismethylene isophthalate]-b-polycaprolactone PCMI-b-PCL copolymer contains an easily crystallizable segment, polycaprolactone (PCL), and a difficult to crystallize segment, poly[(1,4-cyclohexane)-bismethylene isophthalate] (PCMI). The influence of segment sizes

and annealing conditions on the crystallizability of the segments
will be presented.

EXPERIMENTAL

Each of the block copolymers contains a 1/1 weight ratio of
α,ω-dihydroxy PCMI to α,ω-dihydroxy PCL segments coupled with
methylenebis(1,4-phenyl isocyanate) (MDI). The general formula of
the block copolymers is

$$[\{CH_2\text{-⬡-}NHCOOROOCHN\text{-⬡-}\}_m\{CH_2\text{-⬡-}NHCONH\text{-⬡-}\}_n]_p$$

$$\text{where} \quad R \text{ is} \quad (CH_2)_5[COO(CH_2)_5]_x \quad \text{or}$$

$$(CH_2\text{-Ⓢ-}CH_2OOC\text{-⬡-}COO)_yCH_2\text{-Ⓢ-}CH_2$$

The α,ω-dihydroxy esters will be referred to as diols. The
molecular weight of the diols associated with each block copolymer
is given in Table I. The PCL diols were used as received from
Union Carbide. The PCMI diols were prepared by a melt condensation
of 1,4-cyclohexanedimethanol and isophthalic acid. The MDI was
obtained from Fisher Scientific.

TABLE I

Composition of Poly[(1,4-Cyclohexane)Bismethylene Isophthalate]-b-Polycaprolactone Copolymers

Polymer	PCL Mol Wt	PCMI Mol Wt
I	1300	1000
II	1300	3800
III	3000	3800
IV*	3000	3800
V**	3000	3800

*Prepared by first chain extending the PCL segments.
**Prepared by first chain extending the PCMI segments.

General Procedure for Preparation of PCMI-b-PCL Random Copolymers

Typically a mixture of the PCMI and PCL diols and chlorobenzene (230 ml per 60g of the combined diol weight) was heated to
reflux with stirring under a nitrogen atmosphere. Approximately
30 ml of the solvent was distilled to provide for anhydrous
conditions. The temperature of the homogeneous reaction mixture

was reduced to ca. 120C and a solution of MDI in 30 ml of dry chlorobenzene was quickly added. Upon completing the addition of MDI, the reaction mixture was maintained at 120C while stirring. After ca. 18 hr, the reaction temperature was lowered to 100C and water (ca. 40 times the molar excess of isocyanate) was added. Reaction temperatures were subsequently maintained at 105C for 3 hr.

The reaction mixture, while still hot, was added to a large excess of methanol to precipitate the block copolymer. After ca. 18 hr the solvent layer was decanted and the precipitated block copolymer triturated with portions of fresh methanol. The reaction product was dried in vacuo at room temperature overnight and finally at 70C for 2 hr @ <1 torr.

General Procedure for Preparation of PCMI-b-PCL Copolymers Wherein One Segment was Chain Extended

The diol to be chain extended and chlorobenzene (110 ml per 30g of diol wt) was stirred under nitrogen and heated to reflux. A quantity of solvent was distilled to provide for anhydrous conditions. After cooling the reaction mixture to 120C, a solution of MDI (1.5 mol per mol of polyester diol) in dry chlorobenzene was added. After 5 hr at 120C, solutions of the remaining MDI (sufficient to provide for 10% molar excess) and the diol to be copolymerized (100 ml of chlorobenzene per 30g of diol) were added respectively. The reaction mixtures were subsequently treated as described previously.

Test Procedures and Sample Preparation

All test specimens were prepared by compression molding at 180C for 3 min and quenching to 15C using running cold tap water or to 110C in air. Within 15 to 30 min after quenching to 15C samples were run in the Rheovibron from -80C or stored at room temperature or 0C for specified periods of time. The dynamic mechanical properties were determined using a Rheovibron Model DDVII, Toyo Measurement instruments. The DSC results were obtained using a DuPont Model 990 DSC at a scanning rate of 20 deg C/min.

RESULTS AND DISCUSSION

These block copolymers have the unusual property that both high molecular weight segments are soluble in poly(vinyl chloride) (PVC) but insoluble in each other. Melt blended mixtures of the block copolymers and PVC exhibited a single Tg which was a linear function of the composition (Figure 1). All of the polymers except

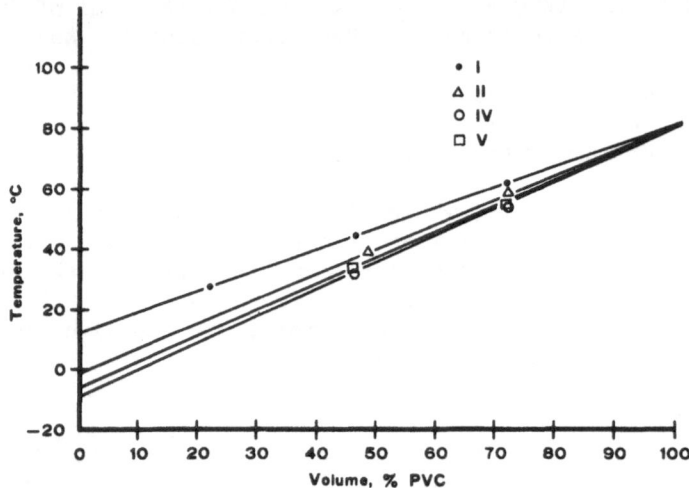

Fig. 1. Tg Of Blends Of PVC And Block Copolymers

TABLE II

Tg's of Quenched Poly[1,4-Cyclohexane)Bismethylene
Isophthalate]-b-Polycaprolactone Copolymers and the
Extrapolated Tg's from Copolymer-PVC Mixtures

Polymer	Extrapolated	3.5 Hz Tg (C) Experimental	MDI w/w%
I	12	9	19.6
II	-1	-9	12.4
III	-7*	-25	7.6
IV	-8	-29	7.6
V	-6	-41	7.6

*Not measured - average of polymers IV and V.

polymer I crystallized from PVC when the mixture contained less
than 50% PVC. Extrapolation of the Tg's of the mixtures to zero
percentage PVC yielded the Tg of the amorphous-compatible block
copolymer. The term compatible is used to signify that the segments
in the block copolymer are mutually soluble. The Tg's of the block
copolymers in the amorphous and compatible state decrease linearly

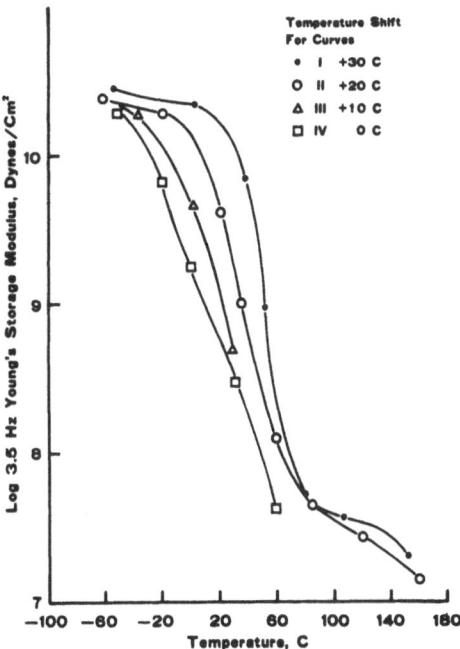

Fig. 2. Modulus vs. Temperature For
Polymers Quenched From Melt

with decreasing MDI content (Table II - Polymers I, II, and IV) as
might be expected from free-volume theory. It has been shown
previously (6) that the Tg's of urethanes decrease linearly with
decreasing weight fraction of [4,4'-methylenebis(cyclohexyl
isocyanate)].

   The difference between the Tg's obtained on the block
copolymers quenched from the melt and the Tg's in the amorphous and
compatible state increases in the order I<II<III<IV<V (Table II).
The breadth of the Tg region increases in the same order (Figure 2).
These results suggest that this is also the order of the increasing
incompatibility of the segments in the block copolymer. The Tg
breadth of polymer V could not be determined because extensive PCL
crystallization occurred during the determination.

Polymer I - Polymer I is judged to be amorphous because no
endotherm could be observed in DSC traces and only an amorphous
halo was observed in wide-angle X-ray diffraction patterns. Based
on the narrow Tg region and the very small difference between the

Tg in the amorphous and compatible state and the Tg in the quenched
state, polymer I is also judged to be compatible. However, the
modulus level in the temperature range of 60–140C indicates that a
low level of PCMI crystallinity may be present, since only
crystalline PCMI could produce this modulus level in this
temperature range (Figure 3). Attempts to crystallize polymer I

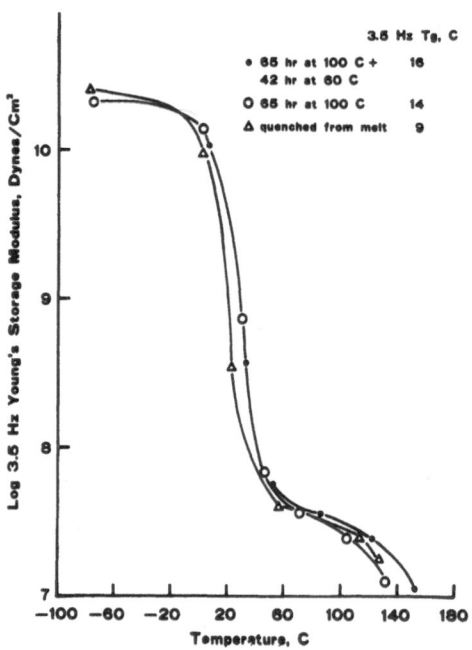

Fig. 3. Modulus vs. Temperature For Polymer I

to any greater degree by various thermal treatments were
unsuccessful as judged by the independence of modulus and thermal
treatment (Figure 3). The reluctance of the PCMI segments to
crystallize most probably arises from their small segment size,
since it will be shown later that larger PCMI segments readily
crystallize under similar thermal treatments. The PCL segments
may not crystallize for the same reason. However, it is well known
that the rate of crystallization of polymers exhibits a maximum as
a function of temperature. At temperatures below this maximum the
crystallization rate is diffusion controlled. At temperatures
above this maximum the crystallization rate is nucleation
controlled. As the difference between Tm (melting temperature) and
Tg decreases, the rate of crystallization will likewise decrease.
A polymer based on the MDI-extended-3000 mol wt-PCL diol has a Tm
of about 50C and a Tg of –60C and crystallizes rapidly at room
temperature. Since the amorphous phase of polymer I is composed of
mutually soluble PCL and PCMI segments, the PCL must crystallize

from this phase which has a Tg of 9C. This effectively reduces
Tm-Tg for the PCL segments and would dramatically decrease the rate
of the PCL crystallization.

Polymer II - The PCMI segments of polymer II could be crystallized
by annealing at 110C for 24 hr as indicated by an endotherm at
140C in DSC traces and the modulus level in the temperature range
of 10 to 160C (Figure 4). This crystallization results in a
decrease of 4 deg C in Tg from the quenched state and 12 deg C
from the amorphous and compatible state. This decrease in Tg

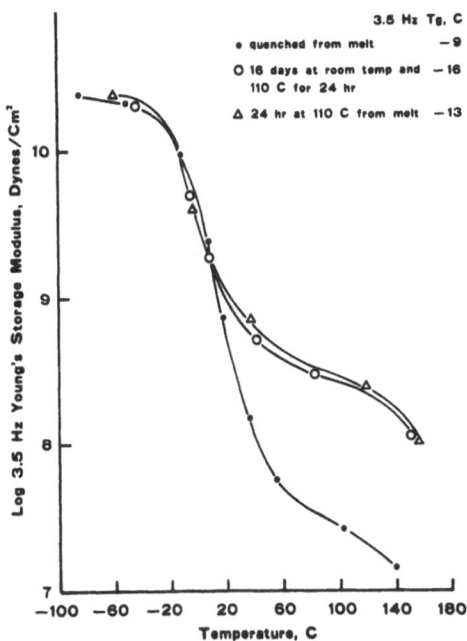

Fig. 4. Modulus vs. Temperature For Polymer II

occurs because PCMI segments (Tg = 90C) are transferred from the
amorphous phase to the crystalline phase. Wilkes and Wildnauer (7)
observed a similar segmental mixing/demixing process using
segmented urethanes. No significant PCMI or PCL crystallization
occurred at room temperature after six days probably because the
Tg (-9C) is too close to room temperature. The PCL segments could
not be crystallized thermally.

A polymer based on the MDI-extended PCMI diol of 3800 mol wt
would not crystallize under the same annealing conditions.
However, it has been observed to crystallize from solution. This
suggests that the PCMI segments in the block copolymer can

crystallize because they are of sufficient size and are plasticized by the PCL segments. These results are similar to those obtained by Merrill and Petrie (1). The reluctance of the PCL segments to crystallize arises either from the proximity of Tg to the PCL crystalline melting point (~50C) and/or the segment size as noted above. Hall, et al (2) obtained similar results using POTM segments with crystallizable polyester segments.

After 3 months at room temperature DSC traces on material annealed at 110C for 24 hr revealed endotherms for both PCL and PCMI crystallinity with enthalpies of fusion of 0.9 and 2.6 cal/g, respectively (Table III). These results indicate that the PCL crystallization is quite slow and even after 3 months has achieved only a small fraction of its potential crystallinity (H = 7.2 cal/g).

TABLE III

Enthalpies of Fusion for Poly[(1,4-Cyclohexane) Bismethylene Isophthalate]-b-Polycaprolactone Copolymers

Polymer	PCMI Cal/g	PCL Cal/g	Thermal Treatment
II	2.6	0.9	110C for 24 hr and 3 months at room temp.
III	2.8	1.6	110C for 24 hr and 3 months at room temp.
III	--	1.3	Quenched from melt to room temp. 3 months at room temp.
IV	--	4.7	Quench from melt and 68 hr at 0C and 50 days at room temp.
IV	2.8	3.0	Quench from melt and 68 hr at 0C and 24 hr at 110C and 50 days at room temp.
IV	3.5	3.5	110C for 24 hr and 50 days at room temp.
V	--	6.2	Quench from melt and 26 hr at 0C and 3 months at room temp.
V	3.3	5.1	Quench from melt and 26 hr at 0C and 24 hr at 110C and 3 months at room temp.
V	3.4	4.9	110C for 24 hr and 30 days at room temp.
VI*	--	14.4	Annealed at room temp. for 3 months.

*Polymer VI is a homopolymer based on PCL diol of 3000 mol.wt.

Polymer III - Polymer III was subjected to the following sequence of thermal treatments: (a) quench to 15C from melt and anneal at room temperature for 40 days; (b) anneal at 110C from the melt for 24 hr followed by (c) annealing at room temperature for 2 days, and (d) an additional 28 days at room temperature. Thermal treatment (a) resulted in a 15 deg C decrease in Tg from the quenched state and a 33 deg C decrease in Tg from the amorphous and compatible state (Figure 5). Both PCL and PCMI segments were crystallized by this thermal treatment. The PCL crystallinity produced an inflection in the modulus/temperature curve between 40 and 60C (Figure 5). The modulus level above 60C indicates the presence of PCMI crystallinity. Above 104C the PCMI segments rapidly undergo further crystallization. Treatment (b) increased the PCMI crystallinity above that obtained by treatment (a) as evidenced by a higher modulus at temperatures above 60C (Figure 5). The PCL crystallinity was eliminated as evidenced by the absence of an inflection in the modulus/temperature curve in the range of 40 to 60C. The Tg was 2 deg C higher after treatment (b) than after treatment (a). This result was unexpected. After treatment (b), the PCMI crystallinity is higher than after treatment (a) and there is no PCL crystallinity. This means that the amorphous regions after treatment (b) should be richer in PCL than after treatment (a) and therefore should have a lower Tg. However, as discussed previously, the segments of polymer III are partially incompatible. Therefore, there are PCL and PCMI rich regions in the polymer. The PCL rich regions which have a lower Tg than the PCMI rich regions, would contribute most to the Tg of the block copolymer (8). The longest segments of each type would be expected to be found in the regions rich in that segment. At room temperature, crystallization from the PCL rich regions would be expected to be most rapid because these regions have the lowest Tg. At 110C the rates of crystallization from both regions would be nucleation controlled and PCMI would be expected to crystallize most readily from the PCMI rich regions because these regions contain the longest PCMI segments and will form the most stable nuclei. Thus although the degree of PCMI crystallinity is higher and the PCL crystallinity lower after treatment (b) than after treatment (a), the amorphous regions contributing to the measured Tg are similar in composition because the crystallizing PCMI segments are not coming from the same amorphous regions.

Treatment (c) reduced the Tg 6 deg C from that obtained after treatment (b). The higher modulus at temperatures above 60C indicates that the PCMI crystallinity has increased from (b) to (c) (Figure 6). The modulus inflection at 20C indicates the presence of some PCL crystallinity of low melting point. The Tg decrease suggests that PCMI segments are crystallizing from the PCL regions at room temperature. The conclusion that further PCMI crystallization can occur at temperatures below 110C after

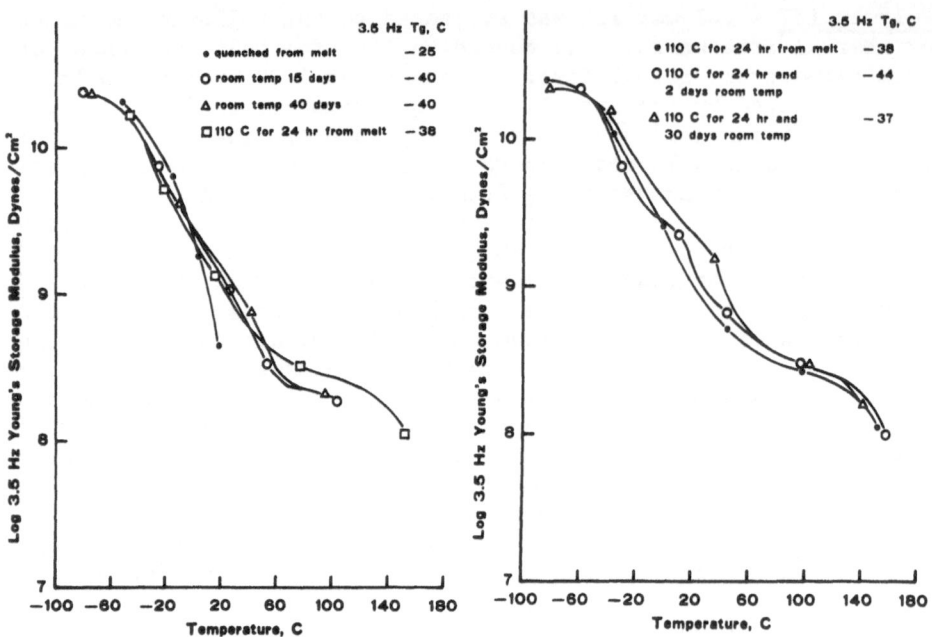

Fig. 5. Modulus vs. Temp.
For Polymer III

Fig. 6. Modulus vs. Temp.
For Polymer III

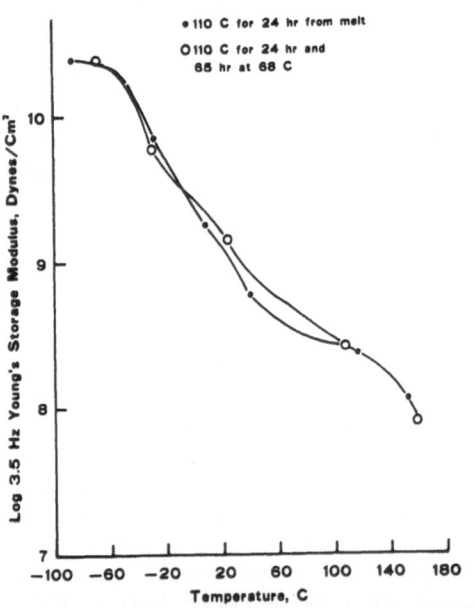

Fig. 7. Modulus vs. Temperature For Polymer III

crystallization at 110C is supported by the observation that annealing polymer III at 110C for 24 hr followed by annealing for 64 hr
at 68C produces an additional DSC endotherm at 80C, increases the
modulus in the temperature range of 0 to 100C (Figure 7), and
decreases the Tg to the level of that obtained after treatment (c).
Annealing at 68C could not produce any PCL crystallinity.

Treatment (d) resulted in no further increase in PCMI
crystallinity as shown by the same modulus levels above 60C as
after treatment (c) (Figure 6). The large change in modulus in the
range of 40 to 60C indicates that PCL crystallinity has been
produced by treatment (d). A Tg increase of 7 deg C accompanies
this increase in PCL crystallinity because PCL is being removed
from the PCL rich amorphous regions.

Enthalpies of fusion obtained after 3 months at room
temperature indicate that significantly more PCL crystallization
occurs in polymer III than polymer II (Table III). In addition,
the level of PCL crystallinity is higher for materials which were
annealed at 110C for 24 hr than for materials quenched to room
temperature. The level of PCMI crystallinity is similar for
polymers II and III (Table III).

These results indicate that polymer III is not homogeneous
and contains PCL and PCMI rich areas. By altering thermal
treatments, PCMI can be selectively crystallized from different
regions. PCMI crystallization from the PCL rich areas occurs at
room temperature and from the PCMI rich areas at 110C. When the
material is first annealed at 110C and then at room temperature,
higher levels of both PCL and PCMI crystallinity are obtained than
if the material is annealed at room temperature only. For short
annealing times (approximately one day) at room temperature, mostly
PCMI crystallization occurs and only after longer time periods at
room temperature does any significant PCL crystallization occur.

Polymer IV - Polymer IV was subjected to the following sequence of
thermal treatments: (a) quench from melt to 15C and anneal at 0C
for 68 hr, (b) anneal at 110C for 24 hr and (c) anneal at room
temperature for 44 days. Thermal treatment (a) results in PCL
crystallization as evidenced by an increase in modulus over that
obtained on the quenched material in the temperature range of -40
to 50C (Figure 8). The Tg after treatment (a) is increased 13 deg
C from the quenched state suggesting that PCL segments are transferred from the amorphous PCL rich regions to the crystalline PCL
regions. Very little PCMI crystallization occurs as evidenced by
the modulus level at 50C. Above 56C a rapid PCMI crystallization
occurs. In contrast, similar thermal treatments on polymer III
resulted in mainly PCMI crystallization. The Tg of the quenched
state of IV is only 4 deg C lower than the Tg of the quenched state

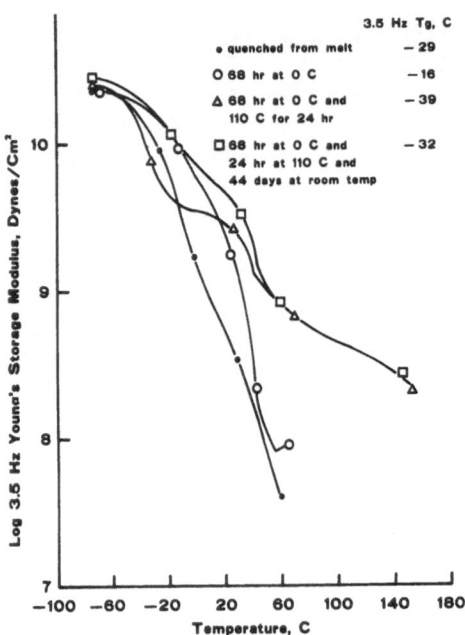

Fig. 8. Modulus vs. Temperature For Polymer IV

of III. However, the modulus/temperature curve of the quenched
state of IV exhibits an inflection in the range of OC which is not
observed in the modulus/temperature curve of III (Figure 2). This
implies that the aggregation into domains after quenching is more
distinct in IV than in III. This would be expected since the PCL
and PCMI segments are larger in IV than in III. In addition,
further changes in the state of aggregation of IV may occur with
time in the quenched state as Wilkes and Wildnauer (6) have
observed in polyurethanes. These differences in the quenched
state could account for the observed differences in the rate of
PCL crystallization.

Thermal treatment (b) results in a decrease of 10 deg C in Tg
from the quenched state (Figure 8). The modulus level above 60C
indicates the presence of PCMI crystallinity. These results
indicate that PCMI segments are transferred from the amorphous to
the crystalline phase. A rapid crystallization of PCL occurs
during the Rheovibron run beginning at approximately -10C, in
contrast to quenched IV or polymer III. This suggests that the
PCL rich areas have been purified sufficiently by the PCMI
crystallization to increase the rate of PCL crystallization.
However, the Tg's of III and IV are nearly identical after
annealing at 110C indicating the same amorphous-phase composition.
The modulus of IV is higher than III above 60C, but this is not

due to increased PCMI crystallinity as judged by enthalpies of fusion of 2.8 cal/g for both polymers. These data indicate a difference in physical structure between III and IV that may also be responsible for the increased rate of PCL crystallization.

Thermal treatment (c) increases Tg by 7 deg C from thermal treatment (b) because the PCL has crystallized which decreases the PCL level in the amorphous phase. Identical modulus results were obtained on a sample annealed at room temperature for 2 days after annealing at 110C for 24 hr directly from the melt. Thus, the PCL crystallization is nearly complete after 2 days. The degree of PCL crystallinity is greater for IV than III as indicated by enthalpies of fusion of 3.5 and 1.6 cal/g, respectively.

Based on enthalpies of fusion the highest level of PCL crystallinity is obtained in IV by annealing at 0C for 68 hr directly from the melt and then at room temperature (Table III). The lowest level of PCL crystallinity is obtained by annealing at 0C, followed by 24 hr at 110C and room temperature. The highest level of PCMI crystallinity is obtained by annealing at 110C for 24 hr directly from the melt.

Polymer V - The rate of polycaprolactone crystallization is much greater for V than for any of the other polymers, as evidenced by the occurrence of PCL crystallization during DSC and Rheovibron runs on the quenched material. This results from a greater degree of phase separation. The Tg after annealing at 0C for 26 hr is 29 deg C higher than that obtained on the quenched sample and is similar to that obtained for IV with a comparable thermal treatment. However, the increase in Tg is greater for V. The degree of PCL crystallinity is greater for V than IV as indicated by enthalpies of fusion of 6.2 and 4.7, respectively (Table III). This is consistent with higher modulus values for V than IV in the temperature range of -40 to 50C (Figure 9). The Tg of V after annealing at 0C and then 110C for 24 hr, is very close to the Tg for the MDI-extended PCL segments, suggesting that pure PCL is present in this material. Annealing at 110C directly from the melt produces a Tg 10 deg C higher, also suggesting that crystallizing PCL first produces a purer PCL phase.

The enthalpies of fusion associated with PCMI melting in IV and V are nearly identical. However, the modulus of V is higher and more temperature dependent than IV in the temperature range of 60 to 140C (Figures 8 and 9). This suggests that the physical arrangement in V is different from the arrangement in IV and most probably results from the greater phase separation in V. After annealing at room temperature for 30 or more days, the Tg's of all of the samples of polymer V annealed at 110C are in the range of -30C.

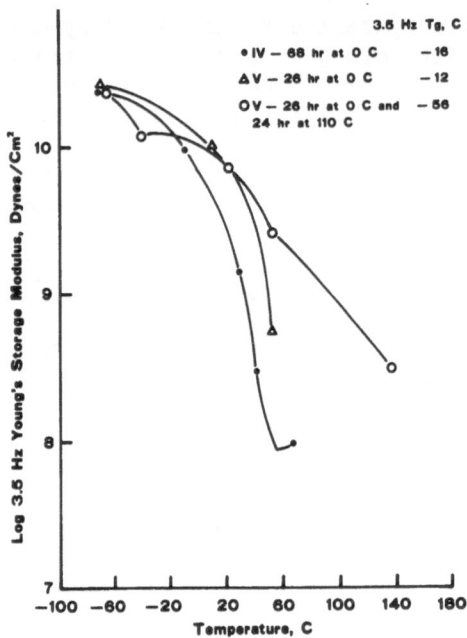

Fig. 9, Modulus vs, Temperature For Polymers IV & V

CONCLUSIONS

Random coupling of a PCL diol of 1300 mol wt and a PCMI diol of 1000 mol wt produces a compatible block copolymer which cannot be thermally induced to crystallize. The aggregation of the segments into domains and their thermally induced crystallizability are enhanced with increasing diol molecular weight and segment size.

The Tg of the phase containing the crystallizable segments controls the ability of the segments to crystallize. Raising the Tg decreases the crystallizability of the segments while lowering the Tg increases the crystallizability of the segments. The changes in Tg accompanying crystallization are correlated with the transfer of segments from the amorphous phase to the crystalline phase. Multiple PCMI crystalline melting points can be observed by sequentially reducing the annealing temperature.

Block copolymers based on PCL and PCMI segments are miscible with poly(vinyl chloride).

REFERENCES

1. S. H. Merrill and S. E. Petrie, J. Polym. Sci., 3, 2189-203 (1965).

2. I. V. Hall, A. Ghaffar, and I. Goodman, Br. Polym. J., 5, 315-25 (1973).
3. R. J. Cella, J. Polym. Sci. Symp. No. 42, 727-40 (1973).
4. R. W. Seymour, J. R. Overton, and L. S. Corley, Macromolecules, 8, 331-5 (1975).
5. M. J. Huet and E. Marechal, Eur. Polym. J., 10 (9), 771-82 (1974).
6. J. L. Work, Polym. Sci. and Tech., 4, 213-23 (1974).
7. G. L. Wilkes and R. E. Wildnauer, J. Appl. Phys., 46, 4148-52
8. J. L. Work, Macromolecules, 9 (5), 759 (1976).

STRUCTURE-PROPERTY RELATIONSHIPS OF POLYURETHANES BASED ON TOLUENE

DIISOCYANATE

C.S. Paik Sung and N.S. Schneider*

Department of Materials Science and Engineering

M.I.T., Cambridge, Mass. 02139

INTRODUCTION

The thermoplastic polyurethanes are linear segmented copolymers which consist of alternating soft and hard segment units. The soft segment is commonly a low molecular weight polyether or polyester whereas the hard segment generally consists of an aromatic diisocyanate condensed with a low molecular weight diol. It is now generally accepted that the properties of these materials are primarily due to the phase segregation of soft and hard segments leading to the formation of hard segment domains which are dispersed in the rubbery matrix. There are a wide variety of compositional variables which can affect the degree of phase segregation and hard segment organization and, accordingly, the sample properties. In this paper we report the results obtained on an extensive series of polyurethanes based on 2,4-toluene diisocyanate (2,4 TDI) and 2,6-toluene diisocyanate (2,6 TDI). In 2,4 TDI polymers, the asymmetric placement of the isocyanate residues with respect to the methyl group can result in some head-to-tail isomerization in the hard segment. In 2,6 TDI polymers, this problem is absent since the molecule is symmetrical. Other compositional variables examined included polyether vs. polyester soft segment, the molecular weight of the soft segment and the length of hard segment. Extensive physico-chemical techniques such as thermal analysis, wide and low angle x-ray studies and IR spectroscopic analyses have been employed to provide detailed information on the thermal transition behavior, structural organization and the properties.

* Polymer and Chemistry Division, Army Materials and Mechanics Research Center, Watertown, Mass. 02172.

EXPERIMENTAL

Preparation of the polymers was carried out by the two step method described previously (1). 2,4 TDI (Aldrich Chemical, 97%) and 2,6 TDI (Kipers Laboratory, 89%) were vacuum distilled. Anhydrous 1,4-butanediol (GAF), polytetramethylene oxide (Quaker Oats) and polybutylene adipate (Hooker Chemical) were used as received. The molar ratio of TDI, butanediol and polyether (PTMO) or polyester (PBA) was varied in five equal steps from 2:1:1 to 6:5:1 with 5% molar excess of diisocyanate. In sample designation, the integer following the TDI isomer indicates the number of moles of diisocyanate per mole of polyether or polyester. Chart I represents the chain structure of the polymers. All the samples used for the study of thermal transition were compression molded into 15 mil thick films with a pressure of 100 bar at 180°C. Differential scanning calorimetry was carried out at heating rate of 10°C/min with a sample weight of approximately 10 mg using the Perkin Elmer DSC II. Thermomechanical analysis (TMA) was conducted with the Perkin-Elmer TMS-1 equipped with the UU-1 temperature programmer at a load of 5 to 10 grams and a heating rate of 20°C/min. The infrared analyses were carried out on films cast directly on sodium chloride plates from 2% solution in dimethyl formamide. The films were dried in a vacuum oven at 50°C and scanned on a Beckman-12 infrared spectrophotometer. The temperature dependence of IR was determined using a variable temperature unit (Wilks model #19) connected to the temperature controller (Wilks model #37). For wide angle x-ray diffraction studies, a Warhus camera was used, while for small angle x-ray studies a Kratky camera equipped with automatic step scanning, counting and printing devices was employed. Radiation was supplied by a graphite monochromated Norelco copper fine-focus tube operated at 50 KV and 30 MA.

RESULTS AND DISCUSSION

Polyurethanes with polyether soft segment

PTMO-1000 Polymers. The two series of polymers in this category are designated as 2,4 TDI-PTMO 1000 and 2,6 TDI-PTMO 1000. Major differences in the properties of the 2,4 TDI and 2,6 TDI polyurethanes were evident as indicated in the top portion of Table I. All of the 2,4 TDI samples were transparent and exhibited a progressive change in properties with composition from soft and tacky to rubbery and finally plastic with increasing urethane concentration. All of the 2,6 TDI samples except the first were opaque, white materials which were hard but tough and proved to be elastic in thin sections. The sample of lowest urethane composition, 2,6 TDI-2, was not used in further studies since it could not be molded into a coherent film due to the low molecular weight of the sample.

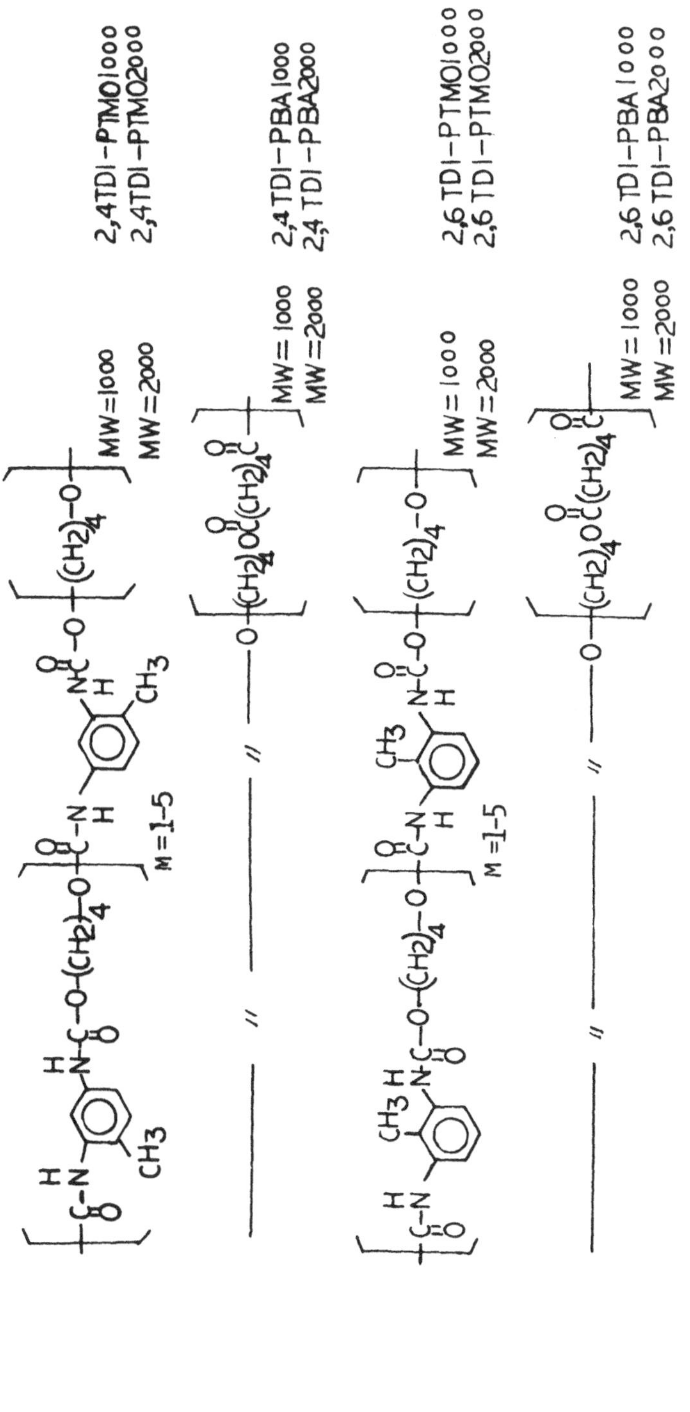

2,4 TDI–PTMO 1000
2,4 TDI–PTMO 2000

2,4 TDI–PBA 1000
2,4 TDI–PBA 2000

2,6 TDI–PTMO 1000
2,6 TDI–PTMO 2000

2,6 TDI–PBA 1000
2,6 TDI–PBA 2000

MW = 1000
MW = 2000

MW = 1000
MW = 2000

MW = 1000
MW = 2000

MW = 1000
MW = 2000

M = 1–5

M = 1–5

Chart I

TABLE 1. Properties and Thermal Transitions of TDI Based Polyurethanes

SAMPLE	URETHANE WR%	PROPERTIES (2,4 TDI Polymers)	T_g (°C)	T_2 (°C)	T_M (°C)	PROPERTIES (2,6 TDI Polymers)	T_g (°C)	T_2 (°C)	T_M (°C)
PTMO 1200									
2	31	CLEAR, SOFT STICKY RUBBER	-36	18	-	CLEAR, WAXY	-	-	-
3	42	TOUGH, LIVE RUBBER	-14	33	-		-60	-	135
4	50	ELASTIC, NOT SNAPPY	1	60	-	OPAQUE, HARD	-60		155
5	56	FLEXIBLE, BUT BOARDY	14	70	-	AND TOUGH	-63		160
6	61	ALMOST PLASTIC	23	80	(160)[1]		-61		165
PTMO 2000									
2	19	CLEAR SNAPPY RUBBER	-67	-	-	CLEAR, WEAK RUBBER	-73	-	-
3	27	CLEAR SOFT RUBBER	-65	50	-	TRANSLUCENT RUBBER	-74	61	-
4	33	TRANSLUCENT	-69	45	-	OPAQUE, SLIGHTLY	-74	71	171
5	39	RUBBERY	-72	40	-	RUBBERY, MODERATELY	-77	72	210
6	44	MODERATELY TOUGH	-70	60	-	TOUGH		-	166-210
PBA 1030									
2	31	TRANSLUCENT, RUBBERY	-13	39	-	TRANSLUCENT	-20	70	103
3	42	CLEAR, WEAK RUBBER	1	50	-	HIGH MODULUS RUBBER	-23	70	184
4	50	CLEAR, SLIGHTLY ELASTIC	14	63	-	OPAQUE,	-22	60	195
5	56	CLEAR, SOMEWHAT STIFF	25	75	-	SLIGHTLY ELASTIC	-14	141	181
6	61	CLEAR, RIGID	36	78	-	TOUGH	-14	138	195
PBA 2000									
2	19	OPAQUE, MODERATELY STIFF	-40	-	-	TRANSLUCENT RUBBER	-	-	-
3	27	STRETCHES, DRAWS	-31	12	-	TRANSLUCENT, SLIGHTLY	-51	85	167
4	33	OPAQUE, SOMEWHAT FLEXIBLE	-23	20	-	ELASTIC, MODERATELY TOUGH	-51	67	177
5	39		-15	45	-	OPAQUE, STIFF, TOUGH	-49	67	194
6	44		-6	64	-		-50	67	195

(1) TRANSITION APPEAR IN THE INITIAL HEATING BUT NOT IN THE SECOND RUN

Wide angle x-ray patterns of the 2,4 TDI samples displayed only a broad amorphous ring and gave no indication of crystallinity. The three 2,6 TDI samples of highest urethane content exhibited five crystalline reflections. None of the spacings corresponded to the lines expected for poly (tetramethylene oxide) indicating that the diffraction was due to hard segment crystallinity. In view of the amorphous nature of 2,4 TDI samples, the small angle x-ray scattering was carried out to characterize the structure of amorphous hard segment domains. The angular dependence of intensity obtained at room temperature is shown in Fig. 1. It is clear that the three polymers with longer hard segment exhibit the type of angular dependence which implies the presence of domain structure. It is also to be noted that total area under the curve is strongly depend-ent on the urethane content. Since the integrated scattering inten-sity is proportional to the product of three terms representing the electron density difference and the weight fractions of the dispersed and the continuous phase, the results imply that the extent of phase segregation increases with increasing urethane content perhaps accompanied by some improving domain organization. No attempt was made to carry out a quantitative analysis of the data along the lines suggested by Bonart (2.3)

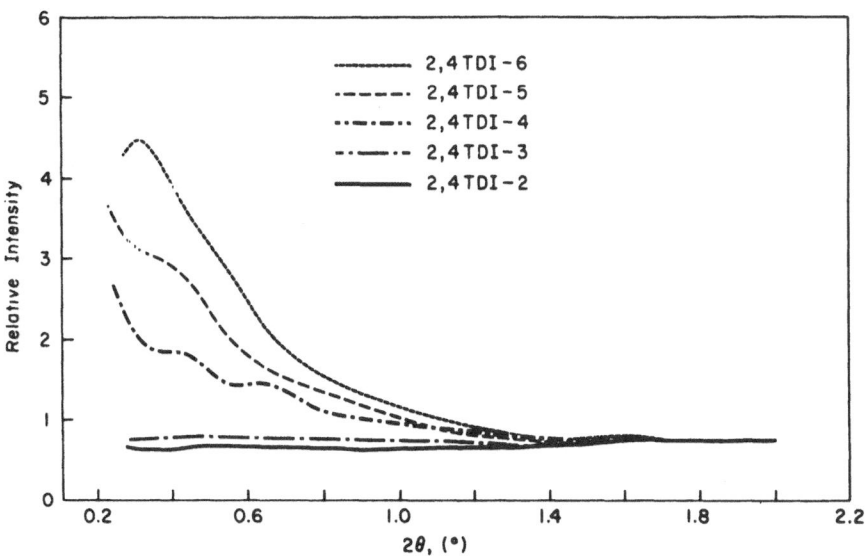

Fig. 1. Angular Dependence of SAXS Intensity in 2.4-TDI
         Polyurethanes

Thermal analyses were carried out using both DSC and TMA to probe the details of the polyurethanae transition of their relation to the structure. In the 2,4 TDI samples, the glass transition temperature of the soft segment phase, Tg, was found to be a strong function of urethane content as shown in Fig 2 and the data of Table 1.

Tg varies from -36° to 23°C with increasing urethane concentration. This corresponds to an elevation from Tg_1 value of the free soft segment (-85°C), in the range of 49° to 108°. The increase in Tg explains the progressive change observed in properties of 2,4 TDI polymers. Similar behavior was shown for an intermediate temperature transition, T_2, observed by TMA analysis. The T_2 transition in 2,4 TDI polymers ranges from 18°C to 80°C, as shown in Fig. 2. This transition is interpreted as the glass transition temperature of the amorphous hard segment domains whose structure improves with increasing hard segment length, which is proportional to the urethane content. Small angle x-ray scattering studies provide independent evidence of domain structure in the three samples of higher urethane content. However, the absence of small angle scattering in 2,4 TDI-2 and 2,4 TDI-3 does not necessarily imply that domain structure is absent since the T_2 transitions for these two polymers occur close to room temperature. A higher temperature transition T_3 was detected only in the samples of highest urethane content and then only on the initial heating. This T_3 transition is believed to result from some allophonate or biuret bonding which arises from the small excess of diisocyanate in the polymerization recipe.

In semicrystalline 2,6 TDI polymers, as listed in Table 1, the glass transition temperature of the soft segment phase was generally independent of urethane concentration and the elevation of T_g above that of the free soft segment is 25°C, which is much smaller than the value for 2,4 TDI polymers. No T_2 transition was observed with 2,6 TDI polymers. Instead, the TMA scans revealed repeatable T_3 transitions in the range of 130°C to 170°C as shown in Fig. 3. The DSC scans showed an endothermal peak at T_3 indicative of crystalline melting. Both the increase in T_3 and change in melting peak appearance indicated improvement in crystalline order with increasing urethane content. These samples readily supercooled and then crystallized during the heating cycle. On further heating, melting occurred followed by recrystallization at a higher temperature and then a final melting process.

Based on the x-ray results and thermal analysis, it can be concluded that the progressive increase in T_g in the 2,4 TDI samples is a consequence of the amorphous domain structure which leads to extensive hard segment-soft segment mixing. In 2,6 TDI polymers, the driving force of hard segment crystallinity leads to improved phase separation with little hard segment-soft segment mixing.

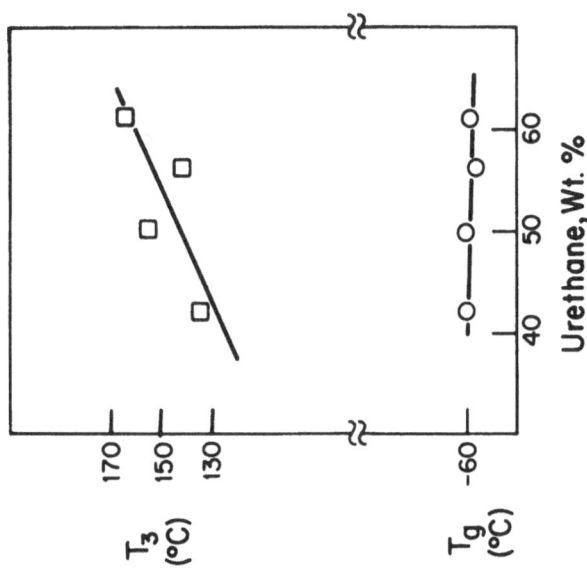

Fig. 3. Glass Transition Temperature (Tg) and Melting Temperature (T_3) as a Function of Urethane Contents in 2,6 TDI-PTMO 1000.

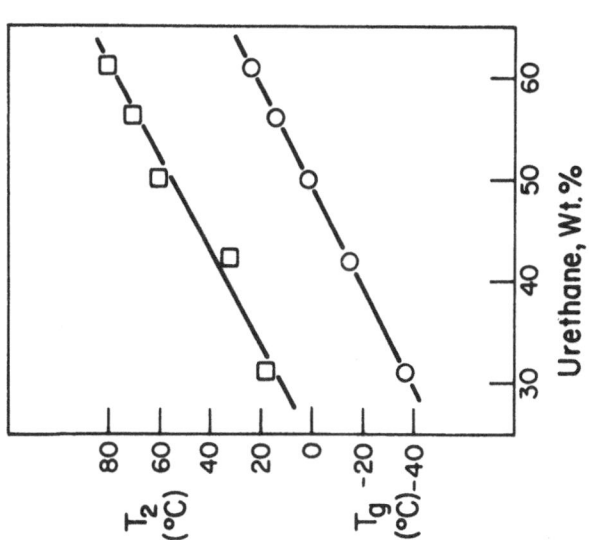

Fig. 2. Glass Transition Temperature (Tg) and Intermediate Transition Temperature (T_2) as a Function of Urethane Content in 2,4 TDI-PTMO 1000.

A more detailed understanding of the relation between composition, transition behavior and properties requires a quantitative
measure of the extent of phase mixing. The analysis of absolute low
angle x-ray scattering provides one approach for estimating the
extent of phase segregation. However, this method depends on a
knowledge of the electron density of the dispersed and matrix phases,
which in turn depends upon the degree of hard segment-soft segment
mixing in these phases. In view of this difficulty, preference was
given to the approach which relies on a quantitative analysis of
the infrared spectra.

The infrared analysis depends on the resolution of the
urethane NH and the carbonyl band into the bonded and non bonded
components. If the only proton acceptors are the carbonyl group
of the urethane hard segment, and the oxygen of the polyether
soft segment, then the fraction of bonded carbonyl can be taken as
a measure of the extent of phase segregation. In turn, the fraction
of NH groups bonded to ether, determined by difference, indicates
the degree of hard segment-soft segment mixing. Fig. 4 illustrates
the IR spectra in the regions of NH stretching and carbonyl stretching. It is well known that a band near 3300 cm^{-1} is caused by the
H-bonded NH groups while the free NH group appears near 3460 cm^{-1}.
MacKnight and Yang have reported an integrated extinction coefficient
Ef=3.44 x 10^3 1/mole-cm for the free NH group and Eb = 1.19 x 10^4
1/mole-cm for the H-bonded NH groups from measurements of a model
compound (3). Based on these extinction coefficients, the fraction
of H-bonded NH groups was calculated by the curve resolving technique (4). The results indicate that for both 2,4 TDI and 2,6 TDI
polymers, 95% of all NH groups are hydrogens bonded in the solic state
at room temperature. The carbonyl region shows spoltting of the absorption band into two peaks at 1740 cm^{-1} and 1720 cm^{-1} for 2,4 TDI
polymers and at 1740 cm^{-1} and 1700 cm^{-1} for 2,6 TDI polymers. Studies
on model compounds of n-butyldiurethane of 2,4 TDI and 2,6 TDI in
solution indicated that a band at 1740 cm^{-1} is due to free carbonyl
and that at 1720 cm^{-1} or 1700 cm^{-1} is a result of hydrogen bonded
carbonyl for 2,4 TDI and 2,6 TDI, respectively.

It can be assumed from other studies that the extinction coefficients for both bonded and free carbonyl bands are approximately
the same. Since the bonded and free carbonyl peaks in 2,4 TDI polymers showed wevere overlapping, the spectra were recorded with four-
fold scale expansion and fitted with multiple Lorentzian peaks using a
du Pont curve resolver. The fraction of bonded carbonyl thus obtained
is listed in the third column of Table 2. These results indicate that
approximately 46-60% of the urethane carbonyl groups of 2,4 TDI polyurethane samples and 80% for 2,6 TDI samples are bonded to the urethane NH group. The remainder of NH groups are necessarily bonded witl
other proton acceptors.

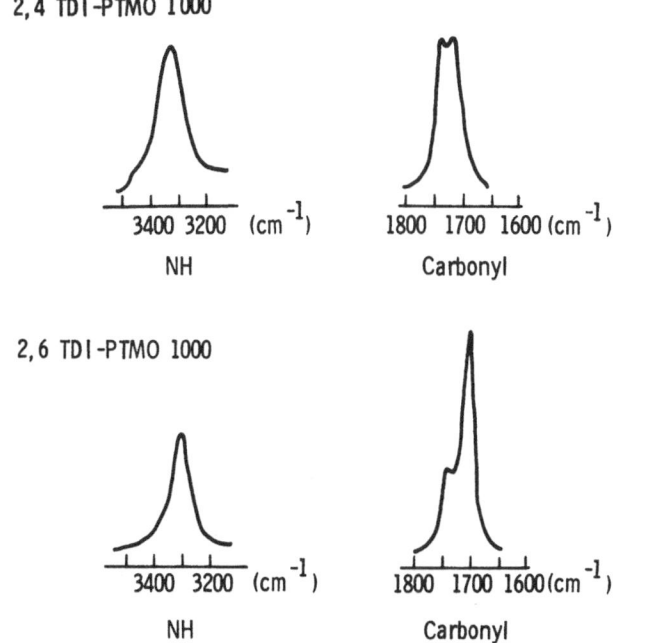

Fig. 4. IR spectra in the NH and carbonyl regions.

TABLE 2

ELEVATION OF T_g (DEG KELVIN) BASED ON

COPOLYMER EQUATION 2,4 TDI-

AND 2,6 TDI - PTMO 1000 SAMPLES

SAMPLES	URETHANE* WT %	T_g, °K (OBS.)	% H-BONDED CARBONYL	W_2+	T_g, °K (CALTD)
2,4 TDI-2	31	237	46	0.20	209
3	42	259	50	0.25	215
4	50	274	60	0.29	220
5	56	287	60	0.34	227
6	61	296	66	0.36	230
2,6 TDI-3	42	213	78	0.13	202
4	50	213	78	0.16	207
5	56	210	80	0.20	209
6	61	212	80	0.20	214

* CALCULATED AS TDI PLUS BD

+ WEIGHT FRACTION OF HARD SEGMENT UNITS (TDI-BD) IN MIXED SOFT SEGMENT PHASE, FROM INFRARED RESULTS.

It has been commonly assumed that the only other proton
acceptor is the ether oxygen of the soft segment. As indicated
earlier, such data then provide a simple indication of hard segment-
soft segment interactions and, by implication, the degree of hard
segment-soft segment mixing. However, in examining several model
systems, it has been found that the situation can be more complicated.
The IR spectra on these copolymers which contained only the appropriate diisocyanate (2,4 TDI, 2,6 TDI and MDI) and butanediol indicated that while NH is almost completely hydrogen bonded, a significant fraction (10-30%) of the carbonyl groups remain in the
free state. The only additional proton acceptor in these copolymers
is the urethane alkoxyl oxygen. It is not known to what degree this
evidence of NH bonding to urethane alkoxyl oxygen in the copolymer
applies to the corresponding hard segments in segmented polyurethanes.
If the alkoxy oxygen is also involved in hydrogen bonding in the segmented polyurethanes, then the fraction of hydrogen bonded carbonyl
underestimates the degree of hard-hard segment bonding and the values
calculated from the difference provide, at best, an estimate of the
maximum amount of intermixing of hard with soft segment. Another
uncertainty in interpreting the IR data as a measure of phase segregation comes from the assumption that the extent of phase mixing is
given solely by the fraction of free carbonyl in excess of the free
NH, which implies that all interurethane bonded hard segment species
are excluded from the soft segment phase. This assumption is probably
valid when there is a strong basis for selectivity, such as the
crystallization driving force in 2,6 TDI polyurethanes. However, it
appears to be an artificial restriction with the 2,4 TDI samples
since the infrared results in Table II indicate that as much as 50%
of the urethane is mixed with the soft segment phase. In these cases,
the amount of interurethane bonding would overestimate the degree of
phase segregation. As we shall see, it will be necessary to take
account of the implication of these two sources of uncertainty in
arriving at an interpretation of the infrared data.

Since the glass transition behavior in 2,4 TDI polymers is believed to be the result of extensive mixing of hard and soft segment
regions, it appeared useful to inquire to what degree the IR evidence
on hard segment-soft segment interaction could be used to explain the
glass transition behavior. One approach is to treat the infrared
data in a simple form of the copolymer equation and then account
for any remaining differences between the predicted and observed T_g
values.

T_g values were calculated using the following relation:

$$\frac{1}{T_g} = \frac{W_1}{T_{g_1}} + \frac{W_2}{T_{g_2}}$$

where W_1 and W_2 are the weight fractions of the soft segment and hard segment in the mixed soft segment phase estimated from IR data, $Tg_1=188°K$ for PTMO, $Tg_2=380°K$ for the 2,4 TDI-butanediol hard segment and $Tg_2=374°K$ for the 2,6 TDI-butanediol analog. The calculated values are listed in the last column of Table 2. It is evident that the observed values of Tg for the 2,4 TDI samples are substantially higher than the calculated values and the difference, ΔTg, increases progressively with the increasing urethane content. Even if the Tg values are calculated assuming complete mixing, using the hard segment concentration which corresponds to the sample composition, the results are 15 to 25% lower than the observed Tg values. On the other hand, for the 2,6 TDI samples there is rather good accord between the observed and calculated results.

For 2,4 TDI polymers, the additional contribution to the increase in Tg, above that predicted by the copolymer equation, can be attributed to hydrogen bonding between hard and soft segments which acts as an effective crosslink. The following empirical equation was employed to fit the experimental results,

$$\Delta Tg = 820 \ Xc/(4.6 - Xc)$$

where Xc, the fraction of ether groups which are hydrogen bonded, is equal to the effective degree of crosslinking. The equation has the same form as the relation between Tg and the degree of crosslinking proposed by DiBenedetto and DiMarzio (5.6).

TABLE 3. Additional Contribution to the Increase
in T_g, 2,4 TDI-PTMO 1000 Samples

SAMPLES	ΔT_g (obs)	% FREE CARBONYL	X_c^*	ΔT_g^+ (calc)
2,4 TDI-2	28	54	0.17	32
3	44	46	0.22	44
4	54	40	0.26	49
5	60	40	0.32	61
6	66	34	0.33	63

* FRACTION OF HYDROGEN BONDED ETHER GROUPS, EQUAL TO THE EFFECTIVE DEGREE OF CROSSLINKING

+ $\Delta T_g = 820 \ X_c/(4.6-X_c)$

The calculated Tg values based on the modified equation are listed in Table 3. A comparison between ΔTg calculated and ΔTg observed shows that, in combination with the copolymer equation, the observed values of Tg can now be predicted within 5°C from the infrared data.

It should be recognized that this suggested relation between ΔTg and the effective degree of crosslinking, Xc, must incorporate several effects discussed earlier, including the contribution of hydrogen bonded interurethane segments which are mixed with the soft segment phase, any limitation in the copolymer equation, as well as the specific contribution of the urethane to ether hydrogen bonding treated as an effective crosslink. In fact, the difference in agreement between the observed and calculated values of Tg based on the infrared data and copolymer equation for 2,4 TDI and 2,6 TDI samples in some degree, must reflect the fundamental differences in the extent of phase segregation. In the 2,4 TDI samples extensive phase mixing occurs, indicated directly by the infrared results, whereas in 2,6 TDI samples hard segment crystallization provides a strong driving force for phase segregation. Thus it is possible that the discrepancy between the calculated and the observed Tg values in the 2,4 TDI polyurethanes is due, at least in part to the phase mixing of interurethane bonded segments. As a result of expected differences in this type of behavior, no songle relation can be expected which is equally applicable to both weak and strongly phase segregated polyurethanes.

Some insight concerning the effect of hydrogen bonding on the higher thermal transitions has been obtained by analyzing the temperature dependence of the infrared spectra in the NH and carbonyl regions using simple procedures for resolving hydrogen bonded and non bonded components of the two absorption bands (7). The change in the NH band as a function of temperature is illustrated in Figure 5(a) for a representative 2,6 TDI polymer. Analogous spectra for the carbonyl band in Fig. 6(a) indicates that extensive disruption of hydrogen bonding occurs within the hard segment domains. This behavior is clarified when the fraction of bonded NH and carbonyl is plotted as a function of temperature as shown in Figure 7. A common onset temperature for the dissociation of hydrogen bonding appears in both the NH and carbonyl results. The onset temperature occurs at about 65°C for all 2,6 TDI samples and the dissociation energy ΔH is about 4 kcal/mole independent of urethane content. These results indicate that the dissociation of hydrogen bonding can occur well below the melting point (130 to 170°C). Perhaps the onset temperature represents the glass transition temperature of amorphous hard segment regions not otherwise detectable by thermal analytical methods (see also results on 2,6 TDI-PTMO-2000 samples), but the fact that dissociation of interurethane hydrogen bonding occurs to levels of 50% or more

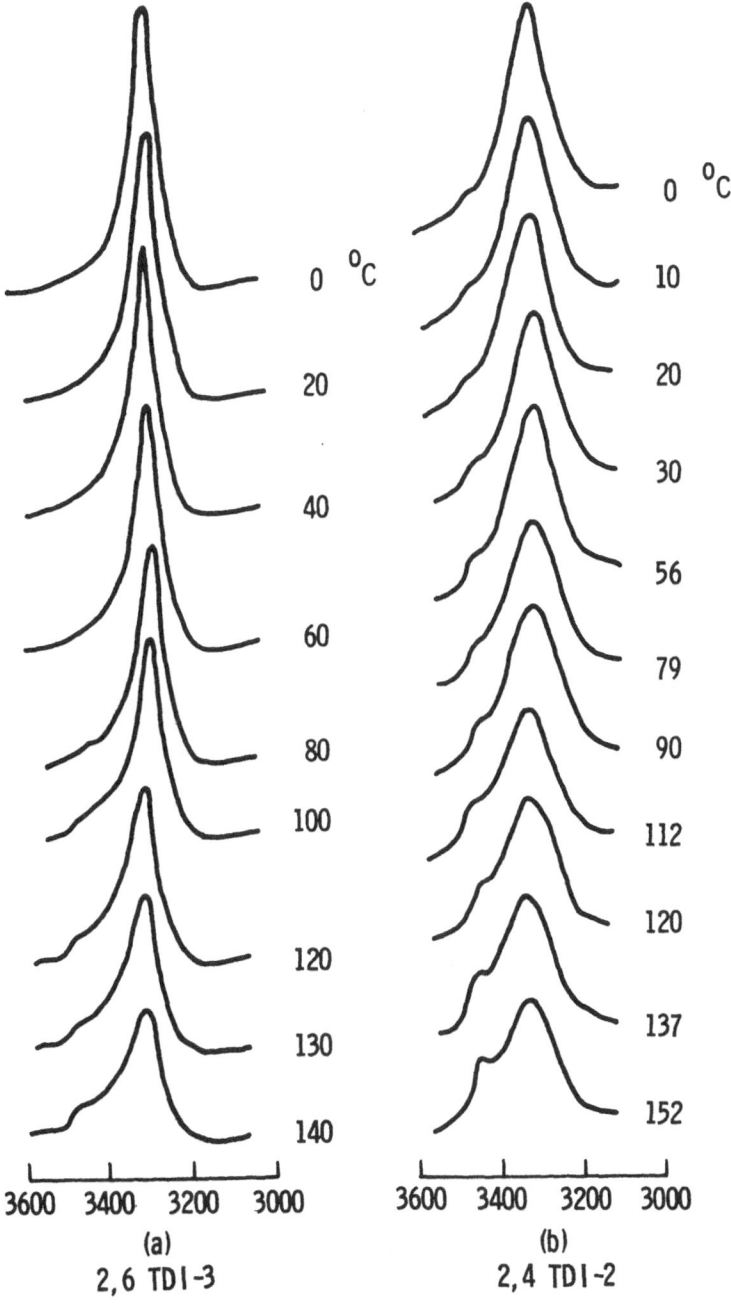

Fig. 5. Changes in NH Region as a Function of Temperature in
2,6 TDI-PTMO 1000 and 2,4 TDI-PTMO 1000 Polyurethanes

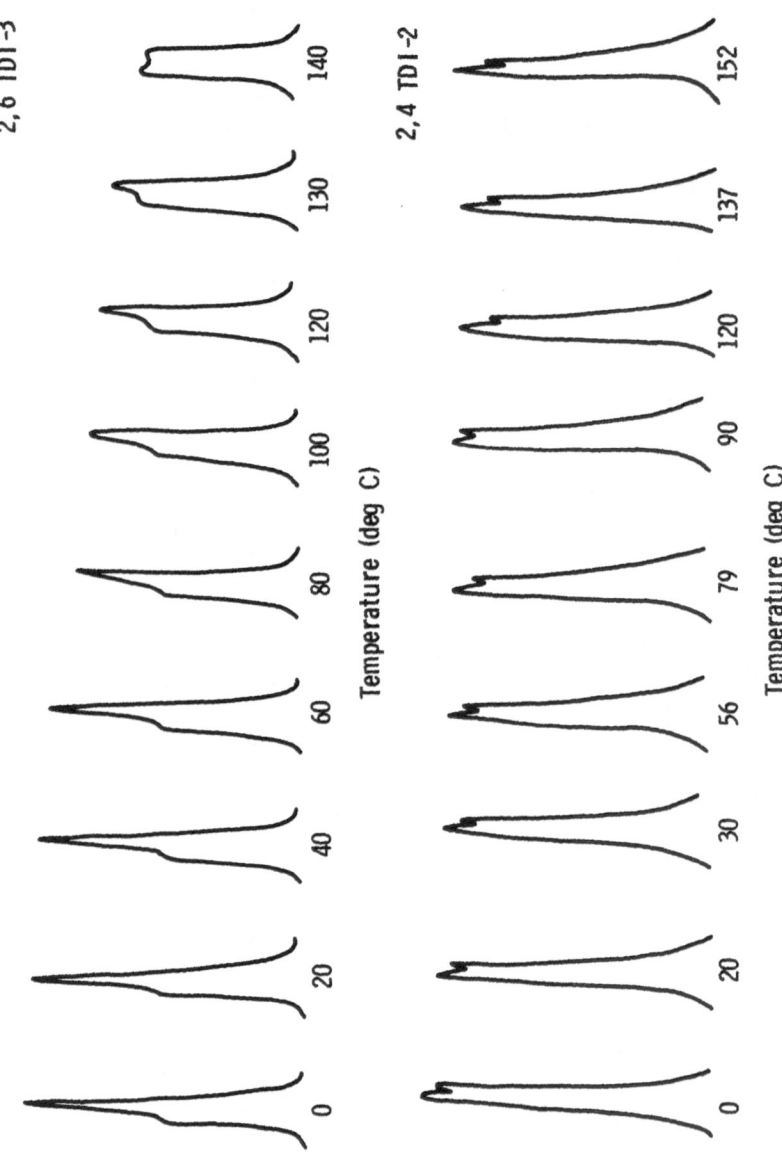

Fig. 6.   Changes in the Carbonyl Region as a Function of Temperature
           in 2,6 TDI-PTMO 1000 and 2,4 TDI-PTMO 1000.

at temperatures which are still well below the melting point
indicates that extensive disruption of hydrogen bonding can occur
within the crystalline structure. The small and nearly constant
difference in the degree of hydrogen bonded NH as compared to
carbonyl in the curves of Fig. 7 could be a reflection of NY bonding
to alkoxy oxygen within the hard segment domains. This would imply
of Table 2 for 2,6 TDI polyurethanes.
mixing which, it should be noted, is also suggested by the results
of Table 2 for 2,6 TDI polyurethanes.

    For 2,4 TDI polymers, the behavior of NH and carbonyl bands
are shown in Fig. 5(b) and Fig. 6 (b), respectively. The onset
temperature for dissociation of hydrogen bonded NH, as shown in
Fig. 8 is variable and occurs at 40° to 60°C, which is close to
the glass transition temperature observed by thermo mechanical
analysis for amorphous hard segment domain structure. Surprisingly,
there is little change in the carbonyl region up to the highest
temperature studied (150°C) as in Fig. 6 (b), implying that the
dissociation of NH is due almost entirely to the disruption of
urethane to ether bonding. The observation that the onset of
dissociation occurs well above the soft segment glass transition
temperature and is possibly related to the domain transition is
difficult to explain at this time.

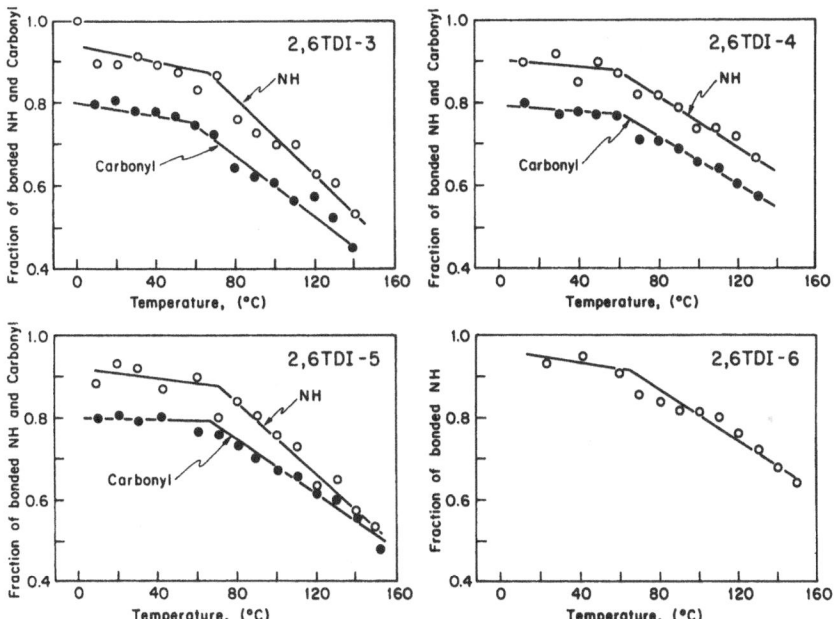

Fig. 7.   FRACTION OF BONDED NH GROUPS AND CARBONYL IN 2,6 TDI
          POLYMERS AS A FUNCTION OF TEMPERATURE.

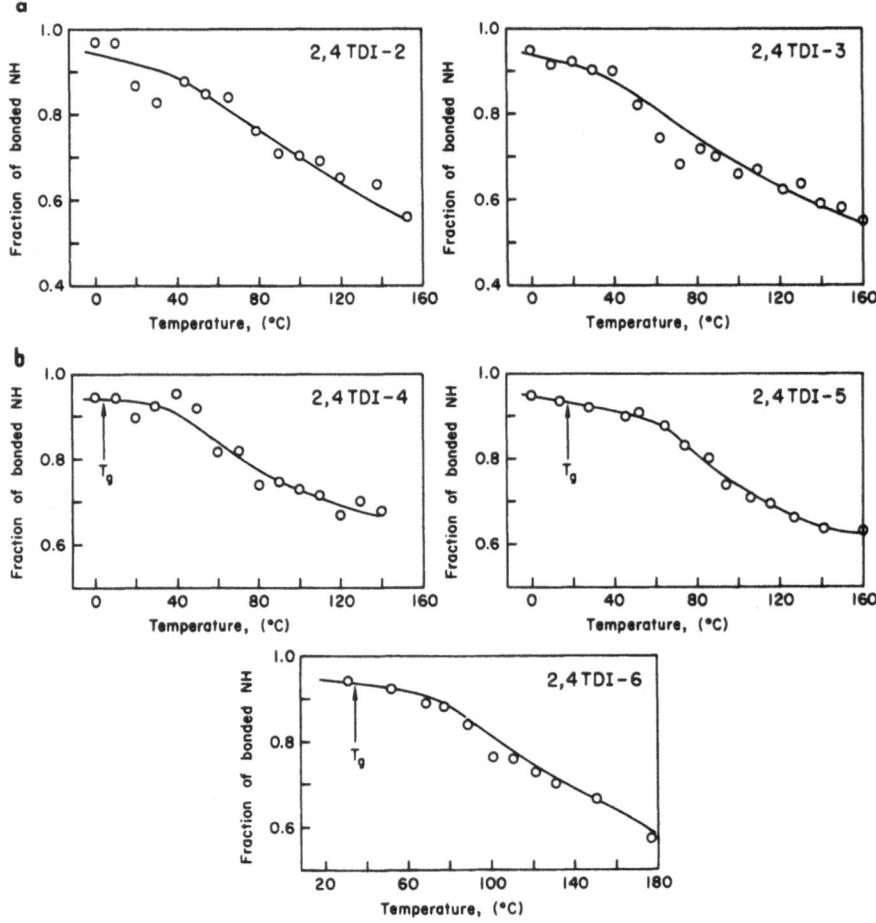

Fig. 8.   Fraction of Bonded NH Groups in 2,4 TDI Polymers as a
          Function of Temperature.

A surprising observation is the relative stability of the interurethane bonding in 2,4 TDI samples by comparison with the behavior in the 2,6 TDI samples. This could reflect the greater stability of hydrogen bonding at the 4 position in the 2,4 TDI ring due to smaller steric hindrance than the 2 or 6 position in the 2,6 TDI ring.

It should be noted here that several of the observations presented here are similar to results reported by Seymour and Cooper for MDI segmented polyurethanes (8). They demonstrated that the onset temperature for hydrogen bond dissociation was insensitive to annealing procedures which resulted in improvement in the hard segment organization as evidenced by the higher domain transition temperatures. Overall, the results obtained here and the important work of Seymour and Cooper indicate that onset of hydrogen bond dissociation bears little relation to the state of structural organization. Seymour and Cooper also reported the persistence of a substantial degree of interurethane hydrogen bonding at relatively high temperatures but concluded that such hydrogen bonding would have little effect on properties due to the rapid interchange of hydrogen bond.

PTMO-2000 Polymers. Increasing the molecular weight of the polyether soft segment from 1000 to 2000 produces a drastic improvement in phase segregation in 2,4 TDI polymers (9). As shown in Fig. 9 there is a clearly defined low temperature glass transition for the soft segment but no melting endotherm, indicating the absence of any soft segment crystallinity in these samples. A small endotherm at 61°C in the first run and at 50°C in the second, appears as the glass transition of the amorphous hard segment domain. Table 1 lists the properties and the thermal transitions of 2,4 TDI samples containing PTMO 2000. The soft segment Tg and the glass transition of amorphous domain for these samples are generally independent of urethane content. This contrasts with the behavior of the 2,4 TDI-PTMO 1000 samples where the soft segment Tg and glass transition of domains exhibited a steep dependence on urethane content. As a result, 2,4 TDI-PTMO 2000 polymers provide good low temperature flexibility in addition to moderate toughness as rubbers. Results from the analysis of the infrared spectra in several of the 2,4 TDI-PTMO 2000 samples are summarized in Table 4. These samples fall intermediate between the 2,4 TDI-PTMO 1000 and 2,6 TDI-PTMO 1000 samples in terms of the extent of phase segregation as measured by W_2, the weight fraction of hard segment in the mixed soft segment phase. The observed value of T_g can be approximated from the data by use of the copolymer equation alone as illustrated in Table 4. These results, like the nearly constant T_g values with increasing urethane content, support the conclusion that the degree of phase selectivity in these samples is close to that of the 2,6 TDI-PTMO 1000 samples.

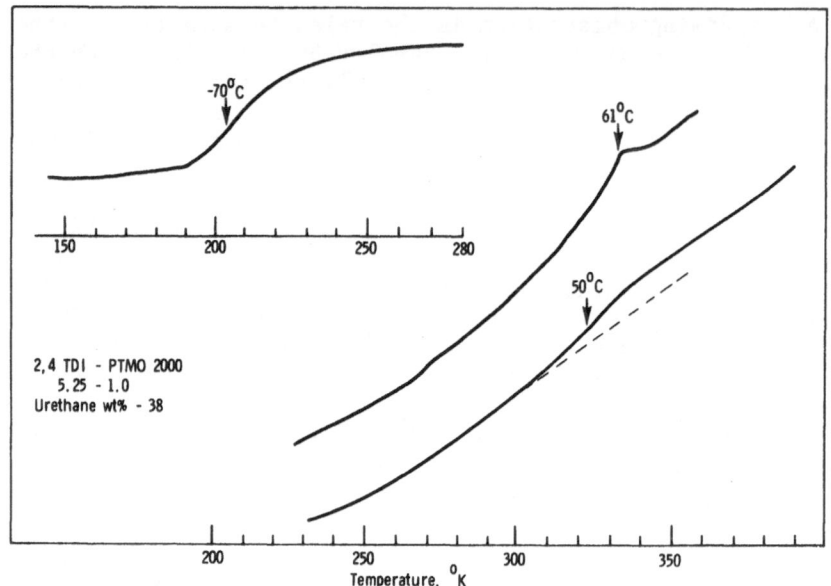

Fig. 9.   DSC Scans, 2,4 TDI-PTMO 2000.

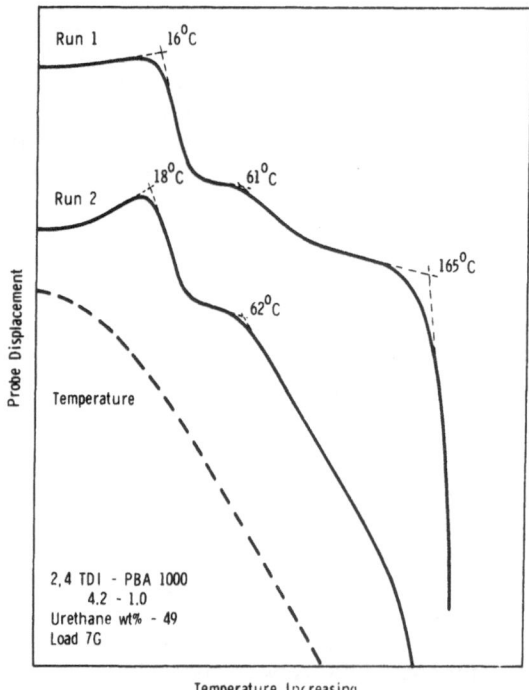

Fig. 10.   Thermomechanical Analysis,   2,4 TDI-PBA 1000 Polymer.

TABLE 4

ELEVATION OF T_g BASED ON THE COPOLYMER EQUATION FOR
2,4 TDI-PTMO 2000 POLYURETHANES

SAMPLES	URETHANE WT %	T_g, K (OBS.)	FREE CARBONYL	W_2*	T_g, °K[†]
2,4 TDI-2	19	206	54	0.11	199
3	27	207			
4	33	204	40	0.16	204
5	39	202			
6	44	202	32	0.19	208

* WEIGHT FRACTION OF HARD SEGMENT UNITS (TDI-BD) IN MIXED
SOFT SEGMENT PHASE FROM INFRARED RESULTS.

† CALCULATED FROM THE COPOLYMER EQUATION.

   In 2,6 TDI-PTMO 2000, soft segment Tg is even lower than in
2,4 TDI-PTMO 2000 samples and appears to decrease somewhat with
increasing urethane content as listed in Table 1. A hard segment
Tg is observed both in DSC and TMA scans on these samples but was
absent from the 2,6 TDI-PTMO 1000 samples. This indicates some improvement in phase segregation due to the increased soft segment
molecular weight with the result that hard segment species which
might otherwise be mixed with the soft segment phase now contribute
to glassy hard segment regions.

Polyurethanes with Polyester Soft Segment

   PBA-1000 Polymers. The 2,4 TDI-PBA 1000 samples like the
2,4 TDI-PTMO samples show a progresseve change in properties from
rubbery to plastic with increasing urethane content (Table 1).
Therefore it is expected that the soft segment Tg would show an
equally marked dependence on the urethane content. This, in fact,
is observed in the results of Table 1. A typical TMA scan for one
of these samples is shown in Fig. 10. Three well marked transitions
are observed in the initial run. The two lower transitions representing the soft segment Tg and hard segment T_2, repeat in the second
run but the upper transition is replaced by a diffuse change in
slope which occurs at a much lower temperature. This upper temperature transition is probably related to allophonate and biuret
bonding due to the 5% excess of diisocyanate. The DSC scan shows
only the soft segment glass transition temperature.

As indicated in Table 1, 2,6 TDI-PBA 100 samples show increased toughness compared to 2,4 TDI-PBA 1000 samples. The several
DSC scans in Fig. 11, show a clearly defined soft segment glass
transition which shifts by about 30°C between the first run and
the third run, after heating above the melting point of hard segment crystallites. This suggests that the original conditions of
compression molding have considerably enhanced phase segregation,
probably as a result of increasing the degree of hard segment crystallization by the application of pressure during the molding and
cooling cycles. The results summarized in Table 1 show that the

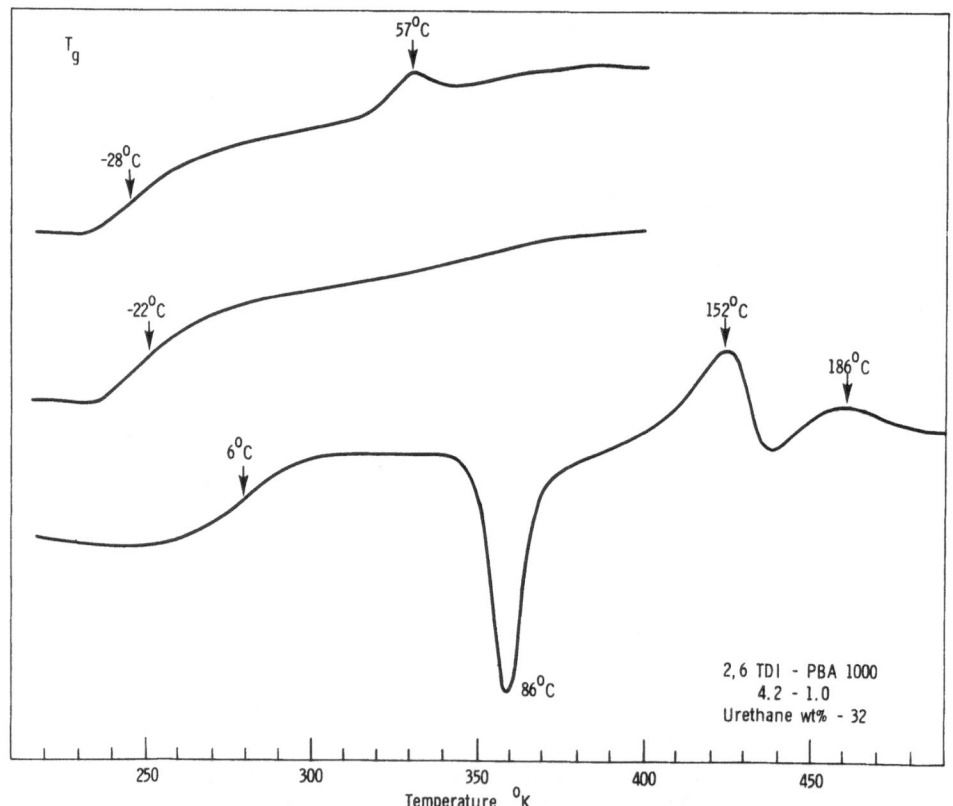

Fig. 11.   DSC SCAN, 2,6 TDI-PBA 1000 POLYMER

initial values of Tg are only slightly dependent on composition but those determined after melting exhibit a marked dependence on urethane content. There is some evidence of a hard segment glass transition in the first but not in the second melting run which again is consistent with greater phase segregation in the original compression molded sample. There are also various transitions associated with the crystalline hard segment structure.

PBA-2000 Samples. In Fig. 12, the DSC scan of 2,4 TDI-PBA 2000 samples show a clearly defined soft segment transition as well as the melting of soft segment polyester crystallinity. Results listed in Table 1 indicate that the soft segment and hard segment Tg are again dependent on urethane content. Both types of behavior

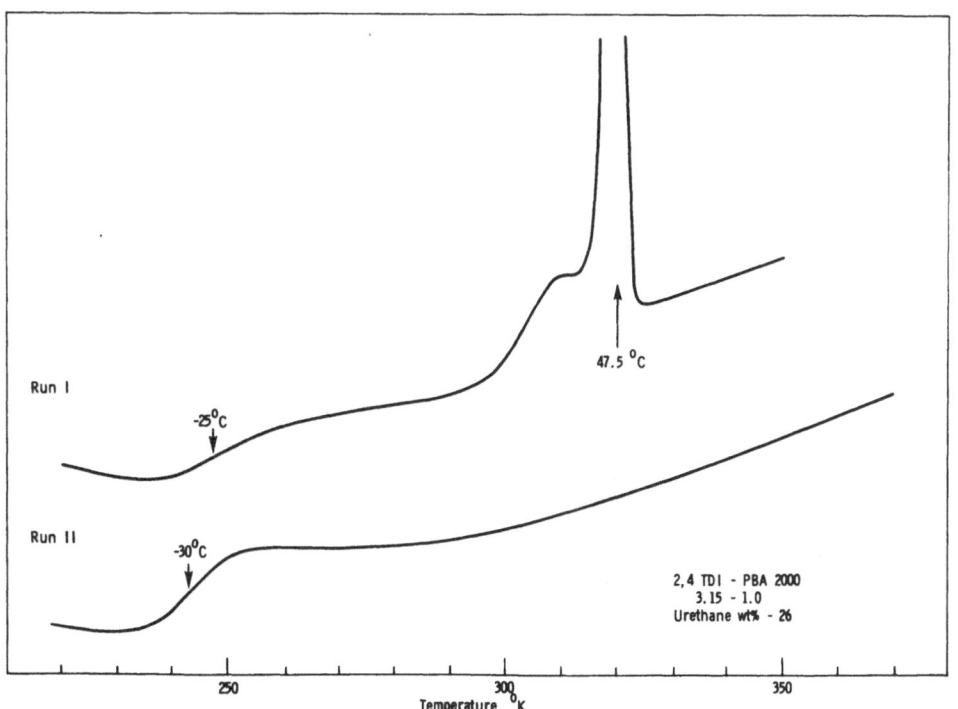

Fig. 12. DSC Scan, 2,4 TDI-PBA 2000 Polymer

suggest that the two-fold increase in polyester molecular weight has not produced a significant improvement in the degree of phase segregation. It is worth noting that the analogous increase in PTMO molecular weight produced a marked improvement in phase segregation in 2,4 TDI–PTMO 2000 samples.

The behavior of the 2,6 TDI–PBA 2000 samples are completely analogous to the 2,6 TDI–PTMO 2000 samples with the appearance of the soft segment Tg, crystallization and melting regions as well as hard segment glass transition and melting temperatures. The data are listed in Table 1.

Comparison of Transition Behavior

The glass transition temperatures plotted as a function of urethane content are shown for the various 2,4 TDI samples in Fig. 13 and for the various 2,6 TDI samples in Fig. 14. The ordinate represents the increase in the soft segment Tg above the value for the free soft segment and is plotted in this fashion to compensate for differences in Tg for PBA and PTMO.

Fig. 13 for the 2,4 TDI samples shows that the elevation of Tg for PBA 2000 as well as PBA 1000 falls close to the curve for PTMO 1000 samples indicating that the degree of phase mixing in both series of polyester samples is comparable to that in the PTMO 1000 samples. The dramatic change in the Tg behavior of the PTMO 2000 samples brought about by the two-fold increase in soft segment molecular weight is emphasized by the contrast with the other samples.

Fig. 14 for the 2,6 TDI samples reveals that extensive phase mixing occurs in the PBA 1000 samples despite the driving force for phase segregation provided by hard segment crystallization. Even the 2,6 TDI–PBA 2000 samples show an elevation, $\Delta Tg=20°$, suggesting that some phase mixing occurs despite the apparent insensitivity of Tg to urethane content. With the 2,6 TDI–PTMO 2000 samples the increase in Tg is limited to a value of about 10°C. This result indicates that the effect of anchoring the ends of the soft segment by attachment to the urethane hard segment is only responsible for a small increase in Tg. Nonetheless, it has sometimes been argued that this anchoring effect represents a major contribution to the increase in soft segment Tg. Seefried and coworkers have carried out a careful examination of the transition behavior in a series of polyurethanes at a fixed 2:1:1 mole ratio of MDI:butanediol:PCL as a function of increasing polycaprolactone (PCL) molecular weight (10). They concluded that the increase in Tg (–43 to 28°C) with reduction in PCL molecular weight (3130 to 340) is due to the increasing contribution of the chain anchoring effect with the lower molecular

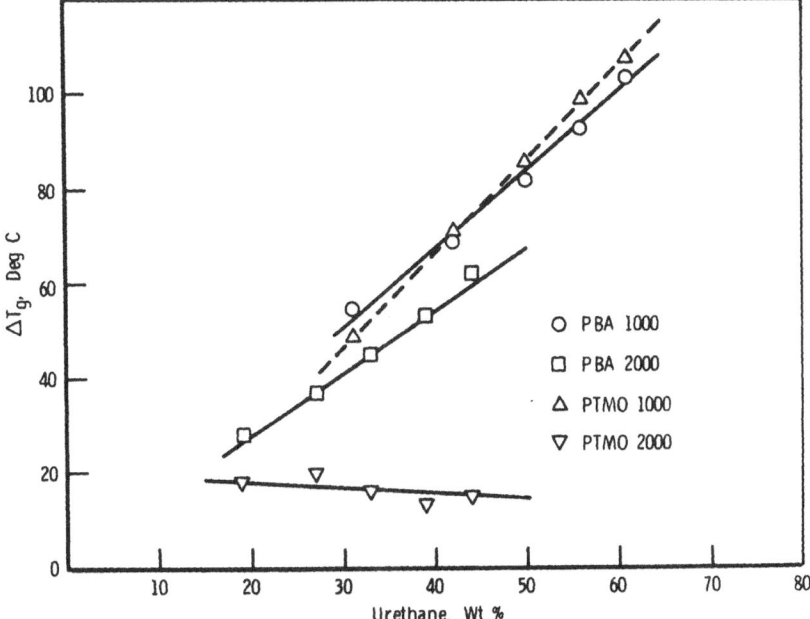

Fig. 13. ΔT_g Values for 2,4 TDI Polyurethanes.

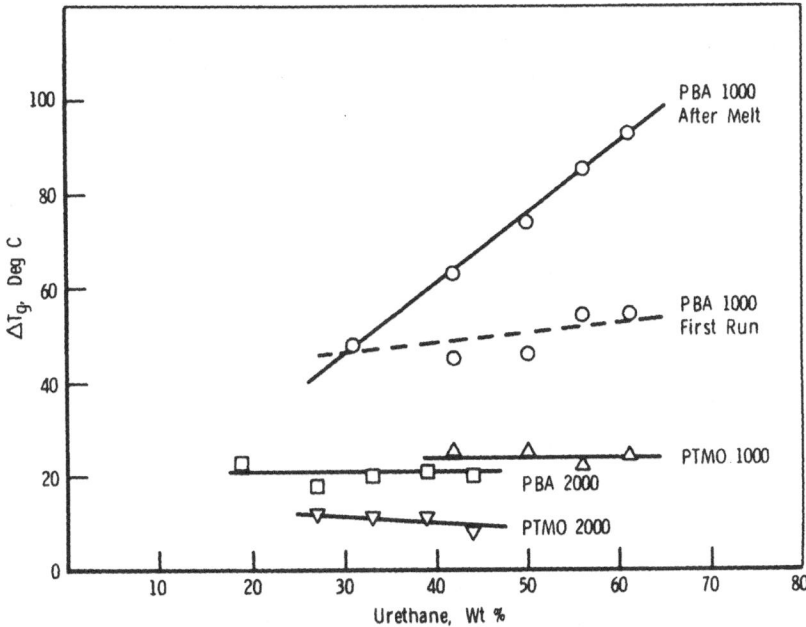

Fig. 14. ΔT_g Values for 2,6 TDI Polyurethanes.

weight soft segment. However, their results can also be interpreted
as the consequence of changes in the extent of phase mixing. In fact,
later torsion pendulum measurements as a function of temperature
show evidence of a marked rubbery plateau in the samples with
higher PCL molecular weight but not in the lower PCL molecular
weight samples. This indicates increasing phase segregation, as expected, with increasing soft segment molecular weight.

Consideration of the chain anchoring effect shows that it
should consist of two factors; (1) the loss of the free volume
associated with the loss of mobility of the chain ends; (2) the
effect of replacing the ether group with the urethane group. The
first is equivalent to increasing Tg to the asymptotic value achieved
with the high molecular weight polymer. The second results in little
restriction to rotational motion since the urethane group is expected to have a low barrier to rotation about the linkage to the alkoxy
oxygen, approximating the situation in aliphatic polyesters where
the melting point decreases with increasing concentration of the
ester group. These considerations suggest that the incorporation of
a soft segment with a molecular weight of 1000 or 2000 into the
polyurethane structure should have very little effect on its Tg, in
keeping with the observed results for the 2,6 TDI-PTMO 2000 samples.

CONCLUSIONS

The influence of an asymmetric diisocyanate structure, which
introduces the possibility of head-to-tail type isomerism and inter-
feres with structural regularity of the hard segment arrangement, has
been shown to have a dramatic effect on phase segregation, domain
organization and polyurethane properties by way of contrast to
polyurethanes based on the symmetric diisocyanate structure. All of
the 2,4 TDI polyurethanes have amorphous domain structure with hard
segment glass transitions below 100°C while all of the 2,6 TDI polymers exhibit semicrystalline hard segment structure with melting
temperatures in the range of 130 to 200°C. Apparently structural
regularity is a requirement for the occurrence of short range order
in hard segment domains and the interurethane hydrogen bonding is
not sufficient in itself to stabilize the hard segment domains or
to assure that extensive phase segregation will occur. Other conditions can also contribute to phase incompatibility such as the increase in the soft segment molecular weight, as demonstrated by
the results presented here, or by the use of a less polar soft
segment, exemplified by the hydroxy terminated polybutadienes which
are now available.

When one compares the effects of polyether and polyester soft
segment, it becomes obvious that the PBA soft segment generally introduces greater phase mixing than the PTMO soft segment. PBA soft

segment even competes with segregation and crystallization of the 2,6 TDI-butanediol hard segment. This solubilization of 2,6 TDI hard segments in the soft segment phase occurs at the expense of the heat of fusion for hard segment crystallites. It is doubtful that this could occur without the specific intervention of hydrogen bonding between the urethane NH and the ester group. Therefore, the difference in extent of phase segregation indicated by the differences in the glass transition behavior in the 2,6 TDI-PTMO and 2,6 TDI-PBA polyurethanes suggests that the hydrogen bonding to ester is stronger than to ether. The net effect of the phase mixing is to raise the glass transition temperature of the soft segment. Thus, it is clear that where the low temperature properties of polyurethanes are important, the choice of polyether soft segment and the use of a 2000 rather than a 1000 molecular weight soft segment is preferred.

It seems reasonable to expect that hydrogen bonding between hard and soft segment units will make a specific contribution to the increase in soft segment Tg beyond the additive effect of mixing described by the copolymer equation. However, the contribution of such specific interactions appears to be important only when the extent of phase mixing is substantial as in 2,4 TDI-PTMO 1000. For 2,4 TDI-PTMO 2000 and 2,6 TDI-PTMO 1000 polyurethanes, where the mixing is limited, one can account for Tg quite well, although not exactly, by applying the copolymer equation to the infrared results. This difference in behavior between poorly and strongly phase segregated materials indicates that there are differences in the details of phase segregation which must be taken into account in developing a model for these systems. There is also some direct evidence that hydrogen bonding interaction can interfere with soft segment crystallization. This is indicated by the fact that soft segment crystallization does not occur in the 2,4 TDI-PTMO 2000 samples, whereas with the slight improvement in phase segregation which occurs in the 2,6 TDI-PTMO 2000 samples as indicated by the further reduction in T_g, the soft segment crystallizes freely. Somewhat the same comparison holds for the PBA 2000 samples. Extensive experimental evidence for both 2,6 TDI and MDI segmented polyurethane samples shows that hydrogen bond dissociation can occur within ordered domain structures at temperatures well below the domain transition temperature and with little regard for improvement in the degree of order. Thus, the thermal behavior of hydrogen bonding appears to be virtually independent of the structural organization in such materials. It remains to be determined to what extent the interurethane hydrogen bonding interactions contribute to the exceptional toughness, abrasion resistance and related properties which are the hallmark of the best polyurethane elastomers.

Finally, the glass transition temperature deserves recognition as a sensitive index of phase mixing, perhaps almost as useful as infrared analysis and low angle x-ray scattering which are also not free of ambiguity. The copolymer equation will continue to serve

as a useful approximation for estimating the extent of phase segregation in strongly phase separated systems. In weakly phase
separated systems, the empirical relation developed here for the
2,4 TDI-PTMO 1000 samples in combination with the copolymer equation
could serve as a useful approximation for relating Tg to the extent
of phase segregation.

REFERENCES

1. N.S. Schneider, C.S. Paik Sung, R.W. Matton and J.L. Illinger,
   Macromolecules, 8, 62 (1975)

2. R. Bonart and E.H. Muller, J. Macromol. Sci.-Phys., B10, 177
   (1974) and ibid, 345 (1974)

3. W.J. MacKnight and M. Yang, J. Polym.Sci., Part C, No.42, 817
   (1973)

4. C.S. Paik Sung and N.S. Schneider, Macromolecules, 8, 68 (1976)

5. A.T. DiBenedetto, unpublished results (see L.E. Nielsen, J.
   Macromol. Sci., Rev., Macromol. Chem. 3, 69 (1969(

6. E.A. Di Marzio, J. Res. Nat. Bur. Stand., Sec. A, 68, 611 (1964)

7. C.S. Paik Sung and N.S. Schneider, Macromolecules, in press.

8. R.W. Seymour and S.L. Cooper, Macromolecules, 6, 48 (1973)

9. N.S. Schneider and C.S. Paik Sing. Polym. Eng. and Sci., 17, 73
   (1977)

10. C.G. Seefried, Jr., J.V. Koleske and F.E. Critchfield, J. Appl.
    Polym. Sci., 19, 2493 (1975)

EFFECT OF ANNEALING ON THE MORPHOLOGY AND PROPERTIES OF

THERMOPLASTIC POLYURETHANES

C. H. M. Jacques

General Motors Research Laboratories

Warren, Michigan 48090

INTRODUCTION

Thermoplastic polyurethanes are essentially linear block co-polymers which have many of the properties of crosslinked elastomers at ordinary use temperatures but may be processed by injection molding at elevated temperatures. Many of the unique properties of these materials are attributed to a microphase separation of the segments into "hard" urethane domains which are surrounded by a "soft" polyester or polyether matrix. Because of their commercial importance, the nature of this domain structure has been extensively studied (1-12). Several investigators have demonstrated that thermal history has a pro-nounced effect on the morphology of the segmented urethanes (4-10). Changes in crystallinity of the hard segment as well as rearrang-ment of the domain structure has been reported to occur with suitable heat treatments (4). Most of these studies, however, have been limited to annealing temperatures of 150°C or lower. In addition, the effects of changes in morphology with annealing on the mechanical properties have not been well characterized. The purpose of this investigation, therefore, was to study changes in morphology which occur during extensive annealing at temperatures of 150°C to 250°C and to relate these changes in morphology to changes in the physical properties of annealed samples.

EXPERIMENTAL

Three thermoplastic polyurethanes supplied by the B. F. Goodrich Company were studied: Estane 58109, Estane 58116, and Estane 58111. These materials are typical polyester urethanes with Shore hardnesses ranging from 90A for Estane 58109 to 55D for Estane 58111. Estane 58116 has properties and composition intermediate between Estane 58109 and 58111.

The effects of thermal treatments on the melting points and heats of fusion of the resins were characterized by differential scanning calorimetry (DSC) using a DuPont 990 Thermal Analyzer. Heating rates of 5 to 80°C/minute were examined, but most of the measurements were performed at 20°C/minute to minimize melting and recrystallization during the heating scan.

Infrared (IR) spectra of solution cast films approximately 0.005 mm thick were recorded using a Perkin-Elmer Model 621 Grating Spectrophotometer. Changes in the infrared spectra of the films were followed as a function of time and/or temperature in a specially constructed cell which was continuously flushed with dry nitrogen.

Tensile properties were determined with ASTM D-1822 type L injection molded samples on an Instron operated at a crosshead speed of 508 mm/minute. A gage length of 25.4 mm was used for calculating ultimate elongation. Relative Young's modulus was determined at 5.08 mm/minute. Injection molded samples were annealed in a dry nitrogen atmosphere and subsequently allowed to remain at ambient conditions for about one month prior to testing.

Wide angle X-ray diffraction patterns of injection molded samples were obtained with a Norelco X-ray diffractometer using nickel filtered CuK$_\alpha$ radiation. Dynamic mechanical measurements on injection molded samples were performed at a frequency of 3.5 Hz using a Model DDV-III-C Rheovibron from the Toyo Baldwin Company.

RESULTS

DSC Analysis

The effect of thermal treatment on melting endotherms present in DSC heating curves was initially studied by annealing samples for two minutes in the DSC at controlled heating and cooling rates. Typical heating curves for Estane 58111 are shown in Figure 1. The material as received exhibited endotherms at 150, 204, and 233°C and a small exotherm near 215°C. The exotherm and subsequent endotherm are a result of crystallization and melting during the heating scan and may be removed by scanning at faster heating rates. Annealing at temperatures of about 191°C or lower results in the disappearance of the 150°C endotherm and intensification and a slight increase in the melting temperature of

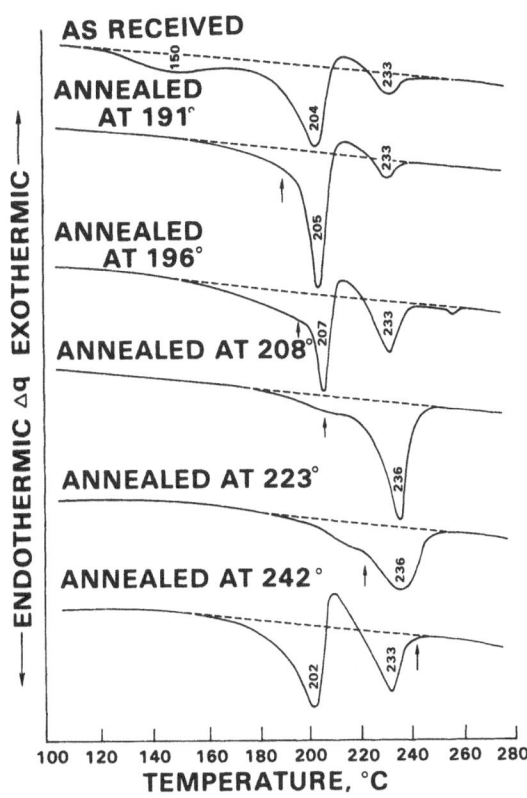

Figure 1   DSC Heating Curves For Estane 58111 Annealed For Two
Minutes in the DSC (Arrows indicate annealing temperatures.)

the endotherm originally at 204°C. The effects of annealing at
these temperatures have been discussed by others (5-8) and are
generally attributed to an improvement in crystalline order of low
molecular weight hard segments. As shown in Figure 1, annealing
at temperatures between 200°C and 240°C results in the disappearance of all lower melting endotherms and the emergence of a
melting peak at 233 to 236°C. The original melting peak near
204°C is restored by heating to temperatures greater than 240°C.
Similiar melting behavior was observed in each of the other
materials.

    Annealing for longer times results in the development of a
single, well defined crystalline melting peak as shown in Figure 2

for Estane 58111 annealed for 20 hours at 180°C and 210°C. Samples annealed at less than about 195°C evidence melting near 210°C while samples annealed between about 195°C and 240°C show a single melting endotherm near 249°C. The heats of fusion for the samples shown in Figure 2 are 6.2 cal/gram for the 210°C peak and 9.5 cal/gram for the 249°C peak.

Extensive annealing of Estane 58109 and 58116 also results in some crystallization of the soft segment. Estane 58109 samples annealed for 908 hours at 150°C exhibited a melting peak at 43°C with a heat of fusion of 4.7 cal/gram. Crystallization of the soft segment in Estane 58111 did not occur with annealing presumably because of the lower soft segment content and possibly shorter soft segment length of this material.

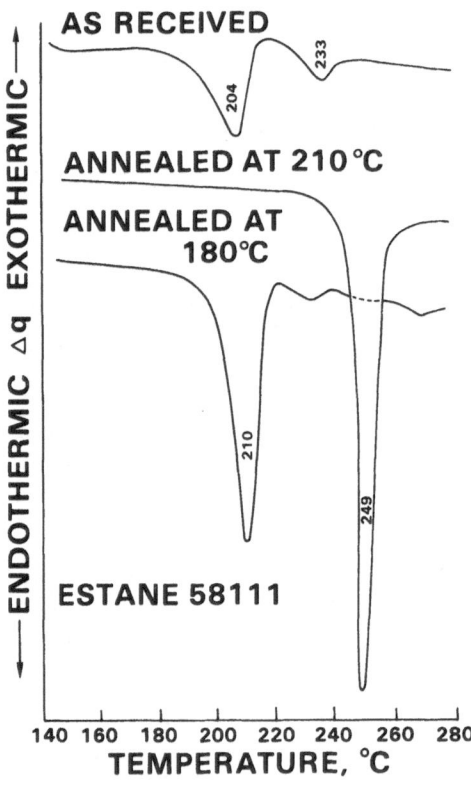

Figure 2 DSC Heating Curves for Estane 58111 as Received and After Annealing for 20 Hours in a Nitrogen Oven at 180°C and 210°C.

Infrared Analysis

No changes were observed in the IR spectra of thin films annealed at temperatures less than about 195°C, even though DSC analysis showed a significant increase in the percent crystallinity of the films. In contrast, several striking changes were observed in the spectra of films annealed at temperatures between 200°C and 240°C. The most dramatic change was observed in the N-H stretching band near 3300 cm^{-1}. Changes in the absorbance and frequency of the N-H stretching band measured at different temperatures are shown in Figure 3. The initial decrease in intensity and increase

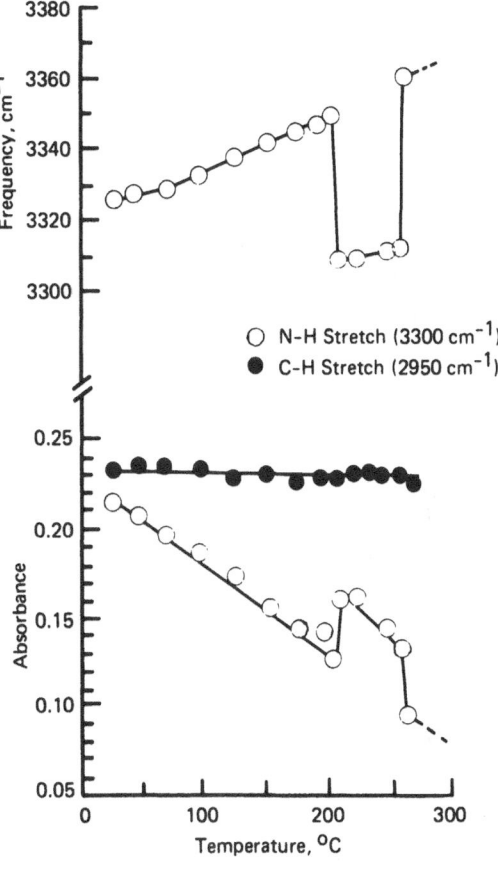

Figure 3 Infrared Analysis of Changes in the N-H Stretching Band of a Solution Cast Estane 58111 Film at Different Temperatures.

in frequency with increasing temperature are primarily attributed to a weakening of intermolecular bonding between N-H and C=O groups [6-11]. Near 200°C an abrupt change is observed in both the intensity and frequency of the N-H band as strong urethane hydrogen bonds are formed. An increase in the ratio of bonded to unbonded C=O groups also occurred as evidenced by an increase in the intensity of the 1700 cm^{-1} band and a decrease in the 1735 cm^{-1} band after annealing. Similar changes were observed at coupled peaks in the amide III region at 1215 cm^{-1} and 1235 cm^{-1} and for peaks at 810 cm^{-1} and 835 cm^{-1}. In addition, the intensities of bands at 1110 cm^{-1}, 1080 cm^{-1} and 745 cm^{-1} increased after annealing and new peaks appeared at 910 cm^{-1} and 775 cm^{-1}.

Changes in the infrared spectra occur at approximately the same temperature at which higher melting endotherms begin to appear in DSC heating curves. The further similarity between the effects of annealing on changes in the IR spectra and changes in DSC heating curves are shown in Figure 4. Both heat of fusion and

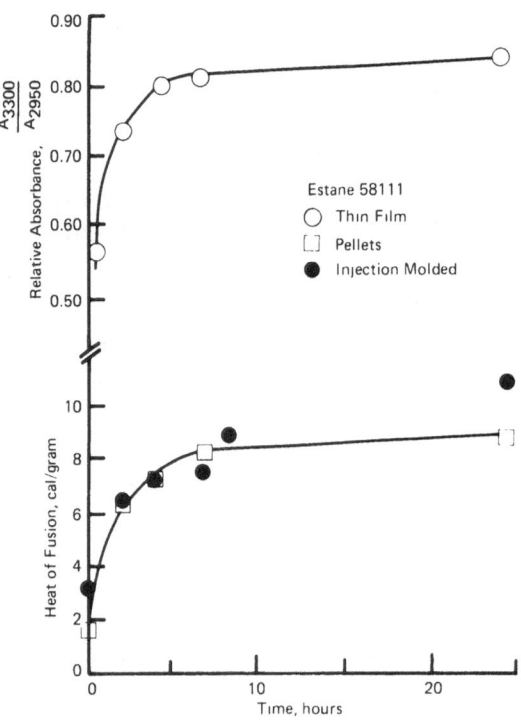

Figure 4 Crystallization Kinetics for Estane 58111 at 210°C Measured by IR Analysis and DSC.

intensity of the N-H stretching band increase at the same relative
rate with annealing at a constant temperature of 210°C.

X-Ray Diffraction

        The relative intensities of X-ray scattering from injection
molded samples of Estane 58116 are shown in Figure 5. All of the
scattering maxima shown are due to order within the hard domains.
These data show that true crystalline order develops with annealing
at 205°C but that only a limited degree of order is present after
extensive annealing at 150°C. However, it is not clear from the
data shown in Figure 5 whether or not two different crystalline
structures are formed under the two different annealing conditions.

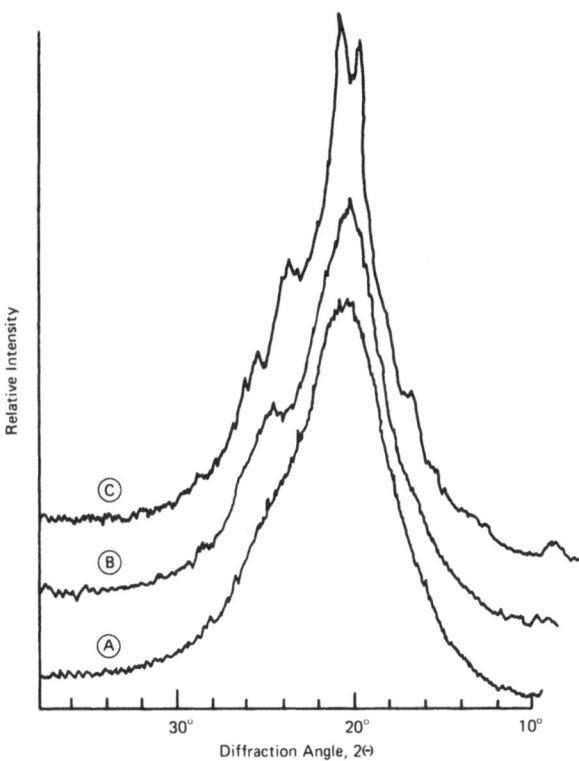

Figure 5   X-Ray Diffraction Patterns of Injection Molded
Estane 58116 (A) As Molded; (B) Annealed at 150°C for 908 Hours;
(C) Annealed at 205°C for 24 Hours.

Tensile and Dynamic Mechanical Properties

Physical properties of injection molded tensile bars of each of the Estanes annealed to constant percent crystallinity at 150°C and 205°C are presented in Table I. Samples annealed for 908 hours at 150°C show relatively little change in physical properties, even though the total heat of fusion nearly doubles for each material. However, annealing for 24 hours at 205°C results in a 100 to 300 percent increase in Young's modulus. Each material also exhibits a loss of ultimate elongation after annealing at 205°C, but even these extensively annealed samples retain between 200 to 510% ultimate elongation.

The effect of annealing on the dynamic mechanical properties of Estane 58116 is shown in Figure 6. The unannealed sample exhibits a single glass transition near +5°C, which suggests that the hard and soft segments are intimately mixed. After annealing a relatively sharp Tan δ peak is observed at -25°C followed by a

TABLE I

PHYSICAL PROPERTIES OF UNANNEALED AND ANNEALED ESTANES

Sample	Total Heat of Fusion (cal/gram)	Ultimate Elongation (percent)	Tensile Strength (MPa)	Relative Young's Modulus (MPa)
As molded				
58111	4.1	380	21.5	61
50/50[1]	3.7	530	20.0	32
58116	3.0	550	23.0	36
58109	2.7	640	18.0	16
Annealed 908 hours at 150°C				
58111	7.7	420	22.0	83
50/50[1]	6.4	570	22.0	34
58116	6.5	610	24.0	39
58109	4.9	660	20.0	14
Annealed 24 hours at 205°C				
58111	11.0	200	22.0	110
50/50[1]	7.7	420	21.5	69
58116	7.7	420	22.0	75
58109	6.2	510	21.5	65

[1] A physical mixture of 50% 58111 and 50% 58109

shoulder at 0°C and another broad peak at 55°C. Similar results were obtained for each of the other materials. It is believed that the -25°C peak corresponds to the Tg of domains composed primarily of soft segments while the 55°C peak is due to hard segment domains. The shoulder at 0°C is probably due to regions where the hard and soft segments are still relatively well mixed after annealing. Annealing also has a significant effect on the storage modulus. Between -40°C and -10°C the annealed sample actually has a lower modulus than the unannealed sample because of the reduced Tg of the soft phase. However, above -10°C the reinforcing effects of the crystallized hard segments become important and the annealed sample exhibits a significantly higher modulus than the unannealed sample.

DISCUSSION

DSC, X-ray diffraction, and infrared analyses clearly show that a distinct change occurs in the morphology of the subject

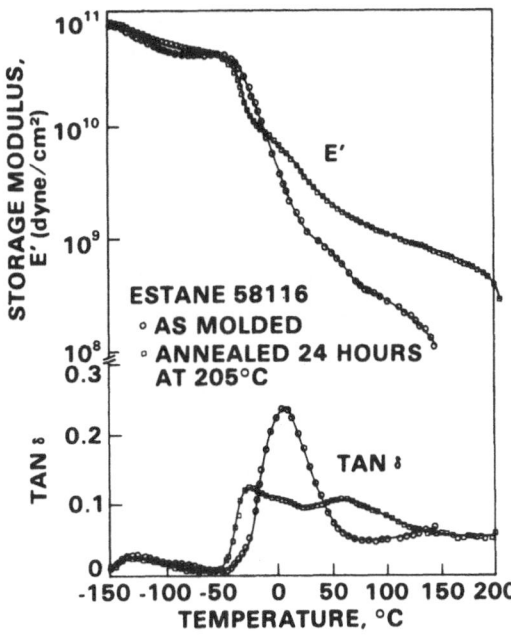

Figure 6 Dynamic Mechanical Properties of Annealed and Unannealed Estane 58116.

urethanes near 200°C. Below this temperature, hydrogen bonding between urethane and ester linkages of the hard and soft segments apparently restricts movements of the polymer chains and thus inhibits crystallization of both phases. Near 200°C, however, IR analysis indicates that dissociation of the ester-urethane bond is sufficiently complete so that more stable urethane-urethane hydrogen bonds may form. This temperature is also sufficiently high to disrupt allophanate bonds between hard segments which may restrict segmental mobility at lower temperatures (12). Specific changes in the IR spectra, the development of multiple glass transitions, and the ability of the soft segments to crystallize after annealing indicate that significant changes occur in domain size and purity during annealing at temperatures from about 200°C to 240°C. These changes in domain structure, along with the concomitant increase in percent crystallinity of the hard (and soft) segments, result in a distinct change in properties of samples annealed at temperatures greater than 200°C. Annealing at lower temperatures increases the order within the existing hard segments but has, in comparison, relatively little effect on domain structure or properties.

DSC and IR analyses also indicate that the changes which occur with annealing are reversible at temperatures of about 250°C. At this temperature, the strong urethane-urethane bonds formed during annealing dissociate, and all crystals formed at lower temperatures melt. Remixing of the hard and soft segments will occur to some extent in the molten polymer (13). Upon cooling, ester-urethane bonds reform, and the morphology which existed before annealing is largely reestablished.

ACKNOWLEDGMENTS

DSC measurements by Dr. S. J. Swarin, X-ray diffraction measurements by F. A. Forster, and instruction concerning the measurement and interpretation of dynamic mechanical properties by Dr. M. G. Wyzgoski are gratefully acknowledged.

REFERENCES

1.   S. B. Clough, N. S. Schneider, A. D. King, J. Macromal Sci., B2, 641 (1968).

2.   C. E. Wilkes and C. S. Yusek, J. Macromol Sci., B1, 157 (1973).

3.   J. A. Koufsky, N. H. Hein, S. L. Cooper, Polymer Letters, 8, 353 (1970).

4.   R. Bonart, L. Morbitzer, G. Hentze, J. Macromal Sci., B3, 337 (1969).

5. D. S. Huh and S. L. Cooper, Poly. Eng. Sci., 11, 369 (1971).

6. R. W. Seymour and S. L. Cooper, Macromol, 6, 48 (1973).

7. H. N. Ng, A. E. Allegrezza, R. W. Seymour, and S. L. Cooper, Polymer, 14, 225 (1973).

8. S. L. Samuels and G. L. Wilkes, J. Poly. Sci., A2, 11, 807 (1973).

9. S. L. Samuels and G. L. Wilkes, J. Poly. Sci., Part C, No. 43, 149 (1973).

10. K. W. Rausch, Jr., and W. J. Farrissey, Jr., J. Elastoplastics, 2, 114 (1970).

11. W. J. Macknight and M. Yang, J. Poly Sci., C, No. 42, 817 (1973).

12. S. B. Clough and N. S. Schneider, J. Macromol Sci., B2, 553 (1968).

13. G. L. Wilkes, S. Bagrodia, W. Humphries, R. Wildnauer, Polym. Preprint, 16, 595 (1975).

THE EFFECT OF CROSS-LINKING ON THE DYNAMIC MECHANICAL PROPERTIES

OF AN INCOMPATIBLE, SEGMENTED POLYURETHANE

J. L. Work, J. E. Herweh, and C. A. Glotfelter

Armstrong Cork Company, Research and Development Center

Lancaster, Pennsylvania 17604

INTRODUCTION

In recent years we have developed an interest in heterogeneous polymeric systems. Of particular interest have been block co-polymers, specifically those prepared by low-temperature condensation techniques and providing for thermoplastic semirigid materials. Coupled with our interest in materials of this type have been studies directed toward determining the effect of various structural modifications on the bulk properties of such block copolymers.

The effects of one such structural modification on the bulk properties of a particular block copolymer as determined by dynamic mechanical spectroscopy is described below. The modification involves the systematic cross-linking of soft segments in an incompatible, segmented polyurethane. In brief the block copolymer is composed of amorphous segments based on a hydroxy-terminated polybutadiene coupled by reaction with a diisocyanate to amorphous segments comprising the urethane from the diisocyanate and a tertiary amine-bearing diol.

EXPERIMENTAL SECTION

Starting Materials — 4,4'Bis(isocyanatocyclohexyl)methane (Dupont Hylene W) and N-methyl diethanol amine (Union Carbide) were used as received. Hydroxy-terminated polybutadiene (Telagen HT 5010, M_n = 3716) containing internal vinyl unsaturation and pendant allyl groups was obtained from General Tire and Rubber Corporation.

Bis(mercaptoethyl) ether (BMEE) was used as received from Evans
Chemetics, Inc. Benzoin isobutyl ether (Vicure 10) obtained from
Stauffer Chemical and T-12 and T-31 catalysts from M&T Corporation
were used as received.

Preparation of Isocyanate-Modified
Hydroxy-Terminated Polybutadiene (A)

A 1:1 solution of hydroxy-terminated polybutadiene (0.05 mol)
in benzene was purged with nitrogen for ca. 15 min and then heated
to reflux under a nitrogen atmosphere. Approximately 50 ml of
solvent was distilled to effect drying. The solution was cooled
to 60C and 4,4'-bis(isocyanatocyclohexyl)methane (0.06 mol) and
T-12 catalyst (2% by wt. of reactants) was added rapidly. The
reaction temperature was maintained at 60C for ca. 24 hr. Chain
extension was effected by adding water (6.6 ml) along with
additional T-12 catalyst (1.5% by wt. of reactants) and maintain-
ing reaction temperatures at 60C for an additional 3 hr. The
reaction product, a rubbery material, was recovered by removing
solvent under vacuum at moderate temperatures and was subsequently
stored under a nitrogen atmosphere.

Preparation of Polyurethane from N-Methyl Diethanol
Amine and 4,4'-Bis(isocyanatocyclohexyl)Methane (B)

The isocyanate (0.1 mol) and N-methyl diethanol amine (0.1
mol) were added to a solution of benzene (180 ml) and DMF (70 ml).
The resulting reaction mixture was heated to reflux with stirring
under a nitrogen atmosphere, and 30 ml of solvent was distilled to
effect anhydrous conditions. T-31 catalyst (0.5% by wt. of
reactants) was added and the reaction mixture was maintained at
75C. After several hours, the reaction mixture became slightly
turbid; DMF (30 ml) was added, but failed to eliminate turbidity.
Heating at 75C was continued; after ca. 24 hr the cooled reaction
mixture was added to excess methanol. The polyurethane precipitated and was triturated with hexane and then water. Following
overnight soaking in water, the polymer was dried in vacuo at 80C
for 2.5 hr.

Preparation of an Incompatible, Segmented Polyurethane (AB)

Hydroxy-terminated polybutadiene (0.01 mol), N-methyl
diethanol amine (0.1025 mol), and 4,4'bis(isocyanatocyclohexyl)
methane (0.1125 mol) were added to a mixture of DMF (100 ml) and
benzene (400 ml). The reaction mixture was rendered anhydrous in
a manner similar to that described above, and T-31 catalyst was

added (ca. 0.5% by wt. of reactants). The clear reaction mixture was maintained at 80C for 6 hr and cooled. Portions of the resulting lacquer were cast into films and the solvent removed in vacuo to provide samples of the block copolymer for characterization.

Photochemical-Initiated Cross-linking of the Isocyanate-Modified Hydroxy-Terminated Polybutadiene (AB)

Solutions (10 w/w%) of the isocyanate-modified polybutadiene and varying levels of bis(2-mercaptoethyl)ether were prepared in 15 w/w% of methanol in benzene. All reaction mixtures contained the photoinitiator benzoin isobutyl ether (1 w/w%). The compositions are given in Table I. Solvent was removed in a forced-air oven at 40C. The resulting films were placed in an inert atmosphere and finally irradiated for 15 min using a Blak Ray lamp[1].

TABLE I

Photochemical-Initiated Cross-linking of the Isocyanate-Modified Hydroxy-Terminated Polybutadiene (A) - Compositions

Polymer No.	A (g)	BMEE[a]		BIE[b]	
		g	w/w%	g	w/w%
A-1	1.0	0.011	1.1	0.01	1.0
A-2		0.030	2.9		
A-3		0.054	5.1		
A-4		0.124	11.0		
A-5		0.253	20.1		
A-6[c]		0.253	20.1		

(a) BMEE = bis(mercaptoethyl)ether
(b) BIE = benzoin isobutyl ether (Vicure 10)
(c) Sample was not exposed to uv

Photochemical-Initiated Cross-linking of the Segmented Polyurethane in Solid State

Solutions of the incompatible, segmented polyurethane, varying levels of BMEE, and photoinitiator were prepared and treated as described above. The compositions are summarized in Table II.

(1) Emits uv in the 365 nm region.

TABLE II

Photochemical-Initiated Cross-linking of the Incompatible,
Segmented Polyurethane (AB) in the Solid State - Compositions

Polymer No.	AB (b)	BMEE		BIE	
		g	w/w%	g	w/w%
AB-1	1.0	0.009	0.9	0.01	1.0
AB-2		0.026	2.6		
AB-3		0.054	5.4		
AB-4		0.119	11.7		
AB-5		0.261	25.5		

Photochemical-Initiated Cross-linking of the Segmented Polyurethane in Solution

Solutions containing 10 w/w% of the segmented polyurethane in
a 30 w/w% mixture of dimethyl acetamide (DMAC) in THF were prepared.
To each solution was added a weighed amount of BMEE and benzoin
isobutyl ether (Table III). In each case 10 g aliquots of the
requisite solution were placed in a vacuum desiccator at 20 mm for
18 hr. The samples were removed from the desiccator and the w/w%
solvent present was determined. The samples were immediately
irradiated under conditions described earlier. They were then
immersed in THF to remove DMAC and after 24 hr were dried an
additional 24 hr in vacuo.

TABLE III

Photochemical-Initiated Cross-linking of the Incompatible,
Segmented Polyurethane (AB) in Solution - Compositions

Polymer No.	AB (g)	Composition of Mixture to be Cross-linked				Solvent w/w%
		BMEE		BIE		
		g	w/w%	g	w/w%	
AB-1S	1.0	0.010	1.0	0.01	1.0	68
AB-2S		0.025	2.5			66
AB-3S		0.053	4.9			67
AB-4S		0.109	9.6			68
AB-5S		0.258	20.5			70

Determination of Swelling Characteristics

Approximately 0.1 g of polymer film was immersed in 10 mls of
THF for 24 hr at room temperature. The film was blotted to remove
surface THF and weighed to determine the level of solvent in the
swollen film. The percent extractables was determined by removing
the THF from the films at 70C in vacuo for 2 hr and also by
evaporating the swelling solvent (Table IV).

TABLE IV

Swelling-Tests Results on Cross-linked Isocyanate-Modified
Hydroxy-Terminated Polybutadiene (A) and the Cross-linked
Incompatible, Segmented Polyurethane (AB)

Polymer No.	w/w% Solvent in Swollen Polymer	w/w% Extractables from	
		Dried Polymer	Solvent Residue
A-1	79.3	5.4	6.2
A-2	69.7	2.2	3.0
A-3	65.2	1.8	2.1
A-4	59.2	1.1	1.3
A-5	42.3	0.9	1.7
A-6	88.8	11.1	15.1
AB-1	88.0	29.8	30.5
AB-2	80.2	11.7	11.7
AB-3	69.6	4.4	3.6
AB-4	58.2	3.0	2.1
AB-5	42.4	4.4	2.1
AB-1S[a]	95.2	26.7	29.6
AB-2S	90.0	12.5	13.7
AB-3S	85.2	9.5	11.5
AB-4S	80.0	13.4	15.9
AB-5S	72.0	13.4	14.1

(a) Test was run for 72 hr rather than 24 hr for polymers
    No. AB-1S through AB-5S.

Dynamic Mechanical Measurements

All measurements were made on 5- to 15-mil films using a
Rheovibron Model DDVII, Toyo Measuring Insts.

DISCUSSION OF RESULTS

Development of a Suitable Cross-linking Process for the Diisocyanate-Extended Hydroxy-Terminated Polybutadiene Segment (A)

Prior to effecting cross-linking of the unsaturated hydrocarbon segment in the requisite block copolymer an effective means of cross-linking was sought. The diisocyanate-modified, water-extended hydroxy-terminated polybutadiene (NCO/OH = 1.2), a rubber, was utilized as a substrate to help determine the effectiveness of various candidate cross-linking reactions. The small amount of urea introduced by water extension caused an aggregation of the polymer into domains. However, the material was soluble in THF and a 15 v/v% solution of methanol in benzene.

Photoinitiated polymerization using 2-ethylhexyl acrylate (2-EHA) and an initiator comprising 50 w/w% N-methyl diethanol amine and benzophenone was examined as a means of cross-linking the polymeric substrate (A). Variations in cure times, initiator levels, film thickness, and level of 2-EHA were tried. Swelling characteristics of the cross-linked polymers were determined. Although optimum conditions for effective cross-linking were found, the use of 2-EHA appeared to significantly reduce the strength and extensibility of the material.

In an attempt to obtain stronger films, bis(2-mercaptoethyl ether) (BMEE) was examined as the cross-linking agent. The reaction was photoinitiated using benzoin isobutyl ether (Vicure 10), and the same procedure established for the 2-EHA cross-linking reactions was followed. A series of samples was prepared (see Experimental Section, Table I) in which the BMEE level was varied. The strengths of all the polymer samples were good. Their extensibility decreased as expected with increasing mercapto ether level. Polymer sample A-6 (Table I) was not exposed to uv irradiation and proved to be the most extensible of all.

Swelling tests were performed on all of the polymer sample films (see Experimental Section, Table IV), using THF as the solvent. The extent of swelling decreased with increasing BMEE content. As swelling decreased, the percent extractables also decreased. The results obtained for Polymer No. A-6 (Table IV) indicate that some cross-linking occurs even when uv exposure is omitted.

In view of the results obtained with BMEE, this reagent was selected as a means to effect cross-linking in the subject block copolymer (AB). Results of these cross-linking reactions will be discussed presently.

Preparation and Characterization of Segmented Polyurethane (AB)

The subject block copolymer has the general formula (I). The

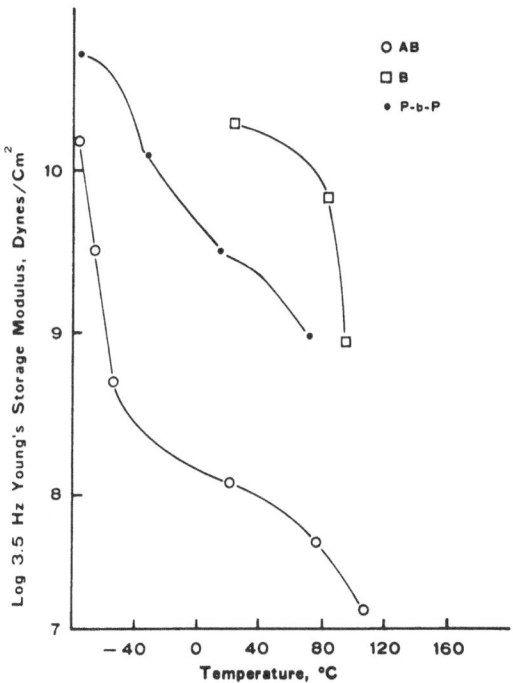

$$R = \{CH_2CH=CHCH_2\}$$
$$R' = \{CH_2\}_2 N(Me)(CH_2)_2$$

wt. % of n units in the block copolymer is 49.6. The polymer was
prepared by a random polycondensation of hydroxy-terminated
polybutadiene and N-methyl diethanol amine with the diisocyanate
at a NCO/OH ratio of 1.0. N-methyl diethanol amine was selected
as a component of the polyurethane hard segment, since experience
had shown that urethanes based on this diol provided for good
phase separation.

The subject block copolymer (AB) exhibited two discrete
phases evidenced by the presence of two distinct Tg's as
determined by the Rheovibron (Fig. 1). The Tg at -74C (3.5 Hz) is
attributed to the isocyanate-modified polybutadiene segments while

Fig. 1. Modulus vs. Temp. For AB, B, polyester-b-polyether (P-b-P)

the Tg in the range of 80C is attributed to the polyurethane
segments. The Tg of the analogous polyurethane homopolymer (B) is
88C (3.5 Hz).

A noteworthy feature of this block copolymer (AB) is its low
storage modulus in the temperature region between the two Tg's (1
to 2 x 10^8 dynes/cm^2) (Fig. 1), when compared to an SBS block
copolymer containing 50 w/w% styrene (5 x 10^9 dynes/cm^2) (1) and
other block copolymers having similar hard/soft segment
compositions (e.g. polyether-b-polyester^2, see Fig. 1) (2). These
results suggest that the morphology of this block copolymer differs
considerably from that of the alternating lamellae attributed to
the SBS block copolymer.

The block copolymer (AB) does not neck down and cold draw as
do a variety of ether-b-ester and ester-b-ester copolymers of
similar hard/soft composition wherein phase separation is rather
diffuse. However, the subject block copolymer does retain
considerable set after deformation, which is typically the case
for the ether-b-ester and ester-b-ester copolymers referred to
above.

Effect on Properties of Cross-linking the Segmented
Polyurethane (AB) in the Solid State

A series of cross-linked block copolymers based on I was
prepared (see Experimental Section, Table II), using the same
procedure as that utilized for the diisocyanate-modified, water-
extended polybutadiene discussed earlier. The extensibility of
the block copolymers decreased as the level of BMEE increased.
At 11.7 w/w% mercapto ether (see Experimental Section, Table II),
the tear strength of the modified block copolymers deteriorates
significantly. Swelling-tests results (see Experimental Section,
Table IV)were comparable to those found earlier for the
isocyanate-modified, water-extended soft segment. As expected,
the density of the cross-linked block copolymers increased as the
level of BMEE increased (Table V).

The low-temperature Tg attributed to the isocyanate-extended
polybutadiene segments increased significantly with increasing
levels of BMEE (Table V). This observation was not anticipated
since the Tg of polyalkylene sulfides is on the order of
-50C; (3,4) consequently, the Tg for the soft segment of the cross-
linked block copolymer would be predicted to be between the limits
of -74 and -50C. A possible explanation for this behavior is that

(2) Coupling of hydroxy terminated polyether and polyester
    segments was effected with a diisocyanate.

TABLE V

Characteristics of the Incompatible, Segmented
Polyurethane (AB) Cross-linked in Solid State

Polymer No.	Density g/cm^3	Tg, $^\circ$C (3.5 Hz)	Unsat./ -SH	w/w% BMEE	
				Copolymer	Polybutadiene
A	1.007	-34	4.6	--	20.1
AB	0.999	-74, 80	--	--	--
AB-1[a]	1.001	-68, 60	122	1.0	1.9
AB-2	1.004	-64, 66	43.9	2.6	5.1
AB-3	1.015	-57, 68	9.6	5.3	10.6
AB-4	1.041	-40, 64	3.8	11.8	23.6
AB-5	1.074	-28,4,62	1.3	23.3	46.6

(a) AB cross-linked in the solid state.

the oxyalkylene sulfide units enter as branches and as a linear comonomer with polybutadiene. This increase in low-temperature Tg is not associated with the block copolymer alone. The isocyanate-modified polyurethane butadiene homopolymer, at a similar BMEE level (Table V), exhibited nearly the same Tg as the block copolymer (Table V). It further appears that the increase in Tg cannot be attributed to increased cross-linking, since the latter would result in a broadening of the Tg region. This is not observed for the isocyanate-modified polybutadiene homopolymer.

At high BMEE levels (Table V) evidence for two Tg's associated with the polybutadiene segments is obtained (Fig. 2). In one case two Tg's are definitely observed; in the other, multiple Tg's appear to be developing, but cannot be resolved using the Rheovibron. It is uncertain at this time what the origin of these multiple Tg's may be.

Following an initial decrease in the Tg of the hard polyurethane phase (ca. 16°C), when cross-linking is introduced in the block copolymer, the Tg remains relatively constant as the level of BMEE increases (Table V). This is indicative of little or no change in the spatial arrangement of the phases in the system with increased cross-link density. The initial decrease in Tg most likely results from incomplete phase separation before major cross-linking reactions begin. It was demonstrated previously, in connection with the isocyanate-modified polybutadiene homopolymer, that some cross-linking occurs during solvent removal. Furthermore, it will be shown subsequently that cross-linking in the presence of solvent reduces the hard-segment Tg with increasing cross-link density.

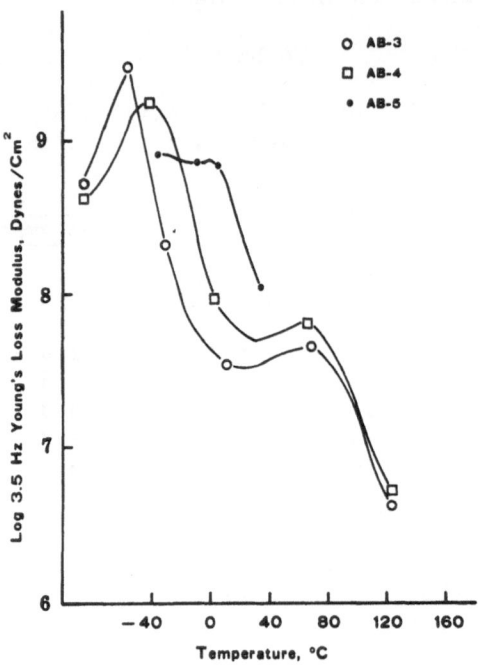

Fig. 2. Loss Modulus vs. Temp. For AB-3, AB-4, AB-5

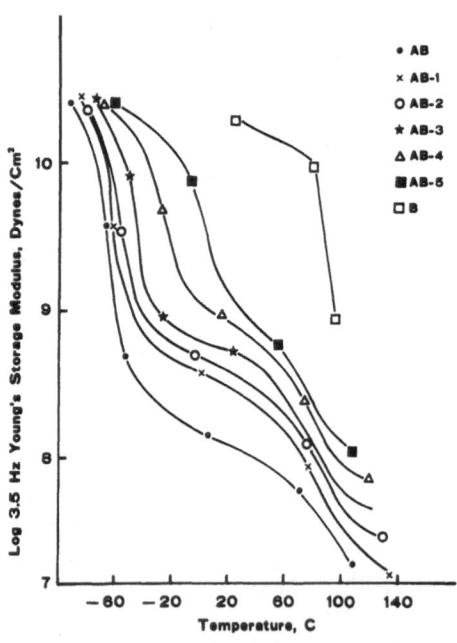

Fig. 3. Modulus vs. Temp. For AB, AB-1, AB-2, AB-3, AB-4, AB-5, & B

The storage modulus between the two Tg's in these block copolymers increases with increased cross-link density (Fig.3). A marked increase in modulus occurs upon the initial formation of cross-links (Fig. 3). Cross-linking accounts for part of this marked increase in modulus. However, partial mixing of soft segments in the hard phase, causing a 16° lowering of the Tg when compared to the polyurethane homopolymer, would also contribute to an increase in modulus. Further increases in cross-link density result in a relatively uniform increase in the storage modulus.

There is no distinct maximum in loss modulus at the Tg of the hard segment domains in the uncross-linked block copolymer. When cross-linking is introduced, a distinct loss modulus maximum occurs (Fig. 4). This suggests that at a given level of strain the hard segment domains are more highly strained in cross-linked materials than in the uncross-linked. Since at large deformations the hard domains undergo plastic deformation, recovery will be very slow in these cross-linked block copolymers unless the temperature is increased or the Tg of the hard domains is lowered. Hence, permanent set which is characteristic of block copolymers having high levels (>25% by wt.) of hard segment is not effectively reduced by cross-linking the soft phase.

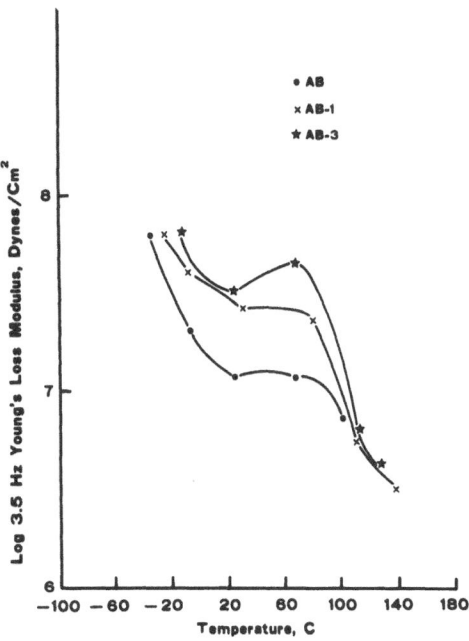

Fig. 4. Loss Modulus vs. Temp. For AB, AB-1, AB-3

Effect on Properties of Cross-linking the
Segmented Polyurethanes (AB) in Solution

In an attempt to further modify the morphology of the block copolymer, cross-linking reactions were carried out in concentrated solution (Table III), where domain morphology is largely absent. The cross-linking reaction was again photoinitiated using BMEE. The swelling data (Table IV) for cross-linked materials show that, at comparable levels of BMEE, less cross-linking occurs than for materials reacted in the absence of solvent.

The characteristics of block copolymers cross-linked in solution are summarized in Table VI. A comparison of these data and accompanying dynamic mechanical spectra, with analogous information for the block copolymer cross-linked in the solid state, reveals several major differences. The most important of these is that the Tg associated with the hard phase decreases with increased cross-link density when the cross-linking reaction is effected in solution vs. relatively little change in solid-state reactions.

TABLE VI

Characteristics of the Incompatible, Segmented
Polyurethane (AB) Cross-linked in Solution

Polymer No.	Density g/cm^3	Tg, ^{o}C (3.5 Hz)	Unsat./ -SH	w/w% BMEE	
				Copolymer	Polybutadiene
A	1.007	-34	4.6	--	20.1
AB	0.999	-74, 80	--	--	--
AB-1S[a]	0.999	-68, 56	122	1.0	1.9
AB-2S	1.006	-62, 44	43.9	2.5	4.1
AB-5S	1.044	--, 22	1.7	20.5	41.0

(a) AB cross-linked in solution.

A logical explanation for this difference is that the hard segments cannot segregate as readily from the cross-linked soft segments as they can from the uncross-linked soft segments. The restrictions placed on the mobility of the soft segments by cross-linking prevents them from assuming conformations that would provide for good phase separation. This leads to a more diffuse interphase between the domains and thus to a greater depth of penetration of the domains into one another. The effect on Tg is predominantly manifested in that of the polyurethane segments, the Tg being decreased from 80 to 22° in the range of cross-linking covered.

A second difference resulting from solution vs. solid-state cross-linking of the block copolymer is that the former materials exhibit a much higher modulus between the Tg's (Fig. 5) at comparable levels of cross-linking. This again is related to the

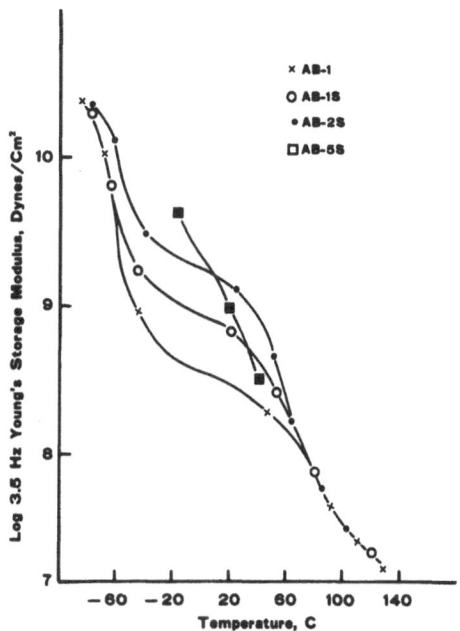

Fig. 5. Modulus vs. Temp. For AB-1, AB-1S, AB-2S, AB-5S

inability of the solution cross-linked block copolymers to aggregate into pure domains. To decrease the Tg of the hard phase, low-Tg material must be solubilized in the hard phase. This will result in an increase in volume of the hard domains and a decrease in volume of the soft domains providing for an increase in modulus. The diffuse interphase produced by this mixing of domains leads to an increased temperature dependence of the modulus between the two Tg's.

A final observation is that the materials cross-linked in solution did not exhibit any gross improvement in recovery from short-term deformations when compared to those materials cross-linked in the solid state.

REFERENCES

1. Gerard Krause, presentation, Sixth Akron Summit Conference, Akron, O., September 1975.

2. J. L. Work, to be published.

3. S. Adamek, B.B.J. Wood, and R. T. Woodhams, _Rubber Age_, _96_, 581 (1965).

4. R. H. Gobran, _Encyclopedia of Polym. Sci. and Tech._ _10_, 333 (1969).

INTERACTION OF WATER WITH HYDROPHILIC POLYETHER POLYURETHANES: DSC
STUDIES OF THE EFFECTS OF POLYETHER VARIATIONS

Joyce L. Illinger

Army Materials & Mechanics Research Center

Watertown, Massachusetts   02172

INTRODUCTION

Chemical composition and structure in segmented polyurethane
materials has a marked effect upon the amount of water which these
polymers will sorb (1,2). These factors will also influence the
states in which the sorbed water is present in the polymer.
Scanning calorimetry has proved valuable for highly hydrated
synthetic and biological polymers in assessing states of water in
several systems (See Reference 3 and its references). One useful
means of interpreting the data is to assume that water which gives
rise to an "endotherm" during the heating scan has the heat of
fusion of pure water. This gives a direct measure of "free" water
and by inference a measure of "bound" water which may exist in one
or more states. Compositional variation of soft segment in a
polyurethane from pure poly(propyleneoxide) (PPO) through a (PEO)-
(PPO)-(PEO) block copolymer to pure poly(ethyleneoxide) (PEO)
produces greater than an order of magnitude change in sorptive
capacity (2). This study is concerned with elucidating the effect
of polyether structure on the state of water in the polymer over
a broad range of water content. The evidence concerning the state

of water is drawn from observations of the shifts in polyurethane soft segment T_g and variations in size of the "melting" endotherm for water.

EXPERIMENTAL

We have studied three segmented polyurethanes synthesized (as described earlier (2)) from MDI, butanediol and a macroglycol in the ratio of 4:3:1. The general structure is illustrated by Figure 1 and polymer variation summarized in Table I.

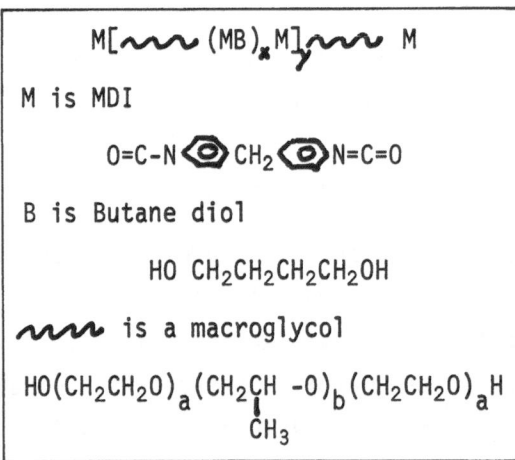

Figure 1. Polyurethane Structure

TABLE I

Polymer Variation

Polymer	Glycol Composition	a	b
10PE33	100% PEO	36	0
5PE33	50/50 PEO/PPO	11	17
OPE33	100% PPO	0	35

The polymer designations are as follows: The first number
refers to the weight percent ethylene oxide in the soft segment,
PE refers to PEO and the final number refers to the weight percent
MDI in the total polymer: thus, 5PE33 is a polymer with 50 weight
percent PEO in the soft segment and 33% MDI by weight in the total
polymer.

Films were cast from DMF solutions and dried under vacuum for
48 hours at 50°C. Samples were prepared from polymer discs of
known dry weight equilibrated with water. Excess water was allowed
to evaporate until desired water content was achieved and gold
sample pans were then hermetically sealed. In addition to the
large, deep pans and lids previously developed (4) a configuration
of large pan nested in large pan was also used. An extended range
of water concentrations in 10PE33 were prepared. Samples were
preconditioned at 10° intervals from 273° to 323°K (T_{eq}) and then
scanned at 20° per minute from 150°K to T_{eq}. The instrument used
was a Perkin-Elmer DSC-2 with subambient accessory cooled with
liquid nitrogen and purged with helium. Samples were quenched at
320° per minute to 150°K before scanning to fix the state of water
distribution determined by equilibration history.

DISCUSSION

T_g of the dry polymer has been attributed to the soft segment
glass transition and is approximately 25° higher then that of pure
soft segment materials. The isolated soft segments at this molecular size of 35 to 39 repeat units all showed a single T_g at
approximately 210°K. The PEO and PPO glass transition temperatures
are sufficiently close to each other that only a single transition
would be expected in the block copolymer. X-ray diffraction
measurements and microscopy (5) have shown that these polymers
display fibrillar spherulitic structure. These structures could

serve as physical cross-links, tying the soft segments down and
increasing T_g. In addition, since the size of the superstructure
is of the order of 5 to 10 microns, these regions must include
soft segment materials. This implies some intermixing of hard and
soft segments which would also serve to increase T_g. IR data show
all NH hydrogen bonded but not all carbonyl, so that there is the
possibility of hard segment hydrogen bonding to the soft segment
which would also increase T_g.

Sorption data on the pure hard segment copolymer $(MDI-Bd)_n$
has shown that no sorption occurs in this material. The assumption
is therefore made that sorption and interaction of water in these
polymers occurs in the soft segment phase.

Previous studies (4) with relatively high amounts of added
water (still lower than the total sorptive capacity of the polymer)
showed that as T_{eq} was decreased the area under the endotherms also
decreased. Assuming this endothermic water to be "non-bound" with
a heat of fusion of bulk water, and subtracting this amount from
total added water, we inferred an increase in "bound" water with
decreasing T_{eq}. A concomittantly greater depression of T_g was
found with decreasing T_{eq}.

In the current work we concentrated on 10PE33 with smaller
increments in amounts of added water in order to study these effects
in more detail. Figure 2 shows DSC scans of this polymer with 48%
added water as T_{eq} was changed. Equilibration times at a given T_{eq}
were lengthened until no further change in endotherm or T_g were seen.
It will be noted that at this high water content on endotherm is
present for all T_{eq}. Also note that at lower T_{eq} there is an
exotherm following the onset of the glass transition which decreases
gradually with increasing T_{eq} The T_g increases slightly and the
character of the endotherm also changes markedly. The onset of

the endotherm is at approximately 260°K and a second peak appears at 273°K at T_{eq} of 303°K. Figure 3 shows the effects of changing water content at a given equilibration temperature, in this case, 273°K. The dry polymer (bottom scan) shows a T_g at approximately 237°K. As the amount of added water increases the T_g markedly decreases. Finally, at an added water content of 38% we see the onset of an endotherm during the scan which was not visible at lower water content. The size of this endotherm is increased in a higher water

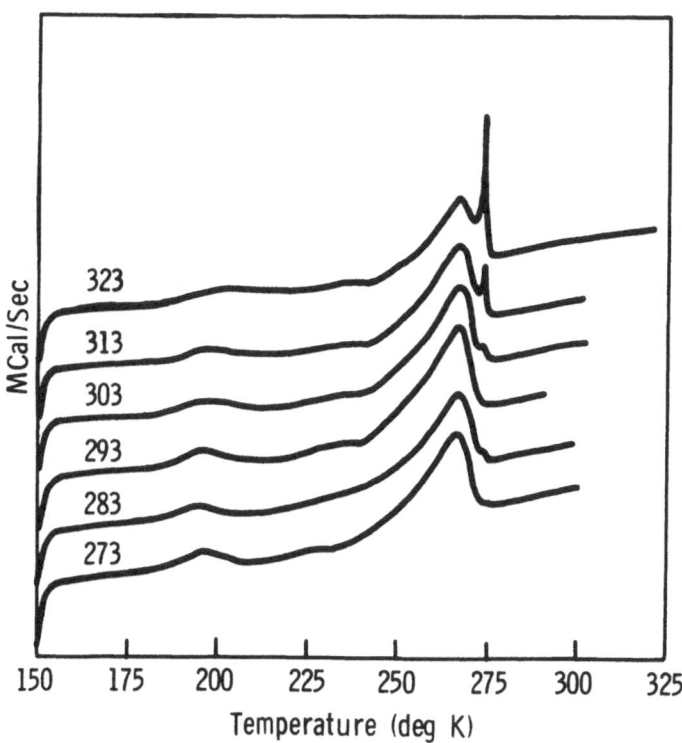

Figure 2. DSC scans of 10PE33 with 48% added water. The individual scans are labelled with T_{eq} (°K). Scans are arbitrarily displaced on the heat capacity axis for clarity.

content polymer. Figure 4 shows similar scans except that the
equilibration temperature is higher, 323°K. The onset of the
endothermic transition is now seen at a 10% lower water content.
Further work with smaller amounts of added water between 19 and
38% added water show that the appearance of an endotherm at this
high T_{eq} is found at 26% added water.

    All of these data would suggest redistribution of water in

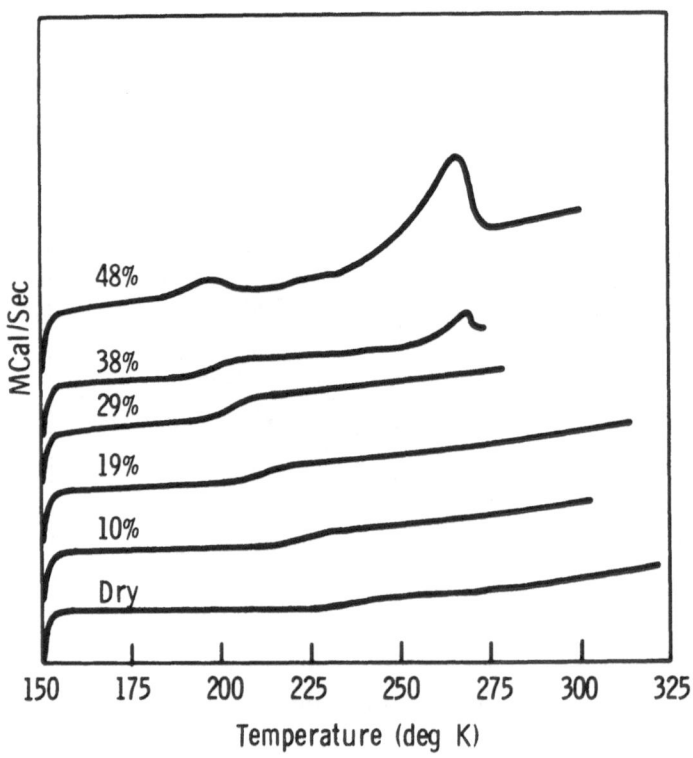

Figure 3. DSC scans of 10PE33 equilibrated at 273°K.
Percentage of added water is shown at each scan.

the polymer depending upon the T_{eq} and time at temperature. In
OPE33 there was a slight antiplasticing effect decreasing as amount
of "bound" water decreased. The plasticizing effect seen in both
polymers with PEO in the soft segment also decreased as "bound"
water decreased. Figure 5 shows these trends assuming all "bound"
water in the soft segment.

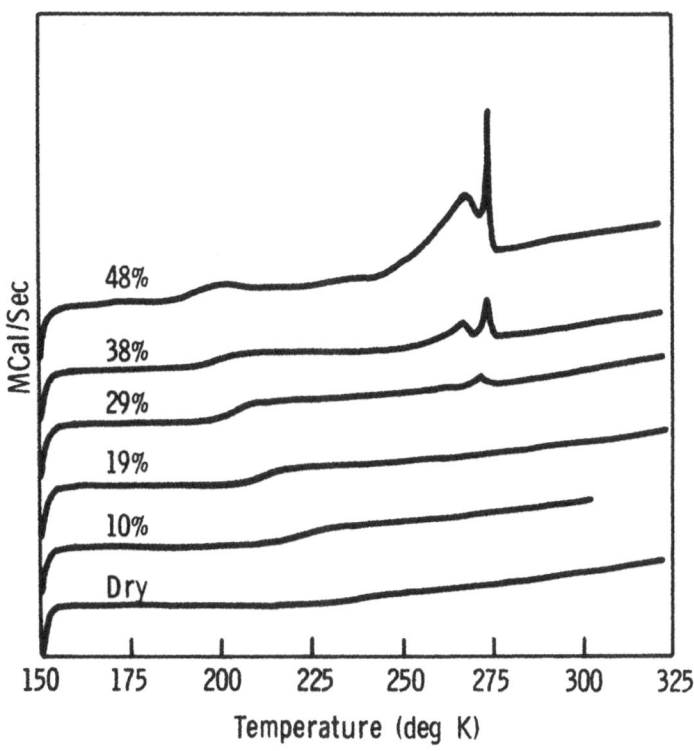

Figure 4. DSC scans of 10PE33 equilibrated at 323°K.

For those polymers in which no endotherm was observed all of the
added water was assumed to be bound to the polymer. These samples
showed no change in T_g with T_{eq}. The positive intercept for the
5PE33 polymer may be indicative of the presence of monomeric or
dimeric species too small to melt and thus not accounted for

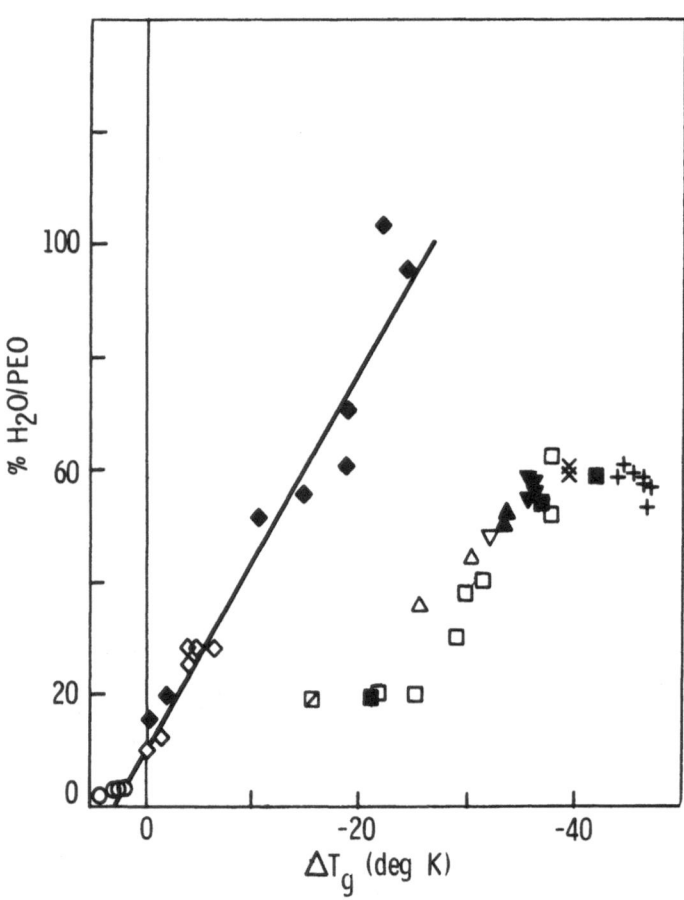

Figure 5. Percent water "bound" to PEO
OPE33, 2% added H_2O (0); 5PE33, 9% (◇), 33% (◆);
10PE33, 10% (◲), 19% (△), 26% (▽), 29% (▲),
34% (▼), 38% (✕), 42% (□), 48% (+), 67% (■).

properly in the calculation of "bound" water. One might alternatively speculate that the specific interactions which lead to antiplasticization in OPE33 are effective at low water concentrations in polymers which have the PPO block in the soft segment. ΔT_g in 1OPE33 is much greater for a given percent "bound" water than it is in 5PE33. If one translates percentage water bound to PEO into molecules of water per EO unit we find that this value in polymer 1OPE33 approaches a limiting value of three. This may be due to the structure in pure polyethylene oxide soft segment itself. NMR measurements by Liu and Parsons (7) have indicated that indeed in pure polyethylene oxide in aqueous solution one finds a limiting value of three molecules of water per ethylene oxide unit in the hydrated material.

Calculations of expected T_g were performed according to free volume theory (8)

$$T_g = \frac{(\alpha_\ell - \alpha_g) v_p \, T_{gp} + \alpha_d \, v_d \, T_{gd}}{(\alpha_\ell - \alpha_g) \, v_p + \alpha_d \, v_d}$$

where:

T_g is the depressed glass transition temperature.

T_{gp}, T_{gd} are the glass transition temperatures of polymer and diluent respectively.

v_p, v_d are volume fractions of polymer and diluent respectively.

$(\alpha_\ell - \alpha_g)$ is the difference in expansion coefficients of polymer in glass and liquid.

α_d is the expansion coefficient of diluent.

T_g values were calculated assuming $(\alpha_l - \alpha_g)$ is 4.8×10^{-4} (Reference 8), α_d is 2.07×10^{-4} for water (See Reference 9) and T_{gd} (water) equals 137°K (Reference 10). Figure 6 is a plot of ΔT_g measured versus ΔT_g calculated. The line is that which obtains

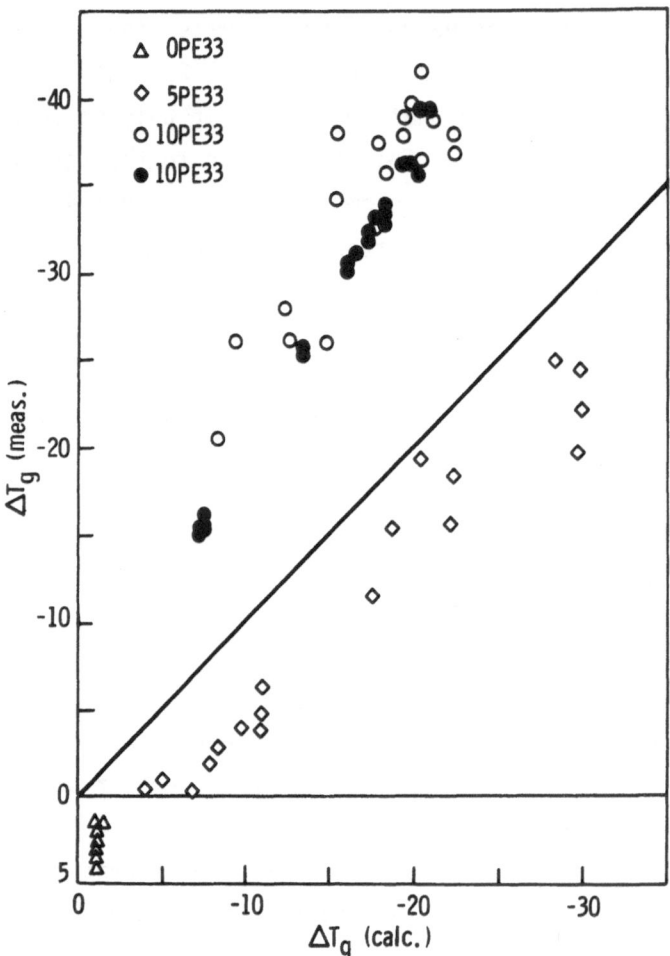

Figure 6. Glass transition temperature depression as
measured vs that calculated from free volume theory.
Open symbols from reference 4, filled circles from
this extended work.

if theory holds. The measured values for the PEO/PPO block copolymer soft segment polymers fall near the theoretical line while both
PPO and PEO polymers deviate markedly, though less so at lower water
contents.

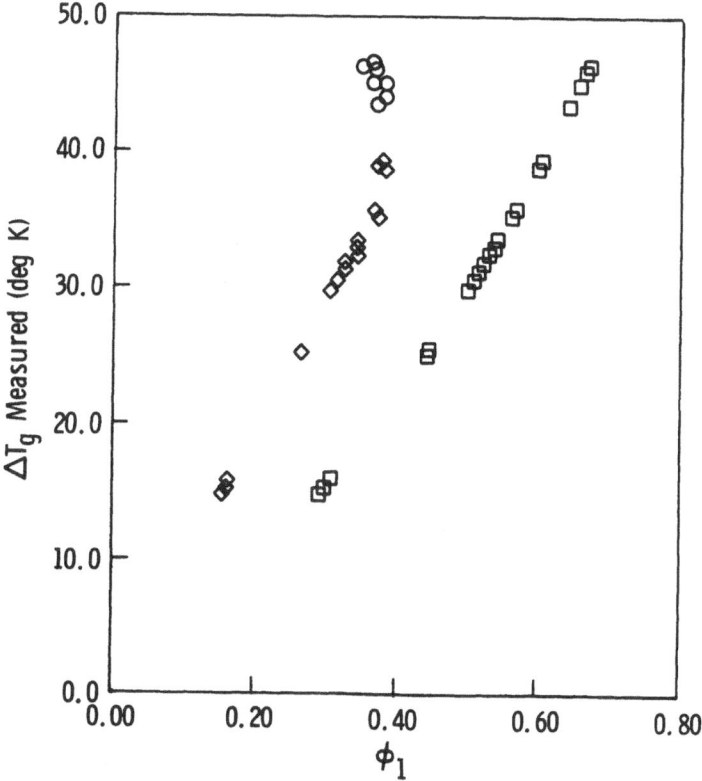

Figure 7. Measured T_g vs volume fraction for 10PE33
              ϕ_1 calculated from
              (◇) water on PEO, no endotherm in DSC trace
              (○) "bound" water on PEO, endotherm in DSC
                  trace
              (□) measured T_g

The recent data using the pan-nested-in-pan configuration show
less scatter than those data obtained using the higher pans. These
data were used to back-calculate the volume fraction of water (ϕ_1)
needed to produce the measured T_g in 1OPE33. Figure 7 shows ΔT_g
measured versus ϕ_1 calculated from "bound" water and from measured
T_g. For samples with all added water bound, there is parallel
behavior, displaced from the back calculated ϕ_1. At higher added
water, additional deviation is seen. The fraction of available
PEO in the polymer needed to interact with the water to give the
calculated volume fraction is constant at approximately 0.45, until
excessive amounts of water are added. From previous microscopic
studies (5) we may infer that the soft segment material in the
spherulites is not as readily accessible to water as is that in the
amorphous areas. It is possible that the PEO available is only
about half of the total PEO in the polymer.

CONCLUSIONS

The structure of the polyether soft segment in these hydrophilic
polyurethanes has a marked effect on the polymer-water interaction,
although the details are not totally clear as yet. We see that in
the polymer with PPO as the soft segment there is an antiplasticizing
effect which might be accounted for by individual water molecules
bridging the soft segments. In the PEO polymer there is a much
greater effect on glass transition temperature than would be
predicted from free volume theory. It is possible that the specific
interactions of the water with PEO and additional water-water-inter-
actions enhance the soft segment mobility. This would further
decrease the glass transition temperature. In the block PEO/PPO/
PEO polymer the competing interactions appear to balance such as
to be almost accounted for by free volume theory.

REFERENCES

1. Schneider, N. S., Dusablon, L. V., Snell, E. W., and Prosser, R. A., J. Macromol. Sci-Phys. <u>B3</u> 623 (1969).

2. Illinger, J. L., Schneider, N. S., and Karasz, F. E., in "Permeability of Plastic Films to Gases, Vapors and Liquids", Hopfenberg, H. B., Ed., Plenum Press., New York, p 183 (1975). Also referenced as Polymer Sci & Tech <u>6</u>, 183 (1975).

3. Kunz, I. Q., and Kauzmann, W., Advances in Protein Chemistry, <u>28</u>,,239 (1974).

4. Illinger, J. L., Schneider, N. S., and Karasz, F. E., submitted to Macromolecules (1977).

5. Schneider, N. S.,Desper, C. R., Illinger, J. L., King, A. O., and Barr, D. B., J. Macromol., Sci.-Phys., <u>B11</u> 527 (1975).

6. Sung, C. P. K., and Schneider, N. S., Macromolecules, <u>8</u> 68 (1975).

7. Liu, K. J., and Parsons, J. L., Macromolecules <u>2</u>, 529, (1969).

8. Kelley, F. N. and Bueche, F., J. Polym. Sci. <u>50</u>, 549 (1961). See also development and references from Meares, "Polymer Structure and Bulk Properties", Van Nostrand Company (1965) Chapter 10.

9. CRC Handbook of Chemistry and Physics, 42nd Edition, (1960) p 2248.

10. Rasmussen, D. H., and MacKenzie, A. P., J. Phys. Chem <u>75</u> 967 (1971).

SOME RECENT DEVELOPMENTS IN RUBBER MODIFIED POLYMERS. INFLUENCE

OF BLOCK AND GRAFT COPOLYMERS ON MORPHOLOGY AND PROPERTIES

G. Riess, S. Marti, J.L. Refregier and M. Schlienger

Ecole Supérieure de Chimie de Mulhouse

3, rue A. Werner, 68093 MULHOUSE Cédex (France)

Polystyrene and similar vinyl polymers, which are brittle materials, are modified with rubber in order to improve their impact strength. Such two phase materials, like HIPS or ABS resins, are generally prepared by polymerization of styrene, or a mixture of styrene-acrylonitrile, in the presence of an elastomer such as polybutadiene (PBut).

The graft copolymer formed in this reaction, acting as an oil in oil emulsifier, has an important influence on the morphology and especially on the particle size of the dispersed phase. Furthermore, by promoting the adhesion between the rubber and the resin phase, it influences directly the mechanical properties of such 2 phase materials (1, 2). A similar effect can also be obtained with block copolymers like poly(styrene-b-butadiene) or poly(styrene-b-isoprene) (3).

In order to specify the influence of the elastomer, as well as that of block and graft copolymers, on the morphology and the mechanical properties, this study has been directed :

- to the formation of *new structure types for the dispersed phase*
- to the use of *rubber latexes in organic solvents* for the preparation of HIPS
- to the formation of graft copolymers based on *stereoblock PBut*

This work has been oriented to approach some actual goals in the domain of rubber modified polymers : improvement of the mechanical properties and the oxidation stability, reduction of the amount of dispersed elastomer, replacement of PBut by saturated elastomers, formation of transparent polyphase materials.

I. CORE-SHELL STRUCTURE

According to **Craig** (4), one can distinguish schematically 3 types of structures for rubber modified polymers (Fig. 1).

Type a is formed by spherical rubber particles dispersed in a resin matrix. This morphology is found in some HIPS and in the case of ABS prepared with rubber latexes.

Type b is a cellular structure where resin occlusions, e.g. PS, can be found inside the rubber phase. This structure of a multiple emulsion is that of most commercial HIPS. It has the advantage to increase the amount of dispersed phase by keeping a low percentage of elastomer.

Type c, the core-shell structure, can be considered as a model system in which the dispersed phase is a resin particle surrounded by a rubber layer. This geometrically more simple structure repre- sents a more practical approach to study the influence of the rubber shell thickness, of the particle size and of the "external" and "internal" adhesion on the mechanical properties (4).

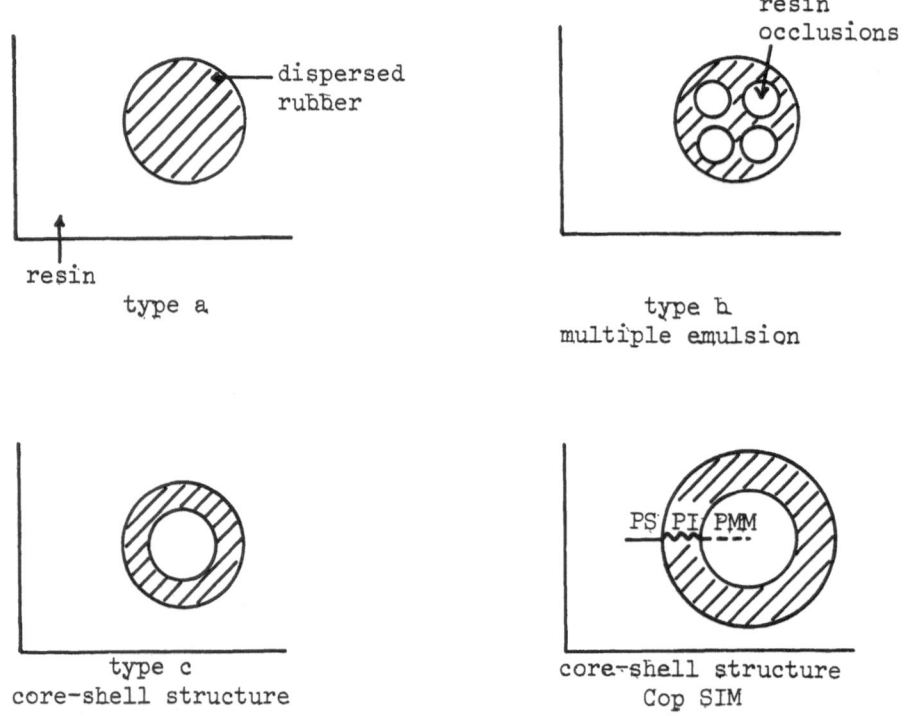

Fig. 1 : Structures of rubber modified resins

Echte (5) has obtained recently such a core-shell structure by polymerization of styrene in presence of PS-PBut block copolymers.

In the present study we have used triblock copolymers containing 3 sequences of different types, as for instance poly(styrene-b-isoprene-b-methylmethacrylate). Such a Cop SIM containing a central polyisoprene block is represented schematically by :

```
————————————wwwwww----------      Cop SIM
    PS          PI       PMM
polystyrene  polyisoprene  polymethylmethacrylate
```

These 3 sequences being incompatible, structures of type c can be obtained either with pure Cop SIM, by adjusting the relative length of the sequences, or with binary blends Cop SIM+PS or PMM, or even with ternary blends Cop SIM+PS+PMM.

1) Preparation of Cop SIM - Emulsifying Properties

Cop SIM are obtained by anionic polymerization in THF and using phenylisopropylpotassium as initiator. Details on synthesis, purification and characterization have been given elsewhere (6). By this technique Cop SIM have been obtained witn various amounts of PS, PI and PMM and with molecular weights ranging from $5 \cdot 10^4$ to $1 \cdot 10^6$.

In a similar manner to our previous results (7), we have shown that Cop SIM are acting as emulsifiers for the system DMF/hexane and cyclohexane-acetonitrile (6).

As for Cop PS-PI and Cop PS-PMM (2, 3), it appeared also that Cop SIM are emulsifiers for binary or ternary blends of the corresponding homopolymers : PS, PI and PMM (6).

2) Polymer Blends based on Cop SIM - Morphology

The morphology of the systems has been examined by electron microscopy. Compression moulded products are ultramicrotomed and the PI phase is selectively stained with OsO_4. The best organization is obtained by slow evaporation of chloroform solutions of Cop SIM, either alone or blended with homopolymers PS, PI, PMM.

For those products containing 3 different phases, one can adjust the relative amounts of PS and PMM in the Cop SIM and/or in the blends, in order to obtain core-shell structures, where PS, respectively PMM, forms the core of the dispersed particles or the matrix (Fig. 1).

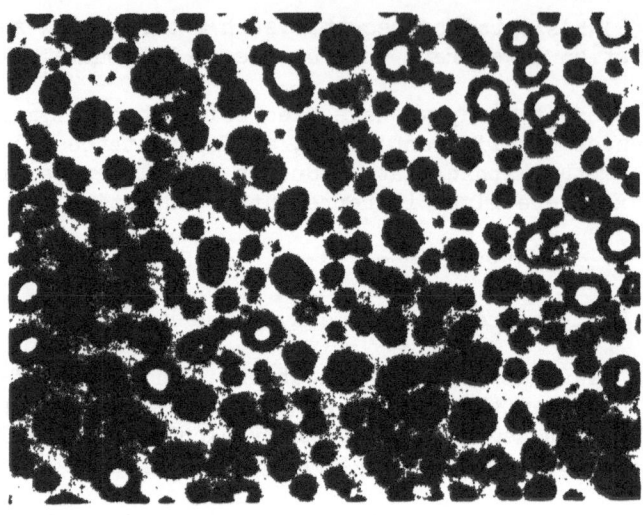

Fig. 2 : Morphology of Cop SIM (n° 8)
         M_n = 150,000
         composition : PS = 72 %
                        PI = 13 %
                        PMM= 15 %
         electron micrography on ultramicrotomed section ;
         magnification 127,500 x

    Fig. 2 shows for instance the core-shell structure obtained
with pure Cop SIM n° 8. For this Cop SIM, containing 72 % PS, PS
forms the continuous phase. The dispersed particles having a diameter of 350-400 Å are formed by a PMM core, surrounded by a PI shell.

    More complex structures (cylinders, lamellas...) have been
obtained by varying the respective sizes of PS, PI and PMM blocks.

    Fig. 3 shows in a similar way core-shell structures obtained
for ternary blends, where Cop SIM is located at the interface of PS
and PMM phase.

                        3) Impact Resistance

    The impact resistance has been determined by the Charpy method on
unnotched compression moulded test bars (50x6x4 mm) for a series of
the previous products. The given impact strength represents therefore only relative values. However, in order to assess the performance of our products, we compared them with 3 commercial HIPS of
varying rubber content.

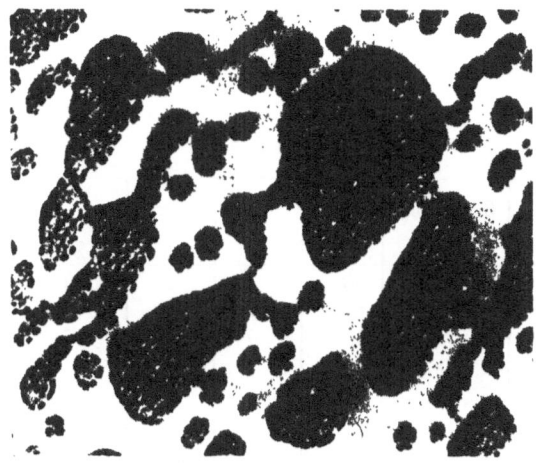

<u>Fig. 3</u> : Morphology of ternary blend
        25 % Cop SIM (n° 12) $\overline{M}_n$ = 330,000
        PS = 27 %   PI = 16 %   PMM = 57 %

        25 % PS $\overline{M}_n$ = 24,000
        50 % PMM $\overline{M}_n$ = 20,000

        electron micrography on ultramicrotomed section ;
        magnification 25,500 x

        The results are given in Table 1. It appears that :

    - the impact strength R increases with increasing amounts of
Cop SIM (samples 1 to 4)
    - at a total amount of elastomer (15.5 % for samples 3 and 7)
R is maximum for binary blends Cop SIM+PMM. In this case all the PI,
coming from the Cop SIM is directly linked to the matrix. This
confirms the importance of the "external" adhesion at the interfaces
on the mechanical properties.

        The comparison of samples 1 and 5 on the one hand (total amount
of PI=3.7 %), and of samples 2 and 6 (total amount of PI=9.25 %) on
the other hand shows that for a same amount of elastomer, binary
blends (Cop SIM+PMM) have better impact properties than ternary
blends (Cop SIM+PMM+PS) for which a part of the occluded PS is not
linked to the rubber phase.

        As postulated by Craig (4), this confirms therefore also the
importance of the "internal" adhesion at the interfaces.

        Although we did not obtain for the moment the performances of commercial HIPS, these model products, with adjusted morphology, have the advantage that the influence of the elastomer on the mechanical properties can be studied, mainly at the interface of 2 resins where the crazes are formed.

Table 1

Sample	Cop SIM %	Homopolymers			PI total %	R
		PMM %	PS %	PI %		
1	10	90	/	/	3.7	1.0
2	25	75	/	/	9.25	2.2
3	42	58	/	/	15.5	3.6
4	100	/	/	/	37	> 5
5	10	60	30	/	3.7	0.9
6	25	50	25	/	9.25	1.6
7	25	68.7	/	6.3	15.5	2.0
8	/	77.5	7	15.5	15.5	0.8
30-0200			98		2 *	5.9
300-0200			97		3 *	6.7
340-0200			94.5		5.5*	8.0

Impact strength of polymer blends based on Cop SIM

Cop SIM n° 16   PS=21 % ; PI=37 % ; PMM=42 % $\overline{M}_n$ 430,000
PMM n° 3 $\overline{M}_n$ = 130,000 (polymer prepared by anionic polymerization)
PS n° 22 $\overline{M}_n$ = 83,000   "   "   "   "   "
PI $\overline{M}_n$ = 100,000   "   "   "   "   "

Samples 30-0200   300-0200   340-0200 are commercial HIPS
* amount of polybutadiene
R = impact strength (Charpy) (relative values)

II. POLYBLENDS BASED ON RUBBER LATEXES IN ORGANIC MEDIUM

In a second approach, we prepared rubber modified polymers of the HIPS type, by using rubber latexes in organic medium, stabilized by block copolymers. This method has the advantage that rubber particles of given size can be prepared. Furthermore, if the Cop A-B (A elastomer sequence, B sequence soluble in the organic solvent), which stabilizes the latex, is well selected by adjusting its composition and molecular weight, it shall be at the interface, e.g. at the surface of the rubber particles.

This latex, which can be crosslinked, can then be blended with various amounts of resin B which is soluble in the organic solvent.

By this procedure, polyblends can be prepared containing rubber types which are difficult to obtain as aqueous latexes, as for instance polyisobutylene or EPDM.

This problem was approached with the model system : polyiso-prene (PI) latex in methylethylketone (MEK) or in dimethylformamide (DMF) stabilized with Cop PS-PI, where the PS sequence is selecti-vely soluble in MEK or DMF. Various amounts of PS homopolymer can be dissolved in this latex.

1) Preparation of Latexes and Polyblends

Cop PS-PI or a blend of Cop PS-PI+PI homopolymer are dissolved in a mixture of THF-MEK or THF-DMF. By distillation of THF, which permits the solution of the PI, one obtains a PI latex in MEK or DMF stabilized by the block copolymer Cop PS-PI.

A systematic study has shown that stable latexes are obtained if the molecular weight of the PI homopolymer is equal or slightly lower than that of the PI sequence of the copolymer. The efficiency of Cop PS-PI for obtaining latexes in MEK is maximum if its molecu-lar weights is higher than 200.000 and if its amount of PI is in the range of 30-70 %.

To this spherical dispersion of PI (1-10 mµ) in MEK, one can add various amounts of PS homopolymer which will form the matrix of the system, after precipitation with methanol.

2) Impact Resistance

These polyblends based on latexes are compression moulded. Their impact resistance is determined as described before. From the results given in Table 2, one can notice :

- that binary or ternary blends containing Cop PS-PI lead to improved impact strength
- that rubber modified resins based on the latex technique give higher values of R than solution blends
- that binary mixtures (Cop PS-PI+PS) are superior to ternary blends (Cop PS-PI+PS+PI)
- that the crosslinking of the PI phase has a favorable influen-ce on R.

III. POLYBLENDS BASED ON STEREOBLOCK POLYBUTADIENES

Classical HIPS is essentially based on polybutadiene (PBut) 1-4 cis imparting its favorable low temperature properties. However the grafting efficiency with PBut 1-4 cis is lower than that of PBut

Table 2

Sample	Method	PS %	Cop %	PI %	R
1	a	100	/	/	1
101	a	83.6	/	16.4	0.7
780	a	81.5	18.5	/	3.8 ± 0.3
780	e (MEK)	81.5	18.5	/	6.9 ± 0.8
781	e (MEK)	82.9	7.1	10	2.1
781	e (DMF)	82.9	7.1	10	2.1
781	f	82.9	7.1	10	4.3

Impact strength of polyblends based on rubber latexes in organic solvents

Total amount of elastomer 16.4 % PI
Cop PS-PI n° 8 $\overline{M}_n$ = 119,000 % PS = 12
PS n° 7 $\overline{M}_n$ = 89,000 (anionic polymerization)
PI n° 1 $\overline{M}_n$ = 30,000   "   "

a) solution blends (dissolution of polymers followed by precipitation in methanol)
e) latex
f) latex (crosslinked) in DMF
R=impact strength (Charpy) (relative values)

vinyl 1-2 (8). Due to the fact that graft copolymer has a great influence on morphology and mechanical properties of HIPS systems, it was interesting to combine the advantages of the 2 rubber types, by using either *rubber blends or stereoblock* PBut :

PBut 1-4 cis PBut vinyl 1-2

A range of such PBut has been prepared according to the method of Antkowiak (9). The new type of HIPS is then obtained by bulk po-lymerization of styrene in the presence of such PBut. After phase inversion, the reaction is continued by means of suspension polyme-rization. A comparative test has been made by using PBut of high cis content (Firestone 55 NFA).

The results are given in Table 3. It appears that by using stereoblock PBut or a blend of this with PBut of high cis content, one obtains an improvement of mechanical properties, particularly of tensile stress limit and of tensile impact energy to break.

In accordance with our previous study (8), we observed also that a smaller particle size (1-5 mμ) for the rubber phase was obtained by using PBut vinyl 1,2 or stereoblock PBut which favors the formation of graft copolymer.

Table 3

Sample	Type PBut	Tensile strength		Impact strength Dynstat kg cm/cm^3	Tensile impact energy kg cm/cm^2
		F kg/cm^2	$\Delta l/l$ %		
10	Firestone 55 NFA	265	48	20	220
12	60 % Firestone 40 % Butalip	267	49	18	278
182	A 24	346	41	19	268
19	85 % Firestone 15 % A 24	349	56	24	268

Firestone 55 NFA   10 % vinyl 1-2 $\overline{M}_n \simeq$ 120,000
Butalip IFP   98 %   "   " $\overline{M}_n$ = 97,000
A 24   29 %   "   " $\overline{M}_n$ = 70,000
1st sequence $\overline{M}_n$ = 45,000 with 8 % vinyl 1-2 content

Total amount of elastomer in PS : 7 %

AKNOWLEDGEMENT

The authors are graciously indebted to DGRST (contracts n° 73.7.1601 and 74.7.0318) and to the "Société Nationale des Pétroles d'Aquitaine" for supporting this work. They wish to thank especially MM. J. Ville and Ph. Hubin.

REFERENCES

1) G.E. Molau and H. Keskkula, J. Polymer Sci. (A 1) 4, 1595 (1966)
2) J. Periard, A. Banderet and G. Riess, Angew. Makromol. Chem. 15, 55 (1971)
3) J. Kohler, G. Riess and A. Banderet, Europ. Polymer J. 4, 187 (1968)
4) T.O. Craig, J. Polymer Sci.(Polymer Chem. Ed.) 12, 2105 (1974)
5) A. Echte, Conf. GDCh Bad Nauheim 29/3/1976
   Angew. Makromol. Chem. under press
6) M. Schlienger, Thesis University Mulhouse 3/3/1976
7) J. Periard and G. Riess, Kolloid Z.-Z. Polymere 251, 97 (1973)
8) J.L. Refregier, Thesis University Mulhouse 1975
9) T.A. Antkowiak, A.E. Oberster, A.F. Halasa and J. Tate, J. Polymer Sci. A 1 10, 1319 (1972)

GRAFTING KINETICS IN THE CASE OF RUBBER MODIFIED POLYMERS.

ABS AND HIPS SYSTEMS

G. Riess, C. Beslin, J.L. Locatelli and J.L. Refregier

Ecole Supérieure de Chimie de Mulhouse

3, rue A. Werner, 68093 MULHOUSE Cédex (France)

High impact polystyrenes (HIPS) and ABS resins (acrylonitrile, butadiene, styrene) are two phase systems combining rigidity and impact strength. HIPS and ABS are generally obtained by polymerization of styrene, or of a styrene-acrylonitrile mixture, in the presence of polybutadiene (PBut). In this reaction, poly(butadiene-g-styrene) graft copolymer, having a PBut backbone and polystyrene grafts (PS_g), is formed simultaneously with non grafted polystyrene (PS_1). These graft copolymers, acting as oil in oil emulsifiers have an important influence on the morphology and on the mechanical properties of such products (1-3).

Grafting reactions for vinyl monomer-elastomer systems have been studied by different authors (4-8). Some contradictions may appear mostly due to the difficulty in separating quantitatively the graft copolymers from the reaction mixture. Furthermore, only grafting on PBut with high 1-4 content has been studied extensively.

Thus it was interesting to perform a systematic study for the grafting on PBut with *various vinyl 1-2 contents*. In order to complete our previous studies on ABS (9-11), we have examined for the HIPS system the influence of the main parameters (initiator concentration, PBut microstructure, molecular weight and concentration, temperature, and conversion) on the characteristics of the grafting reaction (rate of polymerization, grafting efficiency, molecular weight of PS_1 and PS_g). In relation to these parameters, we have examined the reaction medium with respect to the volume fractions and the distribution of the initiator in such two phase systems.

I. EXPERIMENTAL

PBut was prepared by anionic polymerization, according to Antkowiak (12). By this method, products of low polydispersity with vinyl 1-2 contents ranging from 0-80 % could be obtained. Two commercial products : *Cariflex BR 1202 (Shell)* and *Diene 35 NFA (Firestone)*, having respectively 1 % and 10 % vinyl 1-2 content, were also used in this grafting series. The microstructures of PBut were determined by IR and NMR analysis. Their molecular weights were obtained by osmometry and by viscosimetry.

The polymerization of styrene in the presence of PBut was initiated with Bz_2O_2 and was carried out either in a reactor or in a dilatometer. The graft copolymer was isolated from the reaction mixture by the *"reversible gel"* technique (13) or by extraction with a mixture of cyclohexane-acetone (14).

The *grafting efficiency* is given by :

$$r = \frac{PS_g}{PS_1} = \frac{\text{amount of grafted PS}}{\text{amount of non grafted PS}}$$

and the *grafting degree* by $\quad d = \dfrac{PS_g}{PBut}$

After selective degradation of the PBut part of the graft copolymer (14), the average molecular weights and the molecular weight distribution curves of PS_g and PS_1 were determined by viscosimetry and GPC.

II. RESULTS

1) Rate of Polymerization (Rp)

Fig. 1 shows the variation of Rp for such a grafting reaction as a function of the PBut concentration. In order to take into account the variations of the initiator and monomer concentrations with increasing PBut amounts the ratio Rp/Rp_{oc} has been plotted, where Rp_{oc} represents the rate of the corresponding homopolymerization. A *retarding effect*, e.g. a decrease of Rp/Rp_{oc} can be noticed when the PBut concentration reaches a value of about 5 %. A similar decrease of Rp/Rp_{oc} has also been observed by increasing the molecular weight of PBut. This variation as a function of the vinyl 1-2 content of the PBut is shown in Fig. 2. Except for Cariflex (1 % vinyl 1-2), a linear relationship is obtained showing that Rp increases with the vinyl 1-2 content of the PBut. A compensation of the retarding effect can thus be obtained.

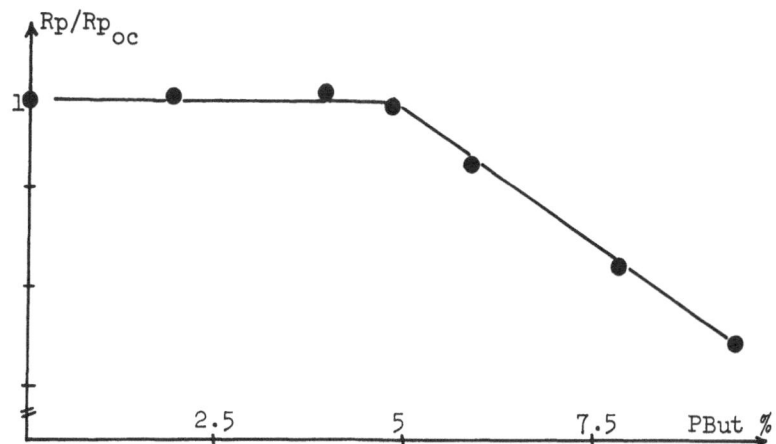

Rp/Rp$_{oc}$ versus PBut concentration (weight %/styrene)

Fig. 1 : PBut : Firestone 35 NFA
10 % vinyl 1-2   35 % 1-4 cis $\overline{M}_n$ 120,000
(Bz_2O_2) = 0.275 weight %
temperature : 70°C   reaction time : 225 minutes

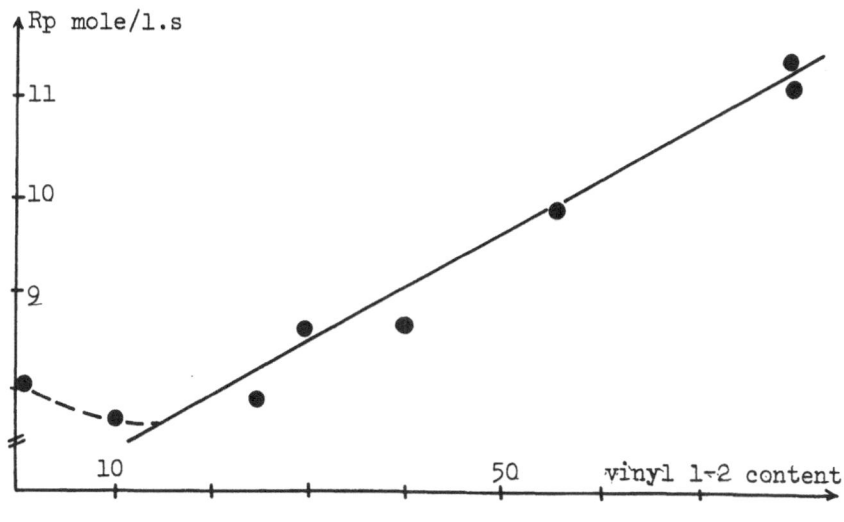

Rp versus vinyl 1-2 content of PBut

Fig. 2 : PBut conc. 8.2 %/styrene
otherwise same conditions as for Fig. 1
Rp$_{oc}$ = 9.352 . 10^{-2} mole/l.s

According to our previous results (9-11), and to those of Rosen et al (15) and Matzuzaki (16), these variations of Rp are the consequence of :

- a reduced efficiency of the initiator in the PBut phase (formation of more stable macroradicals and cage effect resulting from an increased viscosity in this medium)

- a non homogeneous distribution of the peroxide (*preferential solvation effect*)

- a variation of k_t, the termination rate constant of the polymerization.

2) Determination of the Amount of Graft Copolymer

The grafting efficiency as a function of the PBut microstructure is given by Table 1. More especially, it appears that the amount of PS_g increases with the vinyl 1-2 content of the PBut, whereas the amount of PS_1 remains constant in the limits of experimental accuracy. The increase of Rp with vinyl 1-2 content shown previously, can therefore be attributed to the increasing amounts of PS_g.

Table 1

N°	% structure vinyl 1-2	Grafting degree d	Grafting efficiency r	Conversion PS_1 %	PS_g %	Total conversion %
M 4	10	0.36	0.28	10.5	3.0	13.5
M 12	30	0.58	0.45	10.5	4.7	15.2
M 5	42	0.61	0.49	10.2	5.0	15.2
M 13	55	0.94	0.79	9.7	7.7	17.4
M 15	80	1.27	1.13	9.2	10.4	19.6
M 6	80	1.18	0.94	10.3	9.7	20.0

Influence of PBut microstructure on grafting efficiency and conversion
Conditions : PBut : 8.2 % with respect to styrene
        styrene : $(M_o) = 7.735$ mole/l
$(Bz_2O_2) = 9.14 \cdot 10^{-3}$ mole/l (0.275 %/styrene)
Temperature : 70°C   Reaction time : 225 min.

Fig. 3 shows the variation of the grafting efficiency versus the conversion of the styrene polymerization.

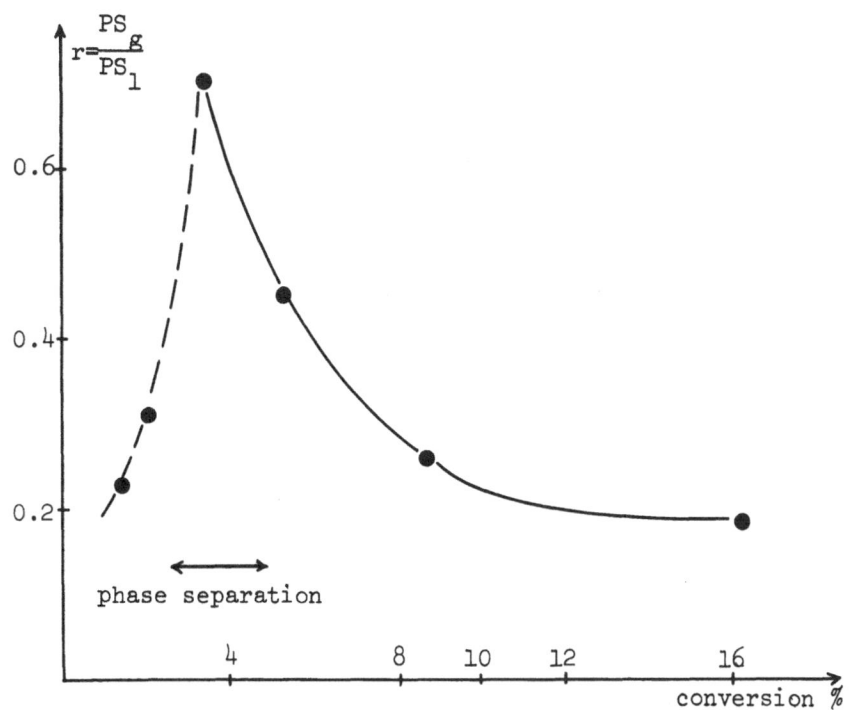

<u>Fig. 3</u> : Grafting efficiency versus conversion
5 % PBut/styrene
PBut : Firestone 35 NFA
(Bz_2O_2) = 0.275 weight %
temperature = 70°C

3) Molecular Weights of PS_1 and PS_g

Table 2 shows the evolution of the molecular weights $(\overline{M}_n, \overline{M}_w)$ for PS_g and PS_1 versus the vinyl 1-2 content of the rubber used in the grafting reaction. One can notice :

- a molecular weight systematically higher for PS_g than for PS_1

- a slight decrease of $\overline{M}_n$ and $\overline{M}_w$, as much for PS_g as for PS_1, with increasing vinyl 1-2 contents of the rubber.

Furthermore it appeared that the ratio

$\overline{M}_n$ of PS_g / $\overline{M}_n$ of PS_1

increases with increasing concentration of PBut and with its molecular weight.

Table 2

N°	% structure vinyl 1-2	PS$_1$		PS$_g$		$\dfrac{M_n\ (PS_g)}{M_n\ (PS_1)}$
		M_n x10^{-3}	M_w x10^{-3}	M_n x10^{-3}	M_w x10^{-3}	
M 4	10	77	172	126	300	1.6
M 12	25	79	162	124	293	1.6
M 13	55	72	148	120	288	1.7
M 15	80	71	152	116	270	1.6

Molecular weights of grafted and non grafted PS (same conditions as table 1)

Finally, we observed that the molecular weights of PS$_1$ are lower than those of homopolystyrene obtained under the same conditions but in the absence of PBut. This result is in agreement with those obtained by Ludwico and Rosen (17). Since the solution of PBut in styrene is a "poorer" solvent than styrene for the growing PS chains, a decrease of the radius of gyration results for these coils thus improving their diffusion and leading to an increase of the chain termination rate constant k_t.

The characteristic values of these grafting reactions are summarized in table 3.

4) Characteristics of the Reaction Medium

The previous results obtained for Rp, the grafting efficiency and for the difference in molecular weight between PS$_1$ and PS$_g$ can be attributed to the fact that phase separation occurs after a given conversion. The reaction proceeds afterwards in 2 media having different characteristics (respective volume fractions, viscosity and distribution of the initiator) (9-11).

For instance, by simulation of a polymerization, e.g. by adding increasing amounts of PS to a PBut solution in styrene containing Bz$_2$O$_2$, one can determine the distribution of the peroxide in the 2 phases. A *surconcentration ratio* can be defined as

$$S = \frac{\text{concentration of Bz}_2\text{O}_2 \text{ in PBut phase}}{\text{concentration of Bz}_2\text{O}_2 \text{ in PS phase}}$$

Table 3

	PBut conc. ↗	Vinyl 1-2 content ↗	PBut mol. weight ↗	Conversion degree ↗	Initiator conc. ↗
Polymerization rate Rp	↘	↗	↘	→ (a)	↗
Grafting efficiency r	↗	↗	↗	∧	--→
$\overline{M}\,PS_1$	--→	↘	↘	→	↘
$\overline{M}\,PS_g$	↗	↘	↗	↗ (step)	↘
V_1/V	↗	--→	↗	↗→	→
S	↗	--→	--→	↗	↘

Summary of the grafting characteristics

V_1/V = volume ratio of PBut phase to total volume V

S = peroxide surconcentration

↗ increasing tendency   ↘ decreasing tendency

→ constant values   --→ no significant change in the examined range

(a) constant values by taking into account the variations of monomer and initiator concentrations

The variation of S versus the PS content, simulating the
conversion degree, is given by Fig. 4. One can notice, that for
these experiments run at room temperature, the concentration of
Bz_2O_2 in the PBut phase is always higher than that in the PS phase
and S varies linearly with the PS content after phase separation.

This preferential solvation effect of the peroxide is therefore similar to that observed in the case of ABS (9-11).

5) Kinetic Scheme

A kinetic scheme can be proposed, taking into account that the
polymerization occurs in 2 different phases, whose characteristics
(viscosity, difference in peroxide concentration) induce modifications of the kinetic constants, especially of the initiator efficiency factor f and the chain termination rate constant k_t (18).

Further, by a stochastic approach of the grafting reaction,
it was possible, by using the *method of Markov chains*, to establish

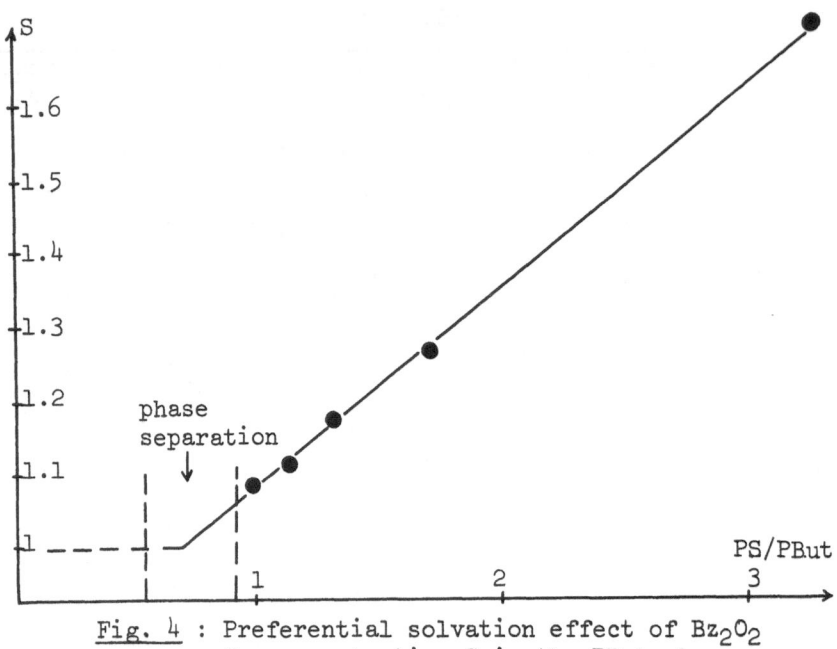

<u>Fig. 4</u> : Preferential solvation effect of Bz_2O_2
Surconcentration S in the PBut phase
Temperature : 25°C

the theoretical distribution curves for the molecular weights of
PS_1, PS_g and $PS_{total}=PS_1+PS_g$. By comparing these model curves to
the experimental distributions obtained by GPC and by using a
suitable non linear regression programme, one can calculate the
parameters of the model (19).

In a first step, in which the preferential solvation effect
of the peroxide has not been considered, we determined by this
Markov chain process the mean values for k_t/k_p^2 at different
temperatures for the homopolymerization of styrene and its polymerization in presence of PBut. In this last case we have considered
the polymerization in the PBut phase and in the medium rich in PS.
The corresponding results are given in table 4.

According to Ludwico and Rosen (17), this table shows an
increase of k_t/k_p^2 for the grafting reaction with respect to the
homopolymerization of styrene. One can also notice a systematic
decrease of k_t/k_p^2 for the polymerization in the PBut phase with
respect to that in the PS phase, which may be attributed to the
higher viscosity of the PBut phase leading to a Trommsdorff effect
(decrease of k_t).

Table 4

| Temp. | Styrene homopolyme-rization | Grafting | |
		PS phase	PBut phase
55°	1,200 ± 50	1,720 ± 30	1,470 ± 30
65°	660 ± 40	850 ± 20	700 ± 40
75°	330 ± 20	420 ± 20	400 ± 20

Kinetic rate constants for PS homopolymerization and grafting on PBut.
Variation of k_t/k_p^2 versus temperature.
5 % PBut Cariflex 1202 ; total conversion = 10 %

CONCLUSION

The particular behavior of these grafting reactions, especial-ly the retarding effect and the difference of PS_g and PS_1 molecular weights, can be attributed to the preferential solvation effect of the peroxide and to the fact that the polymerization occurs in a 2 phase system, leading to modifications of f and k_t.

ACKNOWLEDGMENTS

The authors are graciously indebted to "Société Nationale des Pétroles d'Aquitaine" for supporting this work. They wish to thank especially MM. J. Ville and B. Pflugfelder.

REFERENCES

1) G.E. Molau and H. Keskkula, J. Polymer Sci. (A 1) 4, 1595 (1966)
2) J. Kohler, G. Riess and A. Banderet, Europ.Polym.J. 4, 187 (1968)
3) J. Périard, A. Banderet and G. Riess, Polymer Letters 8, 109 (1970)
4) P. Manaresi, V. Passalacqua and T. Simonazzi, La Chimica e l'In-dustria 51 (4), 351 (1969) ; 52 (3), 234 (1970)
5) A. Brydon, G.M. Burnett and G.G. Cameron, J. Polymer Sci., Poly-mer Chem. Ed. 11, 3255 (1973) ; 12, 1011 (1974)
6) J.P. Fischer, Angew. Makromol. Chem. 33, 35 (1973)
7) Y. Minoura, Y. Mori and M. Imoto, Makromol. Chem. 24, 205 (1957); 25, 1 (1957)
8) J.A. Blanchette and L. Nielsen, J. Polymer Sci. 10, 317 (1956)
9) J.L. Locatelli and G. Riess, Makromol. Chem. 175, 3523 (1974)
10) J.L. Locatelli and G. Riess, Angew. Makromol. Chem. 35, 57 (1974)
11) J.L. Locatelli, Thesis University Mulhouse (1973)

12) T.A. Antkowiak, A.E. Oberster, A.F. Halasa and J. Tate, J. Po-
    lymer Sci. (A 1), 10, 1319 (1972)

13) M.F. Llauro-Darricades, A. Banderet and G. Riess, Makromol.
    Chem. 174, 105 (1973) ; 174, 117 (1973)

14) Ph. Hubin-Eschger, Angew. Makromol. Chem. 26, 107 (1972)

15) W.A. Ludwico and S.L. Rosen, J. Appl. Polymer Sci. 19, 757
    (1975)

16) K. Matzuzaki and S. Nakamura, J. Appl. Polymer Sci. 16, 1339
    (1972)

17) W.A. Ludwico and S.L. Rosen, J. Polymer Sci. 14, 2121 (1976)

18) J.L. Refregier, Thesis University Mulhouse (1975)

19) C. Beslin, Thesis University Mulhouse (1975)

THERMAL GRAFT COPOLYMERIZATION OF STYRENE WITH RUBBER I. KINETICS OF GRAFT COPOLYMERIZATION OF STYRENE WITH POLYBUTADIENE

V. D. Yenalyev, N. A. Noskova, V. I. Melnichenko

O. P. Bovkunenko, O. M. Mezentseva

Donetsk State University, Donetsk, 340055 (U.S.S.R.)

The changing of the overall rate of styrene polymerization and its constituents - homopolymerization rate and styrene grafting to the rubber in the interval of rubber concentration from 2 to 25% and in the temperature range 110-140°C was studied. Rubber works as a polymerization retarding agent when its concentration is at least 6% and with styrene conversion up to 15-20%. Specific degree of grafting polystyrene to rubber related to 1 weight part of rubber in isothermal copolymerization is shown not to depend on rubber concentration. Its limiting value at the investigated temperatures equals 200-220%. The specific degree of grafting can be increased at the stepped temperature-time regimes.

THERMAL GRAFT COPOLYMERIZATION OF STYRENE WITH RUBBER. II. TOPOCHE-

MISTRY OF STYRENE COPOLYMERIZATION WITH POLYBUTADIENE IN BULK AND

ITS CONNECTION WITH HIGH IMPACT POLYSTYRENE MORPHOLOGY

V.D. Yenalyev, N.A. Noskova, V.I. Melnichenko

Donetsk State University, Donetsk, 340055 (U.S.S.R.)

Thermal styrene copolymerization reactions with rubber which took place in the rubber phase were investigated with regard to irregularities of component distribution between phases and continuous change of ratio of their volumes up to the process end. The polymerization rate in terms of rubber phase was found to increase; at styrene conversions of 30-35%, the polymerization rates in both phases were getting equal because of the growth in the rubber phase including both grafting and homopolymerization rate.

Possible diffusion directions of monomer and homopolystyrene from one phase of polymer-polymer emulsion - to another, depending on concrete conditions of polymerization, were studied. The mechanism of rubber particles formation was suggested.

THERMAL GRAFT COPOLYMERIZATION OF STYRENE WITH RUBBER. III. TEM-

PERATURE REGIME INFLUENCE OF STYRENE COPOLYMERIZATION WITH POLY-

BUTADIENE ON HIGH IMPACT POLYSTYRENE MORPHOLOGY

V.D. Yenalyev, N.A. Noskova, S. Hohne

Donetsk State University, Donetsk, 340055 (U.S.S.R.)

The dependence of the change of morphology of HIPS obtained
by means of styrene copolymerization with rubber on the character
of the temperature-time regime was investigated. The relative
volume of the rubber phase was found to increase when the rubber
concentration increased. However the falling of the specific
volume of the rubber phase to 1% of rubber was being diminished.
The specific volume of the rubber phase was lowered with increase
of polymerization temperature as well. These changes are caused
by homopolystyrene diffusion from the rubber phase to polystyrene
as a result of polymer incompatability which is influenced by
polymer molecular weight and medium viscosity. It was shown that
one can change the rubber particle size also after prepolymerization
up to high conversions of about 80-85% due to the stepped temperature-time regimes.

THERMAL GRAFT COPOLYMERIZATION OF STYRENE WITH RUBBER. IV. THE

PECULIARITIES OF THE PROCESS OF PRODUCING HIGH IMPACT POLYSTYRENE

BY MEANS OF CONTINUOUS METHODS.

V. D. Yenalyev, N.A. Noskova, B.V. Kravchenko,
V. A. Gasin

Donetsk State University, Donetsk, 340055 (U.S.S.R.)

Realization of styrene copolymerization with rubber by a continuous method brings diminishing of the rubber efficiency. The
reason for it is deterioration of the product morphology - reducing
of specific volume of the rubber phase, lowering of number of polystyrene occlusions in rubber particles, and increase of dispersion
of rubber particles according to sizes. Change of morphology of
HIPS continuously is caused by introduction of fresh rubber solution into the concentrated polymer to which there are no phase inversion possible: in the lesser volume the degree of grafting of
polystyrene to rubber takes place. Diminishing of the content of
graft copolymer deteriorates conditions of occlusions forming in
the course of the process and the lesser degree of crosslinking of
rubber deteriorates the strength of rubber particles and diminishes
residual internal microtensions in the obtained product.

GRAFTING ISOBUTYLENE FROM SBR

James Oziomek* and Joseph P. Kennedy

Institute of Polymer Science, University

of Akron, Akron, Ohio 44325

INTRODUCTION

Lack of efficient, easily controllable cationic grafting techniques have, until recently, precluded the use of this technique for the preparation of well defined graft copolymers (1). The "grafting onto" approach has been used by Haas and coworkers (1). These authors alkylated poly(p-methoxystyrene) with growing styrene chains using BF_3 and $SnCl_4$. Though copolymer was obtained with polystyrene, no graft formed with vinyl butyl ether, vinyl ethyl ether, or isobutylene. Poly (2,6-dimethoxystyrene-g-styrene) was obtained by Overberger and Burns (2) using $SnCl_4$ in a similar process. The large amounts of homopolymer formed in these experiments indicate that the "grafting onto" technique is inefficient.

The "grafting from" approach has also been used by a number of workers. Kockelberg and Smets (3) obtained low yields of graft copolymer by reacting chloromethylated polystyrene with $AlBr_3$ in the presence of isobutylene. Minoura and coworkers (4) used the chlorobutyl rubber/ $SnCl_4$ system for the graft polymerization of styrene. Optimum grafting efficiency was only 50%. Plesch (5) demonstrated the presence of graft copolymer with solubility data when poly(vinyl chloride) or poly(vinyl

*Visiting Scientist on leave of absence from The Central Research Laboratories, The Firestone Tire and Rubber Company, Akron, Ohio 44317.

chloride-co-vinylidene chloride) with AlCl$_3$ or TiCl$_4$ are
used to initiate the polymerization of styrene, indene,
indole, and trans-stilbene. However, severe degradation
of the PVC was an unavoidable side reaction. Solomon
and coworkers (6) grafted N-vinyl carbazole from a PVC
backbone using AlCl$_3$ with 44% grafting efficiency.

Recently, fundamental studies (7-16) have enhanced
our understanding of initiation in cationic olefin
polymerization. Through these studies efficient cationic
graft copolymerization processes based on macromolecular
halide initiators and alkylaluminum compound coinitiators
have evolved. These include the efficient preparation of
EPM-g-polystyrene (17), butyl rubber-g-polystyrene (18)
and PVC-g-polyisobutylene (18).

Most previous reports of cationic grafting processes
have dealt with grafting from or onto backbones with but
a few mole percent unsaturation. In the course of our
fundamental studies on cationic grafting we became
interested in exploring the possibility of using highly
unsaturated hydrocarbon polymers as backbones. Indeed
the only previous reports of cationic grafting utilizing
unsaturated backbones concern systems wherein the effects
of crosslinking in highly unsaturated systems are minimized, i.e., grafting of isobutylene onto low molecular
weight polydienes (19) and grafting from Neoprene
rubber (20). In the former case gelation can occur only
as a result of extensive crosslinking while in the latter
system the rate of crosslinking is retarded by the presence of the electronegative chlorine on the double bond.

This paper concerns research in which a commercially
available random butadiene/styrene copolymer, poly(butadiene-co-styrene), SBR for brevity, was used as graft
backbone and polyisobutylene, PIB, as branches. More in
particular, this paper describes the synthesis of
poly⟨(butadiene-co-styrene)-g-isobutylene⟩ (SBR-g-PIB),
including kinetic studies and investigations on the
effect of various experimental variables (temperature,
monomer concentration, solvent composition) on the
grafting reaction, the separation of the pure graft by
solvent elution techniques, and the characterization of
the products by molecular weight (osmometry, viscometry),
homogeneity (GPC), and solubility.

Technological significance and experimental convenience led us to select SBR as graft backbone and PIB as
branch. From the technological point of view the examination of the characteristics and physical properties of

a graft copolymer comprising SBR and PIB, both important commercial commodities (SBR the most important synthetic rubber and PIB another significant commercial product on its own right and with isoprene comonomer closely related to butyl rubber), promised to be of great interest. From the experimental point of view, the solubility character- istics of SBR and PIB seemed to be sufficiently different to allow product separation which is of paramount impor- tance for characterization research, and the presence of aromatic moieties in the graft was thought to be con- venient to establish backbone distribution in the reaction mixture by a combination of UV/GPC technique.

EXPERIMENTAL

Materials

Commercially available poly(butadiene-co-styrene) rubber, Stereon 702 of The Firestone Tire and Rubber Company, was used. Isobutylene and ethyl chloride (Linde Co.) were dried by passing the gases through 7 feet columns filled with Drierite and a mixture of BaO and molecular sieves (Linde Co., 4Å, powder). Diethyl- aluminum chloride (Texas Alkyl Co.) was stirred over dry NaCl at $80^{\circ}C$ for 2 hrs and distilled at 10 mm Hg. The liquid was stored over NaCl at -78°.

Various samples of diethylaluminum chloride were used over the course of this study. Though the individual samples gave reproducible results in terms of reaction times, freshly distilled samples appeared to give faster reactions, indicating a deterioration of the sample. This deterioration during storage was also evident by increases in viscosity and turbity of old samples. Very viscous and turbid samples appeared to be totally inactive.

n-Heptane (reagent grade (Fischer Scientific Co.) was refluxed overnight over CaH_2 and fractionated under nitrogen.

The macromolecular initiator was prepared by the allylic chlorination (21) of SBR. Thus 100 g Stereon 702 was dissolved in 2.5 l nitrogen-purged n-heptane. To the vigorously stirred solution kept under a nitrogen blanket were added dropwise over 30 minutes 18 gms t-butylhypo- chlorite dissolved in 250 mls heptane. When the addition was complete, the reaction mixture was irradiated with a sunlamp for an additional 30 minutes. The polymer was then filtered through glass wool and coagulated in acetone.

It was dried for three days in vacuo at ambient temperature. The chlorinated SBR was characterized by infrared,
GPC, viscometry and Cl (turbidimetric) analysis and the
following data were obtained: composition (by infrared,
wt. %): cis-1,4 = 26.9, trans-1,4 = 44.2, 1,2 = 7.6,
total butadiene = 78.7, total styrene = 21.4; $\overline{M}_n$ (by GPC)
84,900, $\overline{M}_w$ (by GPC) = 239,400, $\overline{M}_w/\overline{M}_n$ = 2.82; styrene
content (by GPC, wt. %) = 22.6; intrinsic viscosity
(dl/g) = 1.82; chlorine content (wt. %) = 0.83.

Stock solutions of 5% chlorinated SBR (Cl-SBR) in
n-heptane were subsequently used to initiate graft
copolymerizations. The stock solutions were prepared by
placing 25 g Cl-SBR in an oven dried, cooled 28 oz
beverage bottle. The bottle was capped with a three-
holed crown cap with a hexane-extracted rubber liner.
The bottle was then thoroughly purged with nitrogen and
n-heptane was fractionated as previously described
directly into the bottle using syringe needle adapters
to a total volume of 500 mls. The bottle was then
pressurized to 10 psi with nitrogen and tumbled on a
rotating stirrer until the polymer dissolved. The polymer
solutions were stored in the dark until needed.

Polymerization

Reactions were conducted in a stainless steel
nitrogen purged enclosure (22) at less than 30 ppm atmos-
pheric moisture. The reaction vessel was a 500 ml, three-
necked flask equipped with a thermometer and mechanical
stirrer. The desired quantity of stock polymer solution
was added to the flask and cooled to -20°C. At this point
additional heptane and the required quantities of lique-
fied ethyl chloride and isobutylene were added. The
contents of the flask were cooled to the desired temperature and the desired quantity of diethylaluminum chloride
dissolved in 20 ml n-heptane was added with vigorous
stirring. The onset of grafting was indicated by an
exotherm. The solution remained clear during reaction
and the viscosity did not appear to increase significantly with reaction time. After a few degrees centigrade
exotherm, the polymerization was terminated by the additioı
of 2-3 ml cold methanol. The reaction vessel was removed
from the enclosure and the polymer coagulated in 2 litres
ethanol containing about 1 gm di-t-butyl-p-cresol. The
polymer was dried in vacuo at ambient temperature.

Gel permeation chromatographs, run on a Waters Associates Model 200 instrument equipped with a UV detector operated at 256 mμ, were used to determine grafting efficiency by styrene distribution. Solutions of 1/16 weight % in THF were run at ambient temperature through a series of Styragel columns with pore sizes of 10^6, 10^5, 10^4, and 10^3 Å. at a flow rate of 1 cc/min.

In some instances increased resolution of GPC chromatographs were obtained on 0.5 weight % THF solutions run on a Waters Associates Ana-Prep Chromatograph operated at 1 ml/min flow rate at 25°C. Grafting efficiencies were determined by partial peak resolution. Styragel columns with pore sizes of 7 x 10^5-5 x 10^6, 1.5-7 x 10^5, 5 x 10^4-1.5 x 10^5, 1.5-5 x 10^4, 5 x 10^3-1.5 x 10^4, 5 x 10^3-1.5 x 10^4, 2 x 10^3-5 x 10^3 Å were used.

Number average molecular weights were determined with a Mechrolab Model 502 membrane osmometer using toluene solvent at 37.5°C.

Intrinsic viscosities were measured using Ubbelohde viscometers and THF solutions at 25°C.

Graft copolymer compositions were determined by ir analysis following a modified method of Binder (23) on a Beckman IR-4 spectrometer using carbon disulfide solutions. The following absorption bands for the various components in the polymers were employed:

Polymer Component	Absorption Band, (mμ)
Butadiene cis-1,4	13.7
trans-1,4	10.35
vinyl	10.98
Styrene	14.3
Isobutylene	7.3

Nuclear magnetic resonance spectra were run on 5% solutions in carbon tetrachloride on a Varian Model T-60 spectrometer.

Grafting Efficiency

Two methods were used to determine grafting efficiencies (G.E.).

A.) <u>G.E. by Styrene Distribution</u>. The aromatic
moiety in SBR can be used to determine G.E. by a combination of GPC and UV spectroscopy. Since the styrene
distribution in SBR is random, a plot of weight fraction
of styrene in the polymer versus the cumulative weight
fraction of the polymer should give a horizontal line for
the backbone polymer. Similarly, for a perfectly random
SBR-g-PIB, the plot of weight fraction of styrene in the
graft versus cumulative weight fraction of the graft,
would also be a horizontal line. In contrast, the GPC/UV
scan of a mixture of SBR-g-PIB plus PIB would not show a
horizontal plot because the styrene distribution would not
be uniform across the fractions. Consequently, it is
possible to determine G.E. from the experimentally obtain-
able styrene distribution plot by the following reasoning:

The weight fraction of styrene in SBR, S_{SBR}, is
proportional to the ratio of the UV absorption coefficient
of styrene, ϵ, and the refractive index difference of the
backbone and solvent Δn_{SBR}:

$$S_{SBR} = k \frac{\epsilon}{\Delta n_{SBR}} \tag{1}$$

The weight fraction of styrene in the SBR-g-PIB is:

$$S_G = k \frac{\epsilon f_{SBR}}{\Delta n_{PIB}(f_{gPIB} + f_{hPIB}) + \Delta n_{SBR} f_{SBR}} \tag{2}$$

$$= \frac{S_{SBR} f_{SBR} \Delta n_{SBR}}{\Delta n_{PIB}(f_{gPIB} + f_{hPIB}) + \Delta n_{SBR} f_{SBR}} \tag{3}$$

where Δn_{SBR} and Δn_{PIB} are the refractive index differences for SBR and solvent, and PIB and solvent, respec-
tively; and f_{SBR}, f_{gPIB}, f_{hPIB} denote weight fractions of
SBR, grafted-PIB, and homo-PIB, respectively. By
definition

$$f_{SBR} + f_{gPIB} + f_{hPIB} = 1 \tag{4}$$

Assuming that at high molecular weights (high
cumulative weight fraction) $f^{\infty}_{hPIB} \to 0$ i.e., neglibible
amounts of homo-PIB are present, in other words
$f_{gPIB} + f_{SBR} = 1$, the backbone content of the SBR-g-PIB

can be calculated from the styrene content:

$$S_G = \frac{S_{SBR} f_{SBR} \Delta n_{SBR}}{\Delta n_{PIB} f_{gPIB} + \Delta n_{SBR} f_{SBR}} \qquad (5)$$

or $\quad S_G = \dfrac{S_{SBR} f_{SBR} \cdot X}{1 + (X-1) f_{SBR}} \qquad (6)$

where $X = \Delta n_{SBR} / \Delta n_{PIB}$. The following refractive indices have been used to determine X: $n_{SBR} = 1.5345$, $n_{PIB} = 1.4070$; $X = n_{SBR} - n_{THF} / n_{PIB} - n_{THF} = 1.25$.

The weight fraction of the backbone at any point is then

$$f_{SBR} = \frac{S_G}{X S_{SBR} - (X-1) S_G} \cdot \qquad (7)$$

Since at high molecular weights $f_{hPIB}^{\infty} \approx 0$

$$f_{gPIB}^{\infty} = 1 - \frac{S_G}{X S_{SBR} - (X-1) S_G} = 1 - f_{SBR}^{\infty} \qquad (8)$$

Since grafting is a random process

$$f_{gPIB} = \frac{f_{gPIB}^{\infty}}{f_{SBR}^{\infty}} f_{SBR} \qquad (9)$$

i.e., the ratio of weight fraction grafted-PIB to weight fraction SBR at any point is invariant. Thus

$$f_{hPIB} = 1 - f_{SBR} - f_{gPIB} \qquad (10)$$

$$f_{hPIB} = 1 - \frac{S_G}{X S_{SBR} - (X-1) S_G} - \frac{f_{hPIB}^{\infty}}{f_{SBR}^{\infty}} \frac{S_G}{X S_{SBR} - (X-1) S_G} \qquad (11)$$

$$= 1 - \left(1 + \frac{f_{gPIB}^{\infty}}{f_{SBR}^{\infty}}\right) \frac{S_G}{XS_{SBR} - (X-1)S_G} \qquad (12)$$

Thus by plotting f_{gPIB}, f_{hPIB} and f_{SBR} versus cumulative weight fraction, the areas under the curves will give cumulative weight fractions of the various components. The cumulative weight fraction of homo-PIB divided by the total cumulative weight fraction of graft-PIB and homo-PIB gives G.E.

The calculation procedure will be illuminated by an example (experiment in which we used 35% EtCl at -50°, cf. Table 1):

At 100% cumulative weight fraction (cf. Figure 1) $S_G^{\infty} = 0.104$. From equation 7 we obtain $f_{SBR}^{\infty} = 0.408$ and, consequently, from equation 8, $f_{gPIB}^{\infty} = 0.592$. At a point a, $S_G = 0.046$ and $f_{SBR} = 0.172$ and we obtain from equation 9, $f_{gPIB} = 0.249$ and, subsequently, from equation 10, $f_{hPIB} = 0.579$. Several points were used to calculate the curves shown in Figure 1. The areas A, B, C, and D in Figure 1, respectively, are directly related to the amount of homo-PIB, grafted-PIB, butadiene in SBR and styrene in SBR and can readily be obtained by integration, for exampl by planimetry. In our example G.E. = 100A/(A+B) = 100 x 95/(95+823) = 89.7%.

B) <u>G.E. by Curve Resolution</u>. If homo-PIB can be satisfactorily separated from SBR-g-PIB and an independent method for the determination of overall graft composition is available, G.E. can be calculated directly from GPC elution volume versus differential refractive index curves Thus

$$A_{hPIB} = f_{hPIB} \cdot \Delta n_{PIB} \cdot w \qquad (13$$

$$A_{gPIB} = f_{gPIB} \cdot \Delta n_{PIB} \cdot w \qquad (14$$

$$A_{SBR} = f_{SBR} \cdot \Delta n_{SBR} \cdot w \qquad (15$$

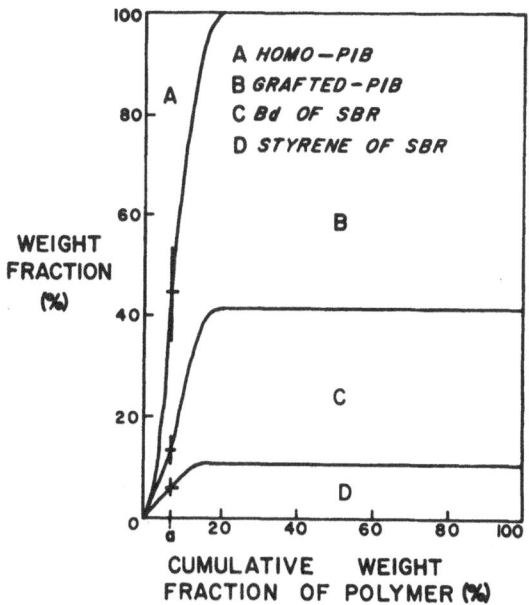

Figure 1.   Polymer Composition as a Function of
            Cumulative Weight Fraction of Polymer

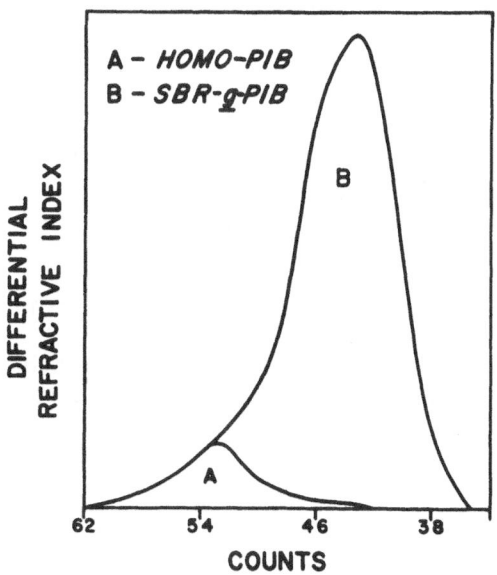

Figure 2.   Partial Resolution GPC Chromatogram

where A_{hPIB}, A_{gPIB} and A_{SBR} are the areas under the curves associated with homo-PIB, grafted-PIB and SBR backbone, respectively, and w is the weight of total polymer sample. By definition:

$$A_T = A_{hPIB} + A_{gPIB} + A_{SBR} \tag{16}$$

Thus

$$G.E. = \frac{A_T - A_{SBR} - A_{hPIB}}{A_T - A_{SBR}} \cdot 100 \tag{17}$$

$$G.E. = \frac{1 - A_{SBR}/A_T - A_{hPIB}/A_T}{1 - A_{SBR}/A_T} \cdot 100 \tag{18}$$

$$\frac{A_{SBR}}{A_T} = \frac{Xf_{SBR}}{f_{gPIB} + f_{hPIB} + Xf_{SBR}} = \frac{Xf_{SBR}}{1 - f_{SBR} + Xf_{SBR}} \tag{19}$$

$$G.E. = \frac{1 - \dfrac{Xf_{SBR}}{1 + (X-1)f_{SBR}} - \dfrac{A_{hPIB}}{A_T}}{1 - \dfrac{Xf_{SBR}}{1 + (X-1)f_{SBR}}} \cdot 100 \tag{20}$$

$$G.E. = \frac{1 - f_{SBR} - \left[1 + (X-1)f_{SBR} \right] \dfrac{A_{hPIB}}{A_T}}{1 - f_{SBR}} \cdot 100 \tag{21}$$

Again an example should illuminate the actual calculation procedure (example using 35% EtCl at $-50°$ in Table I). The area by plainimetry under the homo-PIB curve (Figure 2) was found to be 160 units and the total area was 1542 units Thus from equation 21:

$$G.E. = \frac{1 - 0.46 - (1 + 0.25 \times 0.46)\dfrac{160}{1542}}{0.54} \cdot 100 = 78.6\%$$

C) G.E. by S_2Cl_2 Crosslinking. Grafting efficiencies can also be determined by crosslinking the total polymer with S_2Cl_2 followed by extraction of the essentially unreactive homo-PIB. Thus three grafting experiments were conducted as described below:

	A	B	C
SBR[a] (g)	5	5	4.5
/Polymer-Cl/ x 10^3	4.5	4.5	4.1
Solvent[b]			
n-heptane (mls)		200	
Ethyl Chloride (mls)		60	
/i-C_4H_8/, M		1.67	
/Et_2AlCl/, M		2×10^{-2}	
Temperature, $^\circ$C	$-50 - -35$	$-50 - -45$	-70
% Conversion	100	28	57
% Polyisobutylene[c]	85	61	78

a) 0.96% Cl
b) Total Volume (n-C_7H_{16} + EtCl + i-C_4H_8) = 300 ml
c) % of total mixture

Approximately 2 grams of each polymer mixture (dissolved in 20 mls of hexane) were treated with S_2Cl_2 (0.2 ml) for 24 hours with the following results:

	A	B	C
Sample Weight (g)	2.06	2.08	2.09
Weight after S_2Cl_2	2.40	2.37	2.59
Weight of Sol (PIB by nmr)	.46	0.26	0.16
Weight of Gel	1.86	2.09	2.18
Grafting Efficiency %	74	79	90

However, this method proved tedious and was not used further. But the results did suggest quite high grafting efficiencies were to be expected.

RESULTS AND DISCUSSION

Preliminary Studies

Since the SBR backbone contains olefinic unsaturation and aromatic nuclei, the possibility for carbenium ion attack on these nucleophilic sites during the grafting experiment exists. These grafting-onto reactions may lead to undesirable structures, ill defined networks, etc. An experiment was designed to determine the extent of these potentially disturbing side-reactions and to work out

reaction conditions to avoid them during grafting. To
this end an isobutylene polymerization was carried out in
the presence of SBR under conditions essentially identical
to a grafting experiment except initiation was effected by
a small molecule, allyl chloride, and not by Cl-SBR. Thus
SBR (5 g) was dissolved in a n-heptane-ethyl chloride (110
ml-150 ml) mixture, cooled to -70°, isobutylene (40 ml or
1.6 M) and Et$_2$AlCl (2 x 10^{-2} M, in 20 mls n-heptane) were
added; the polymerization was initiated by the dropwise
addition of allyl chloride (4 x 10^{-3} M). The temperature
rose to -65° within 6 minutes. At this point the reaction
was terminated by the introduction of a few mls of pre-
cooled methanol and subsequent coagulation in excess
methanol yielding 16.7 g of product mixture. The mixture
was extracted with isopentane which removed 69% of the
product. According to nmr spectroscopy this fraction was
pure PIB. GPC showed $\overline{M}_n$ = 77,400 and $\overline{M}_w/\overline{M}_n$ = 2.2.
Selective elution with 3-pentanone (diethyl ketone, DEK)
yielded 32% product; when the DEK soluble fraction was
coagulated into isopentane and the solid analyzed by nmr,
the spectrum indicated less than 3% PIB. In view of this
result, not more than 1% of the PIB formed could be bound
directly to or entangled intimately with the SBR. Grafting
efficiency measured by styrene distribution was 4.3%. An
examination of GPC traces of the SBR before and after
isobutylene polymerization, the SBR/PIB mixture, and PIB
indicated that the SBR remained essentially unchanged
during this experiment. Evidently, then, the grafting
of growing polyisobutylene carbocations onto SBR is a
negligible side reaction under the above conditions.

Detailed Studies

    A. Fractional Extraction Studies. Great efforts
have been made to separate by selective elution mixtures
of SBR, PIB and SBR-g-PIB. While mixtures of SBR and PIB
are readily separable by isopentane (solvent for PIB,
nonsolvent for SBR) and DEK (solvent for SBR, nonsolvent
for PIB), efforts to separate SBR-g-PIB from the respective
polymers by these solvents or solvent mixtures remained
fruitless. Indeed SBR-g-PIB was found to be readily
soluble both in isopentane and DEK. The isopentane solutions were slightly hazy while DEK solutions were more
hazy but polymer did not separate even after 1 hr of
centrifugation. The solubility parameters of SBR and PIB
are quite similar: δ = 8.5 and $\sim$ 8.0 (24).

Fortunately, however, this difficulty of solvent-separating SBR-g-PIB from the component polymers was avoided by developing two convenient GPC/UV methods (cf. Experimental) for the quantitative determination of grafting efficiencies. By the use of these methods, grafting efficiencies can be determined without cumbersome, repeated solvent elutions. Also, as it turned out, reaction conditions could be readily defined under which grafting efficiencies of ∿100% were obtained, i.e., very little if any ungrafted PIB was formed (chain transfer was absent).

B. <u>The Effect of Temperature and Solvent Polarity on Grafting</u>. Table I is a compilation of a large number of grafting experiments designed to show the effect of temperature and solvent polarity on the grafting of iso-butylene from Cl-SBR. The temperature range was from -30 to -70°C whereas the polarity was changing the ethyl chloride/n-heptane volume ratio from 20/80 to 35/65 to 50/50 (cf. columns 1 and 2 of Table I respectively).

All data were obtained on crude, unfractionated products isolated directly from reaction mixtures. Attempts to separate graft from component polymers by selective solvent extraction remained fruitless; for although isopentane and DEK are specific solvents for SBR and PIB, respectively, the products were all completely soluble in isopentane and most were virtually completely soluble in DEK. Evidently, the grafted PIB branches render the graft copolymer soluble in isopentane whereas the SBR backbone makes the graft soluble in DEK.

Grafting efficiencies have been determined by the styrene distribution and curve resolution methods (cf. columns 6 and 7 in Table I). The latter method gave consistently lower values. While the reason for this discrepancy is obscure, it is possible that the sensitivity of these methods is limited by the amount of the UV active styrene in the SBR and the amount of both homo- and grafted PIB.

TABLE I

GRAFTING ISOBUTYLENE FROM SBR

Experimental Conditions[a]			Results			Characterization			
				Grafting Efficiency by				$\overline{M}_n \times 10^{-5}$	
EtCl Vol %	i-C$_4$H$_8$ M	Time[b] Min	Yield g	Sty Dist.	Curve Resol.	% SBR[d]	dl/g	Calc[c]	Osmotic
-30°C									
20	1.6	6[1]	11.8	74.2	51.8	45.9	1.17	--	--
35	"	3[1]	11.0	83.4	58.3	45.5	1.36	1.86	1.10
50	"	.17[3]	29.9	41.5	33.5	25.2	1.03	--	--
50°C									
20	"	15[1]	12.7	87.4	73.8	41.7	1.40	2.03	1.46
35	"	8[1]	11.9	89.7	78.6	46.0	1.29	1.84	1.52
50	"	1.5[2]	20.3	83.5	64.3	31.9	1.44	--	--
70°C									
20	"	8[3]	11.8	100	96.7	45.5	1.62	1.86	1.73
35	"	15[1]	11.9	100	100	44.9	1.47	1.89	2.06
50	"	8[3]	12.3	97.6	100	43.3	1.34	1.96	2.13
50	"	11[3e]	10.6	100	100	48.8	1.45	1.74	1.87
50	"	.84[3f]	18.0	100	89.0	34.5	1.40	2.46	2.16
50	1.1	3[3]	11.5	100	94.7	45.1	1.47	1.88	1.85
50	0.52	13[3]	9.7	91.5	77.9	52.0	1.42	1.63	1.40
50	0.28	17[3]	9.0	90.1	64.2	57.5	1.62	1.47	1.21

a) Total Vol = 300 mls = n-heptane + EtCl + isobutylene; 5 g SBR per charge
   Cl Content of SBR in Solution: 3.9×10^{-3} M. $[Et_2AlCl] = 2 \times 10^{-2}$ M
b) Superscript denotes Et$_2$AlCl sample
c) $\overline{M}_n = \overline{M}_n$ of SBR/(fraction SBR)
d) By ir spectroscopy
e) $[Et_2AlCl] = 1 \times 10^{-2}$ M
f) $[Et_2AlCl] = 3 \times 10^{-2}$ M

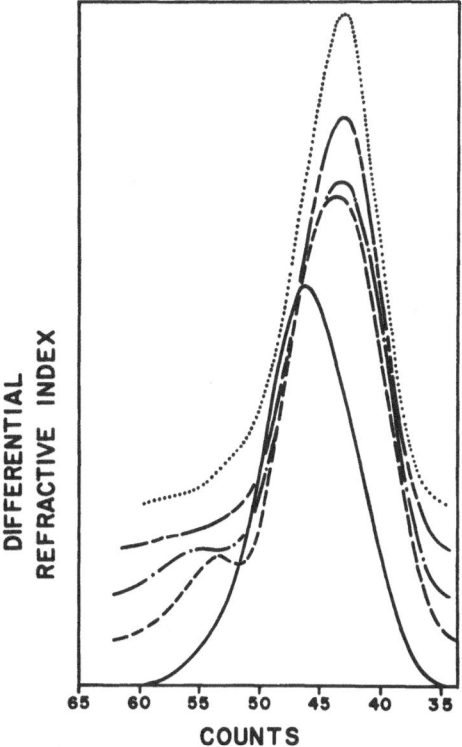

Figure 3.   Effect of Monomer Conversion (Concentration)
            on Grafting Efficiency at -70°C and 50 Vol %
            Ethyl Chloride

	SBR	
. . . .	1.6 M	i-C$_4$H$_8$
— — — —	1.1 M	" $_4$H$_8$
—— · ——	0.52 M	"
— — — — —	0.28 M	"

        Grafting efficiencies (G.E.) increase with decreasing
conversions and decreasing temperatures. By reducing the
monomer concentration, the effect of transfer on grafting
efficiency as a function of conversion is magnified. Such
a picture is presented in Figure 3. Here as initial
monomer concentration is reduced (simulating higher conversion) lower molecular weight polyisobutylene peaks

became apparent. Similarly, lower molecular weight PIB peaks appear as the temperature is raised, Figure 4. Both effects have been observed previously (18) in similar systems, e.g., butyl rubber-g-PSt, poly(vinyl chloride-g-isobutylene). The reasons are always the same: chain transfer to monomer, the most important G.E. determining event, increases with increasing monomer conversions and increasing temperatures. In the present system the effect of temperature on G.E. is particularly striking: At -70 G.E. is ∿ 100% at moderate conversions, i.e., chain transfer is essentially absent.

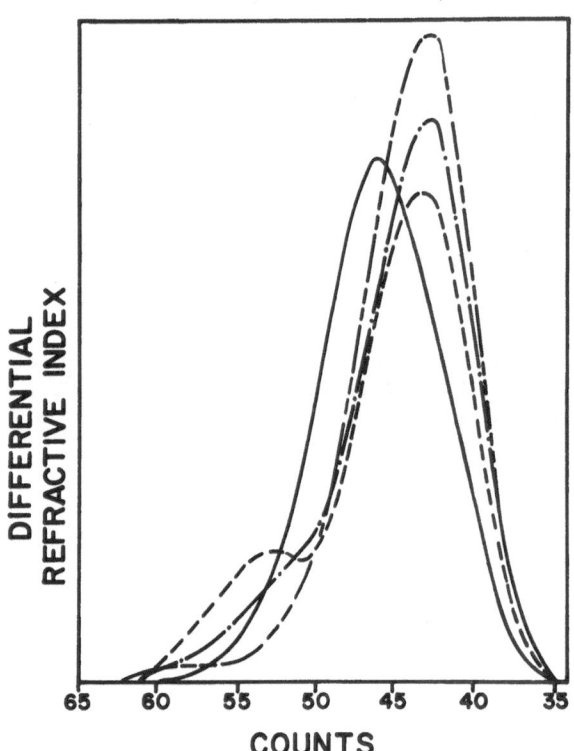

Figure 4.   Effect of Temperature on Grafting Efficiency at 20 Vol % Ethyl Chloride

SBR

-30°C

-50°C

-70°C

The fact that chain transfer is not significant in these experiments is of import not only in the present context of producing pure graft but is also of great significance for the cationic polymerization of olefins in general. Indeed this seems to be the first chemical (nonkinetic) evidence for an isobutylene polymerization in the virtual absence of chain transfer. Kinetic evidence for the virtual absence of chain transfer in cationic isobutylene polymerizations coinitiated by alkylaluminum compounds has already been found by Italian workers (25) and in our laboratories (26).

C. <u>Polymer Characterization</u>. Graft characterization data are also meaningful only with products obtained at 100% G.E. if fractionation is not undertaken. Indeed possibly the strongest evidence for the purity of our grafts and reliability of our analytical methods is the agreement between calculated and experimentally determined $\bar{M}_n$ values (Columns 11 and 12 in Table I). The $\bar{M}_n$ of grafts can readily be calculated from the known $\bar{M}_n$ of the SBR backbone and overall graft compositions available by ir spectroscopy. A plot of reciprocal $\bar{M}_n$ versus percent polyisobutylene for select polymers prepared at -70° and moderate conversion levels further substantiates this point (Figure 5).

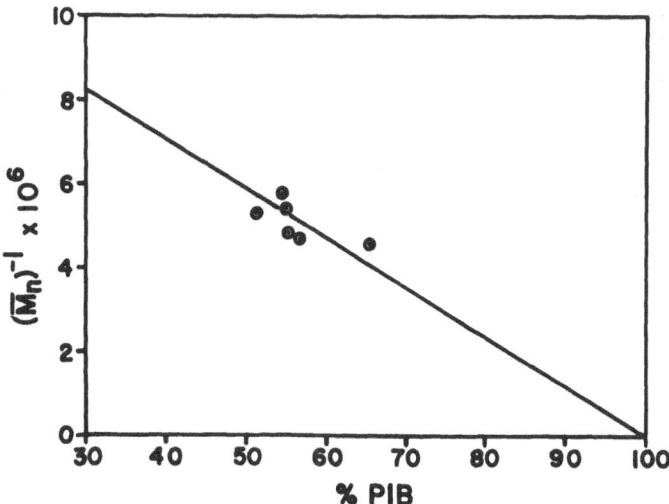

Figure 5.   Comparison Plot of Calculated (line) and Experimental (Osmotic) Molecular Weights as a Function of % PIB in the Polymer

The intrinsic viscosity data are most intriguing.
While η = 1.82 for the SBR backbone, it decreases to
1.3-1.6 after adducting to it 43-65% PIB branches.
Independent $\overline{M}_n$, GPC evidence and control studies indicate
that the SBR backbone has not degraded during grafting.
The reason for this phenomenon of decreasing η upon
grafting is unknown, however, similar effects have been
noted previously by us (27,28). It is interesting to
contemplate that the hydrodynamic volume of the graft
coil with $\overline{M}_n \sim$ 200,000 is smaller than that of the
"naked" backbone with $\overline{M}_n \sim$ 85,000.

Efforts have been made to estimate b/B (branches/
backbone) ratios. We theorized that this information can
be derived by oxidatively destroying the highly unsaturated
backbone using OsO_4 and determining the amount and molecular weight of the surviving saturated PIB branches.
However, control experiments showed that OsO_4 degrades
PIB also. Thus, upon treatment with OsO_4 (at 120° for 10
min in chlorobenzene) of a PIB of $\overline{M}_n$ = 200,000, the $\overline{M}_n$
dropped to 50,000.

Since we could not obtain b/B by direct oxidative
degradation, a very rough estimation of b/B was obtained
by the following reasoning. Knowing the $\overline{M}_n$ of the
SBR-g-PIB ($\sim$ 200,000) and that of the starting SBR
($\sim$ 85,000), the $\overline{M}_n$ of total grafted PIB is 115,000.
Further, we have obtained by extraction a PIB of $\overline{M}_n$ =
77,000 from a homopolymerization experiment under grafting
condition in the presence of SBR (cf. Preliminary Studies).
By this reasoning b/B $\cong$ 1.5.

As expected, DTA showed only one Tg transition for
representative samples of pure SBR-g-PIB. This is not
unexpected since the Tg's of SBR and PIB are ~ -65 and
$\sim -70^\circ$, respectively.

In spite of the fact that DTA showed only one Tg, the
presence of two distinct polymer phases was clearly indicated by transmission electron microscopy. The SBR-g-PIB
sample was obtained in a grafting experiment with G.E.
$\sim$ 100% and contained 65% SBR and 35% PIB by nmr spectroscopy. The unextracted sample was compounded (sulfur,
ZnO, accelerator), lightly crosslinked (30 mins at $153^\circ C$),
microtomed, and stained with OsO_4. The micrograph shown
in Figure 6 indicates phase separation with the existence
of light PIB domains of $\sim$ 300 Å diameter dispersed in the
continuous dark SBR matrix.

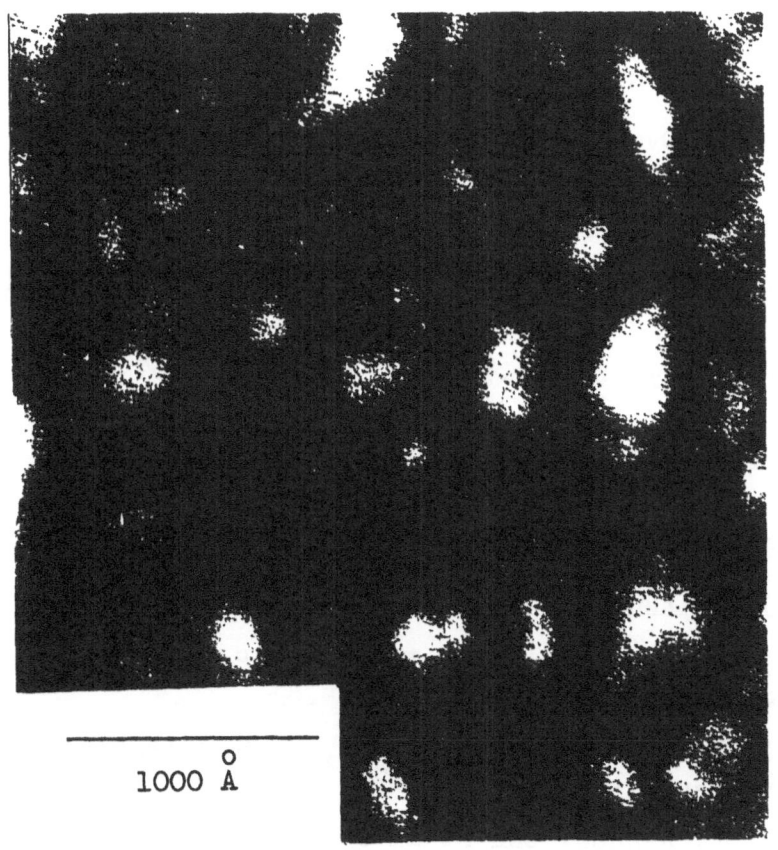

1000 Å

Figure 6.   Transmission Electron Micrograph of an SBR-g-PIB
             (65% SBR/35% PIB). Magnification: 396,000 X.

        A physical blend of SBR and PIB gave PIB domains at
least one order of magnitude larger. Moreover, the gum
vulcanizate of the graft copolymer had a tensile strength
of 3 MPa with 200% elongation while the physical blend
had no measurable tensile strength.

REFERENCES

1. H. C. Haas, P. M. Kamath, and N. W. Schuler, J. Polymer Sci., 24, 85 (1957).

2. C. G. Overberger and C. M. Burns, ibid., A-1, 7, 333 (1969).

3. G. Kockelberg and G. Smets, ibid., 33, 227 (1958).

4. Y. Minoura, T. Hanada, T. Kasaba, Y. Ueno, ibid., A-4, 4, 1665 (1966).

5. P. H. Plesch, Chem. Ind. (London), 954 (1958).

6. O. F. Solomon, M. Dimonie, and C. Ciuciu, J. Polymer Sci., A-1, 8, 777 (1970).

7. J. P. Kennedy, J. Macromol. Sci.-Chem., A3 (5), 861 (1969).

8. J. P. Kennedy, J. Polymer Sci., A-1, 6, 3139 (1968).

9. J. P. Kennedy and M. Ichikawa, Polymer-Eng. and Sci., 14, 322 (1974).

10. J. P. Kennedy, N. V. Desai, S. Sivaram, J. Amer. Chem. Soc. 95, 6386 (1973).

11. J. P. Kennedy and S. Sivaram, J. Org. Chem., 38, 2262 (1973).

12. J. P. Kennedy, J. Org. Chem., 35, 532 (1970).

13. J. P. Kennedy, J. Polymer Sci., A-1, 6, 3139 (1968).

14. J. P. Kennedy, Int. Symp. Macromol. Chem., Tokyo, Kyoto, 1966, Abstract 2.104.

15. J. P. Kennedy, J. Macromol. Sci.-Chem., A3 (5), 861 (1969).

16. J. P. Kennedy, J. Macromol. Sci.-Chem., A7 (4), 969 (1973).

17. J. P. Kennedy and R. R. Smith, Polymer Preprints, 14 (2), 1069 (1973).

18. J. P. Kennedy, J. J. Charles, and D. L. Davidson, ibid., 14 (2), 974 (1973).

19. U.S. Patents 3,458,599; 3,476,831 (1969).

20. J. P. Kennedy and D. K. Metzler, Polymer Preprints, $\underline{16}$ 512 (1975).

21. "Free Radicals in Solution" by Ch. Walling, John Wiley and Sons, Inc. Pub., New York, 1957, p. 386.

22. J. P. Kennedy and R. M. Thomas, Advan. Chem. Ser., $\underline{34}$, 111 (1962).

23. J. L. Binder, Anal. Chem., $\underline{26}$, 1877 (1954).

24. P. A. Small, J. Appl. Chem., $\underline{3}$ 71 (1953).

25. S. Cesca, P. Giusti, P. L. Magagnini and A. Priola, Makromol. Chem., $\underline{176}$, 2319 (1975).

26. J. P. Kennedy and P. Trivedi, paper 1-3-20 presented at the XXIII IUPAC Intl. Symp. on Macromol., Madrid, Spain, Sept. 1974.

27. J. P. Kennedy and A. Vidal, J. Polymer Sci., Chem. Ed., $\underline{13}$, 1765 (1975).

28. J. P. Kennedy and J. J. Charles, to be published.

SYNTHESIS AND CHARACTERIZATION OF POLY(CHLOROPRENE-g-ISOBUTYLENE)

D.K. Metzler* and J.P. Kennedy

Institute of Polymer Science

University of Akron, Akron, Ohio 44325

ABSTRACT

Poly(chloroprene-g-isobutylene) was synthesized by cationic grafting of isobutylene from polychloroprene, using diethyl-aluminum chloride as coinitiator. Pure graft copolymers were isolated by selective solvent extraction. Effects of reaction conditions on grafting efficiency were investigated. Determination of branch molecular weight was carried out by oxidation degradation of graft copolymer backbones, followed by isolation and characterization of residual polyisobutylene. Molecular weight and molecular weight distribution of grafted polyiso-butylene were indistinguishable from corresponding values for ungrafted homopolyisobutylene formed in the same polymerization.

INTRODUCTION

Kennedy and co-workers (1-7) have demonstrated efficient synthesis of graft copolymers by cationic polymerization, using alkylaluminum compounds as coinitiators to graft cationically polymerizeable monomers from halogen-bearing backbone polymers. Backbones employed previously include halogenated EPM (1) and EPDM (2), halogenated SBR (3), chlorobutyl and bromobutyl rubber (4), chlorosulfonated polyethylene (5), and PVC (6,7). Monomers grafted include styrene (1,4,5,7), isobutylene (2,3,6), α-methylstyrene (2), and indene and acenaphthalene (4). In

*Present Address: Celanese Research Co., P.O. Box 1000, Summit
New Jersey   07901

this study a similar process has been employed to graft isobutylene from polychloroprene, also known as neoprene.

Polychloroprene (PCR) has proven an extremely useful backbone for cationic grafting, since labile chlorine structures inherent in the polymer are unusually active initiators for cationic polymerization. In addition, this system was proven to be a valuable model for exploration of cationic grafting processes in general, since direct characterization of grafted polyisobutylene (PIB) branches could be carried out by selective backbone degradation and subsequent PIB characterization.

Cationic polymerization is readily initiated by a dialkyl-aluminum halide or trialkylaluminum coinitiator in conjunction with a suitable organohalide initiator, such as an allyl, benzyl, or tertiary alkyl chloride (8). Other halides (e.g. vinyl, aryl, and primary and secondary alkyl chlorides) do not initiate polymerization. In addition, such alkylaluminum compounds do not initiate polymerization in conjunction with trace impurities such as water (8).

In cationic grafting, a halogen-containing backbone polymer is employed as a macromolecular initiator, resulting in "grafting from" (9) the backbone of cationically polymerizable monomer branches. The required labile halogens may be incorporated in the backbone by halogenation (1-4), or preferably may be inherent in the original polymer (5-7). With polychloroprene, the latter situation prevails.

Commercially available polychloroprene, produced by emulsion polymerization (10,11), consists largely of 1,4-trans monomer en-chainment (I), with small amounts of 1,4-cis (II), 1,2 (III), and 3,4 (IV) structures (12). A recent study has also identified an isomerized 1,2 structure (V) arising from post-polymerization rearrangement of II (13). For a polymer produced at 40°C, as in the commercial polymerization process, the distribution of structures was 89.5% I, 5.2% II, 2.1% III, 1.5% IV, and 1.7% V (13).

I

II

III   IV   V

Vinyl chlorines, I, II, and IV, are inactive as initiators for cationic polymerization, but the substituted allyl chlorides, III and V, are extremely reactive. Studies by Kennedy, et al. (14,15) have demonstrated extremely fast reactions of $\alpha,\alpha-$ and $\gamma,\gamma-$dimethylallyl chloride (model compounds for structures III and V respectively) with coinitiators trivinylaluminum and boron trichloride, both in alkylation and in model initiation reactions. Furthermore, detailed kinetic studies by Vernon (16) have shown that solvolyses of $\alpha\alpha-$ and $-$dimethylallyl chloride (a nucleophilic substitution akin to initiation in cationic polymerization) are $\sim 10^8$ times faster than unsubstituted allyl chloride.

Thus, the mechanism of cationic grafting isobutylene from polychloroprene is proposed to proceed by 1). <u>Cation generation</u>, i.e., abstraction of labile allylic chlorines of structures III or V by the Lewis acid coinitiator:

$$\text{(structure)} \quad \text{Cl} \quad + \quad Et_2AlCl \quad \longrightarrow \quad \text{(cation)} \quad Et_2AlCl_2^{\ominus}$$

2). <u>Initiation</u>, i.e., addition of the first monomer molecule to the allyl cation:

$$\text{(cation)} \oplus \quad Et_2AlCl_2^{\ominus} \quad + \quad \text{(isobutylene)} \quad \longrightarrow \quad \text{(product)} \oplus \quad Et_2AlCl_2^{\ominus}$$

3). <u>Propagation:</u>

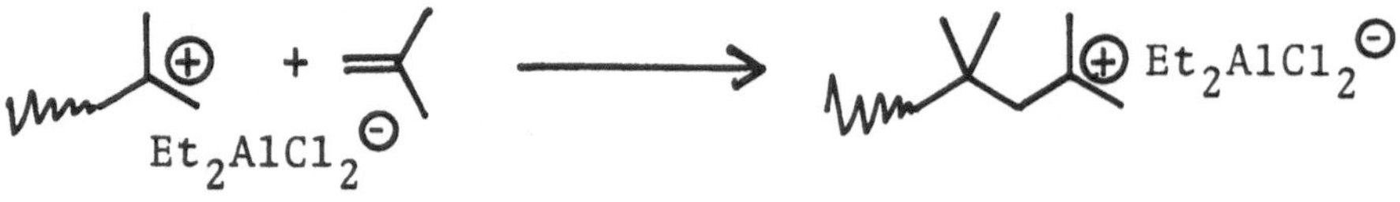

4). <u>Chain transfer to monomer:</u>

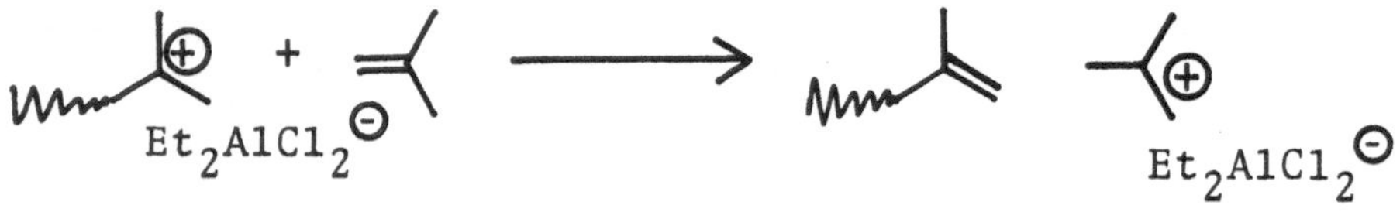

and 5). <u>Termination:</u>

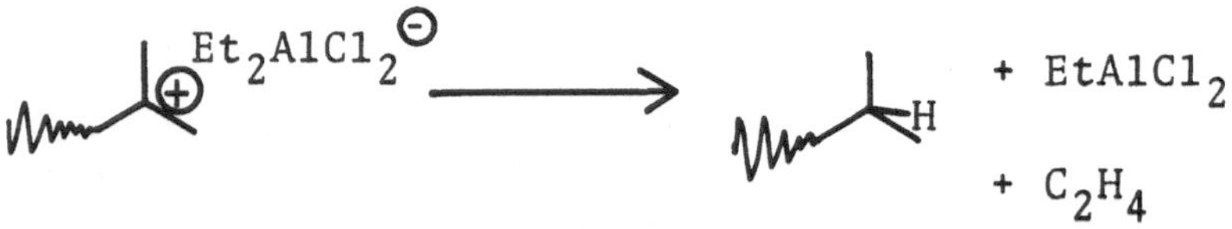

Propagation, chain tranfer to monomer, and termination are as in
the homopolymerization of isobutylene, and have been discussed in
detail (17).

<u>EXPERIMENTAL</u>

<u>Materials</u>

Polychloroprene (Neoprene W, E. I. duPont de Nemours and Co.)
was purified by three precipitations from toluene solution into
methanol, then dried under vacuum at room temperature. Isobutylene (99.9%, Union Carbide Corp.) was dried by passage through
a column of barium oxide and molecular sieves, then condensed
under a nitrogen atmosphere inside a stainless steel enclosure.
Diethylaluminum chloride (Ethyl Corp.) was distilled from NaCl
at reduced pressure under inert atmosphere (bp 90°C/15 mm Hg)
and stored at -78°C. Triethylaluminum and trimethylaluminum
(Ethyl Corp.) were distilled at reduced pressure under inert
atmosphere (bp 90°C/10 mm Hg and 60°C/70 mm Hg, respectively).
n- Pentane (Aldrich Chemical Co.) and methylene chloride (Eastman Organic Chemicals) were distilled from trimethylaluminum.

Solvents used for extraction and precipitation of polymers were used as received. Nitrogen (dry grade, Union Carbide Corp.) was dried by passage through a column of barium oxide and molecular sieves.

Apparatus

All polymerization experiments and associated materials transfer operations were carried out inside a stainless steel drybox enclosure containing a dry nitrogen atmosphere (moisture content 10-50 ppm by electrolytic analysis). This apparatus has been described (18). Glassware was dried overnight at 150°C before use, then cooled under dry nitrogen in the drybox entry port. Most polymerizations were carried out in 1-liter 3-neck round bottom flasks equipped with thermometer and overhead mechanical stirrer. Alternatively, some experiments were conducted in 8 in. test tubes with screwcap closures. Reaction vessels were supported in a n-pentane bath inside the drybox. Temperature control was maintained by liquid nitrogen circulated through copper coils immersed in the bath.

Graft Copolymerizations

A typical grafting experiment was carried out as follows. A 1-liter flask equipped with stirrer and thermometer was charged with methylene chloride (335 ml) and a stock solution of polychloroprene in methylene chloride (5 g in 100 ml). The charge was thermoequilibrated to -50°C and isobutylene (50 ml) was added. Polymerization was initiated by addition of diethylaluminum chloride diluted with methylene chloride (15 ml. 1.0 M solution). After 2 min., polymerization was terminated by addition of 20 ml chilled methanol.

Separation of Products

Pure graft copolymer was isolated from the crude product by the procedure illustrated in Figure 1. Extraneous components, i.e. aluminum alkoxides coinitiator residues, polyisobutylene, ungrafted polychloroprene, and gelled graft copolymer were removed.

After termination, the reaction flask was removed from the drybox and the reaction mixture precipitated into methanol and stirred overnight to dissolve aluminum alkoxides. Precipitated polymer was collected and dried under vacuum at room temperature.

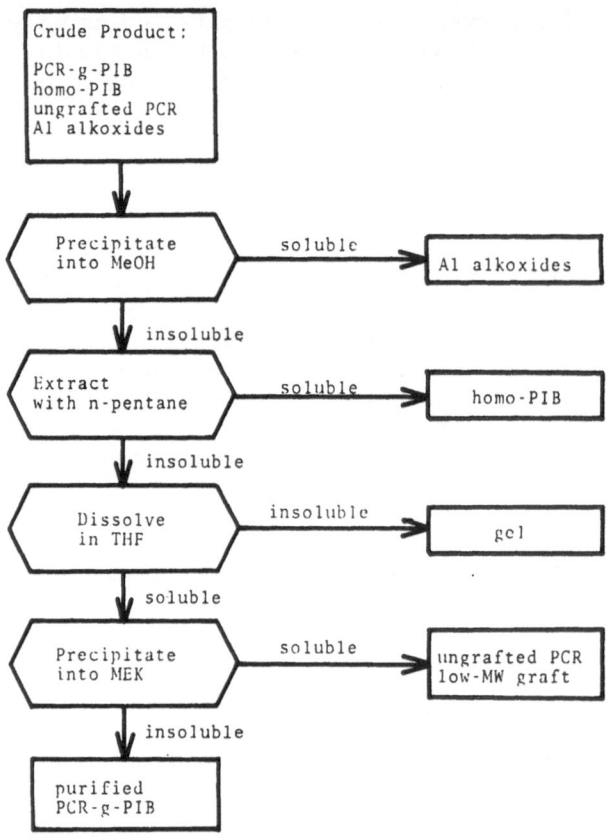

FIGURE 1

Graft Purification Sequence

After weighing to determine conversion, the product was cut in
small pieces, placed in an Erlenmeyer flask, and covered with n-
pentane (20 ml/g polymer). The flask was shaken gently for 16
hrs using a Burrell "Wrist-Action" shaker, then the n-pentane
solution was decanted and replaced with fresh solvent. The extraction was repeated for two additional cycles of 8 hrs each.
By this process, PIB was efficiently removed while graft copolymer and ungrafted polychloroprene remained highly swollen but
insoluble. Soluble and insoluble fractions were dried at room
temperature and weighed. The n-pentane-insoluble fraction
(graft copolymer and ungrafted backbone) was dissolved in THF
and centrifuged to remove gel, then precipitated into MEK.

Ungrafted polychloroprene was isolated from MEK solution using
a rotary evaporator. MEK-insoluble polymer, pure poly(chloroprene-g-isobutylene) was collected and dried under vacuum at room
temperature.

Characterization

Graft copolymer compositions were determined by NMR spectroscopy, using a Varian T-60 NMR spectrometer. Sample solutions
were about 50 g/l in $CDCl_3$. Wt % PIB was calculated from instru-
ment-integrated areas of the methylene multiplet of PCR (2.7 ppm)
and the sum of methyl and methylene singlets of PIB (1.4 and 1.7
ppm) using the relation

$$wt.\% \ PIB \ = \ \frac{A_2 \ x \ 100}{(3.16 \ A_1) \ + \ A_2} \tag{1}$$

in which A_1 and A_2 are the areas of the polychloroprene and polyisobutylene peaks. Composition values determined by NMR and calculated by mass balance generally agreed within 5%,

Molecular weights of graft copolymers were determined by
membrane osometry, using a Mecrolab 503 high-speed membrane osmometer with S&S 08 membranes. Measurements were carried out at
$37^{\circ}C$ using toluene solutions.

Molecular weights of PIB were determined by gel permeation
chromatography using a Waters Associates Ana-Prep instrument
with a seven column analytical set and THF solutions. Molecular
weights were calculated using a PIB calibration curve (courtesy
of Prof. D. McIntyre).

Oxidative Degradation of Graft Copolymer Backbone

Backbone double bonds of graft copolymer samples were epoxidized with m-chloroperbenzoic acid (19), followed by hydrolysis
and cleavage with periodic acid (20). Epoxidation was carried
out for 24 hours at room temperature in chloroform solution,
using a peracid charge equivalent on molar basis to the double
bond content of the polymer sample. Epoxidized polymer was
collected by precipitation into slightly alkaline methanol, containing sufficient dissolved KOH to neutralize the acid charge.
Hydrolysis and cleavage were carried out simultaneously in 90%
aqueous THF at $60^{\circ}C$ for 8 hours, using 50% molar excess of periodic acid. Polyisobutylene precipitated from solution during

the course of reaction was collected from the bottom of the vessel and dried under vacuum.

RESULTS AND DISCUSSION

Poly(chloroprene-g-isobutylene), PCR-g-PIB, graft copolymers
were readily synthesized under a variety of experimental conditions using diethylaluminum chloride (Et$_2$AlCl) coinitiator. Results of polymerizations carried out in 1 liter flasks, cf. experimental) are summarized in Table I.

Reactions were carried out at temperatures ranging from -10
to -70°C, due to constraints imposed by the boiling point of isobutylene (-6°C) and the temperature of precipitation of polychloroprene from the reaction medium (about -80°C, depending on
composition of the solvent mixture). Polymerizations were conducted using methylene chloride, a good solvent for PCR but nonsolvent for PIB, modified by addition of varying proportions of
n-pentane, a good solvent for PIB but nonsolvent for PCR. In
the temperature range of interest, polychloroprene is insoluble
in methylene chloride/n-pentane mixtures of less than about
50 vol % methylene chloride. Accordingly, to insure backbone
solubility, grafting experiments were conducted in mixtures of
60 vol % or more methylene chloride. Monomer concentrations
up to 2.4%M were employed, with most experiments conducted using
1.2M (10 vol. %) isobutylene.

Preliminary grafting experiments carried out in test tubes
demonstrated that polymerizations were generally very fast, and
unless terminated proceeded to 100% conversion. At high
conversion solid plugs of polymer formed, occluding the entire
volume of solvent. These observations mandated use of stirred
reactors rather than test tubes. Reactions carried out in flasks
were also characterized by high rates of polymerization and solid-
ification of the reaction medium at high conversion. However,
with rapid stirring, the medium remained fluid to high conversion
and efficient termination was possible. Polymerization exotherms
were generally controlled to within 5°C of the initial temperature.

Effect of Backbone Type on Rate of Polymerization

Figure 2 shows conversion/time data obtained in three different grafting systems. In addition to PCR/isobutylene, the
systems shown are chlorinated EPM/styrene (1) and PVC/isobutylene
(6). Experimental conditions (indicated in the figure legend)
are sufficiently similar to permit qualitative comparison. The
figure shows the very high relative reactivity of polychloroprene
as a backbone/initiator in cationic grafting.

TABLE I

SYNTHESIS OF POLY(CHLOROPRENE-g-ISOBUTYLENE)

Expt.	PCR (g/l)	i-C$_4$H$_8$ M	Et$_2$AlCl M	Solvent (% CH$_2$Cl$_2$)	T (°C)	t (min)	conv. (%)	X$_b$ (%)	GE (%)
N1	10	1.2	0.030	90	-50	4	92	-	-
N2	10	2.4	0.030	80	-50	3	95	-	-
N3	10	0.60	0.030	95	-50	2	98	-	-
N4	10	1.2	0.030	80	-50	1	97	-	-
N5	10	0.60	0.030	80	-50	1	87	-	-
N6	10	0.60	0.030	60	-50	2	62	-	-
N8	20	0.60	0.020	60	-50	2	100	-	-
N11	10	1.2	0.010	90	-30	2	59	64	40
N12	10	0.60	0.020	90	-30	5	57	56	42
N13	9	0.30	0.020	80	-30	5	96	26	24
N14	10	1.2	0.030	90	-10	1	81	53	21
N15	10	0.60	0.020	90	-10	0.5	74	27	15
N16	20	1.2	0.024	90	-10	0.5	82	24	11
N17	10	1.2	0.030	90	-50	4	70	51	62
N18	10	1.2	0.030	90	-70	1	77	64	76
N20	5	1.35	0.011	90	-70	2	9	50	73
N21	10	1.2	0.030	80	-70	2	94	81	72
N22	10	0.60	0.030	80	-70	2	46	53	71
N26	10	1.2	0.030	90	-70	0.5	75	83	82
N27	10	1.2	0.030	70	-70	1.25	16	52	80
N28	10	1.2	0.030	60	-70	3	76	83	75
N29	10	1.5	0.030	80	-70	0.5	74	81	68
N30	10	0.90	0.030	80	-70	1	55	66	72
N31	10	0.30	0.030	80	-70	3	61	39	64
N32	10	0.15	0.030	80	-70	6	12	9	100
N33	10	1.2	0.030	70	-70	2	51	69	64
N34	10	1.2	0.030	70	-70	1.5	37	65	73
N35	10	1.2	0.030	70	-70	1	2.3	13	100
J36	10	1.2	0.030	70	-70	0.5	0.8	4.5	91

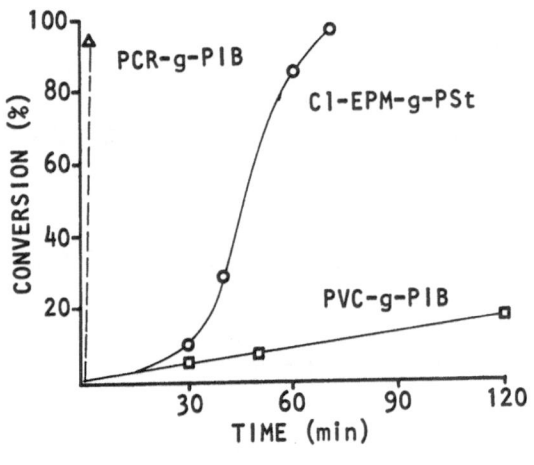

FIGURE 2

Comparative Conversion/Time Data in Three Cationic
Grafting Systems. ○ Cl-EPM-g-PSt at -35°C in 30/70
EtCl/n-hexane (1).   □ PVC-g-PIB at -50°C in CH$_2$ClCH$_2$Cl
(6).   △ PCV-g-PIB at -50°C in CH$_2$Cl$_2$.

This effect is probably due to differences in graft site
structure among the three backbones, and resulting differences
in their rates of carbenium ion generation. In chlorinated EPM,
the postulated graft site is a tertiary alkyl chloride, giving a
tertiary cation upon chlorine abstraction. In polychloroprene,
as discussed above, the graft site is a dialkylallyl chloride,
yielding a tertiary allyl cation. Since the latter cation is
more stable, cation formation should be faster with polychloroprene than with chlorinated EPM, leading to faster overall rates
of polymerization. A similar situation may exist for PVC. However, the nature of labile chlorines in PVC has not been defini-
tively established, which precludes meaningful analysis at this
time.

Effect of Coinitiator Type on Rate of Polymerization

The activities of diethylaluminum chloride (Et$_2$AlCl), triethylaluminum (Et$_3$Al), and trimethylaluminum (Me$_3$Al) in grafting from PCR have been investigated. Although grafting occurred
readily with Et$_2$AlCl under a wide range of conditions (Table I),
experiments with Et$_3$Al and Me$_3$Al did not produce grafts,
Experiments were attempted at temperatures from -30° to -80°C,

with coinitiator concentrations up to 0.2 M, using methylene chloride solvents 1.2 M isobutylene, and 10 g/l polychloroprene. Experiments were run up to 2 hrs. without detectable conversion of isobutylene. These results were unexpected, since Et_3Al and Me_3Al are active coinitiators with other backbones (1, 4, 6), although not as reactive as Et_2AlCl. Since polychloroprene was found to be an unusually active initiator in the presence of Et_2AlCl, measurable activity was also anticipated with Et_3Al or Me_3Al. Failure of these coinitiators to produce grafts thus presented an apparent paradox.

Resolution of this paradox may lie in consideration of the relative stabilities of the initiating and propagating cations. After formation of the initial cation (equation 1), for polymerization to occur initiation (equation 2) must compete with alkylation (analogous to termination):

$$P-Cl \ + \ R_3Al \longrightarrow P^{\oplus} \ R_3AlCl^{\ominus} \xrightarrow{\ M\ } PM^{\oplus} R_3AlCl^{\ominus}$$

$$\downarrow$$

$$P-R \ + \ R_2AlCl$$

In the chlorinated EPM/styrene system (1), initiation results in conversion of a relatively unstable tertiary cation into a more stable secondary styryl cation. In such a system initation should be relatively fast and efficient. However, with polychloroprene/isobutylene the stability relations are reversed. Initiation converts the relatively stable dialkylallyl cation to a less stable tertiary cation. Thus, though carbenium ion formation is very fast with polychloroprene, initiation should be relatively slow and may not compete effectively with alkylation in the presence of a highly nucleophilic counterion such as a trialkylchloroaluminate. The experimental results support the hypothesis that with Et_3Al or Me_3Al polychloroprene is preferentially alkylated without initiating polymerization.

Effect of Polymerization Variables on Grafting Efficiency

Grafting efficiency is a measure of effectiveness of a graft polymerization; it is defined as the percent of polymerized monomer grafted to the backbone, as distinguished from that present as ungrafted homopolymer:

$$GE \ (\%) \ = \ 100 \ x \ W_b/(W_b + W_h)$$

Where W_b is the weight of grafted branches and W_h is the weight of homopolymer. $W_b + W_h$ is thus total yield of PIB. In contrast to grafting efficiency, graft composition is defined as

$$X_b \ (\%) \ = \ 100 \ x \ W_b/(W_b \ + \ W_B)$$

in which W_B is the weight of the backbone. Ideally, grafting efficiencies of 100% are sought. In this system the less-than-100% grafting efficiencies obtained are due largely to effects of chain transfer, since initiation can only occur by the backbone.

As described in the Experimental section, crude products were purified by selective extraction with n-pentane to remove PIB. Grafting efficiencies were calculated from weights of homopolymer, backbone, and total yield:

$$GE \ (\%) \ = \ 100 \ x \ (W_S \ - \ W_B)/(W_T \ - \ W_B) \qquad (2)$$

where W_T is total crude polymer yield. To establish optimum grafting conditions, effects of several reaction variables on grafting efficiency were investigated in detail.

Figure 3 illustrates the effect of temperature on grafting efficiency: Grafting efficiency increased sharply with decreasing temperature in the range investigated (-10° to -70^oC), as chain transfer was "frozen out". This finding is consistent with results in other cationic grafting systems, (1, 4, 6), as well as fundamental kinetic studies (17). Of the variables studied in relation to grafting efficiency, polymerization temperature exhibited the largest effect.

The effect of conversion on grafting efficiency is shown in Figure 4. Grafting efficiencies were nearly 100% at low conversion, then decreased with increased conversion as chain transfer became significant. This result is in line with the "grafting from" mechanism, since initiation by the backbone should give very high grafting efficiencies at low conversions, while "grafting onto" results in low initial grafting efficiencies increasing with conversions (21).

Although overall grafting efficiency decreased with conversion, incorporation of PIB into graft copolymer continues over the entire range studied (Figure 5). Furthermore, even at high conversions grafting efficiencies were in general relatively high. For the 15 experiments at -70^oC in Table I, the minimum grafting efficiency obtained was 64%, and the average was 77%.

The effect of solvent polarity on grafting efficiency was also investigated, by varying proportions of methylene chloride and n-pentane in the reaction medium. Results are shown in Figure 6. Grafting efficiency increased significantly but relatively slightly over the range from 60% to 90% methylene chloride, probably reflecting increased rates of carbenium ion formation and initiation in the more polar media.

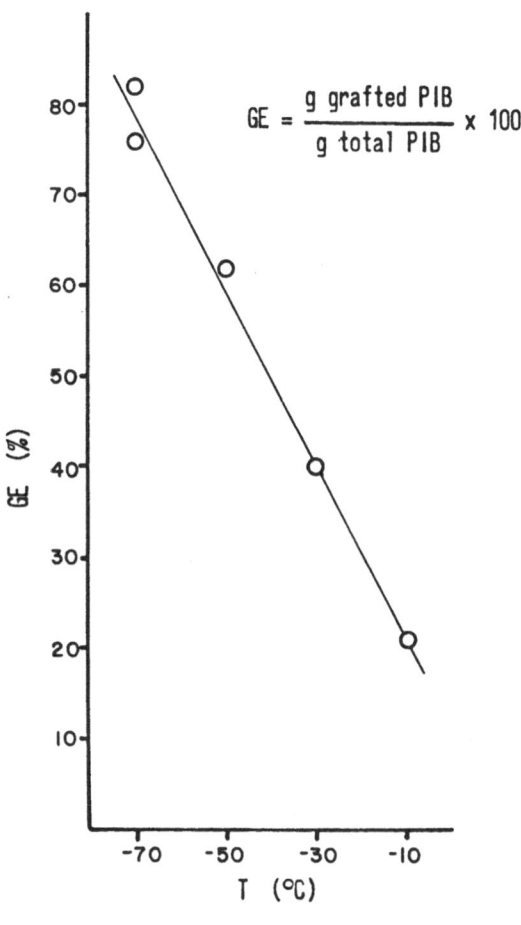

FIGURE 3

Effect of Temperature on Grafting Efficiency. [PCR] = 10 g/1;
[i-C$_4$H$_8$] = 1.2M; [Et$_2$AlCl] = 0.03M; solvent 90% CH$_2$Cl$_2$

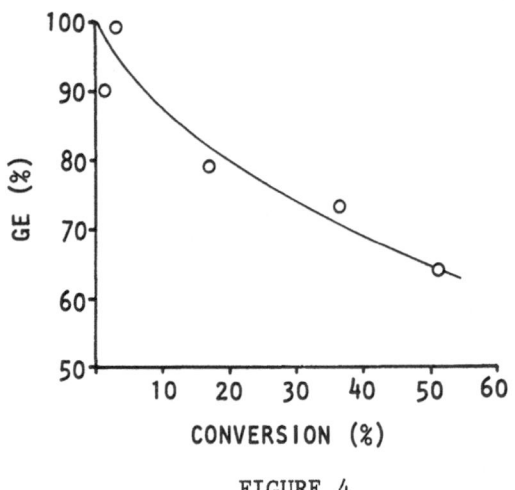

FIGURE 4

Effect of Conversion on Grafting Efficiency. T = -70°C;
$[PCR]$ = 10 g/l; $[i-C_4H_8]$ = 1.2 M; $[Et_2AlCl]$ = 0.03 M;
solvent 70% CH_2Cl_2.

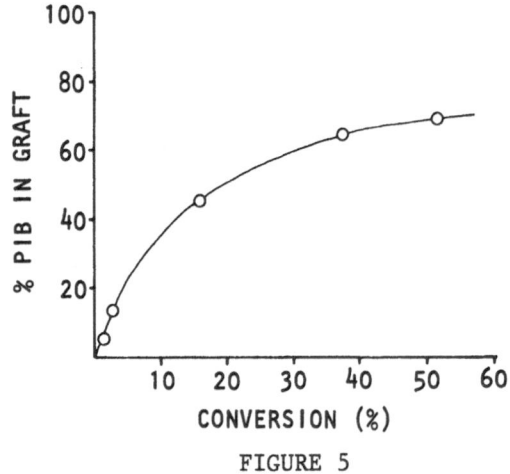

FIGURE 5

Effect of Conversion on Graft Copolymer Composition. T = -70°C;
$[PCR]$ = 10 g/l; $[i-C_4H_8]$ = 1.2 M; $[Et_2AlCl]$ = 0.03 M;
solvent 70% CH_2Cl_2.

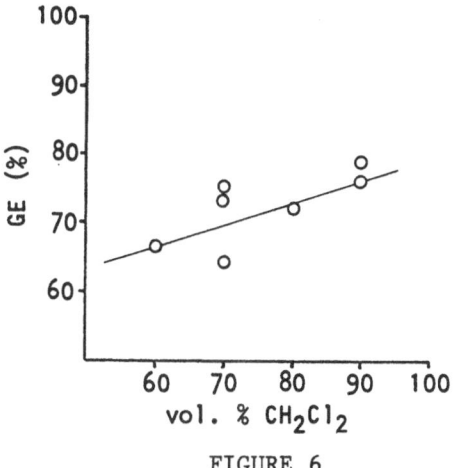

FIGURE 6

Effect of Solvent Composition on Grafting Efficiency. T = -70°C;
$[PCR]$ = 10 g/l; $[i-C_4H_8]$ = 1.2 M; $[Et_2AlCl]$ = 0.03 M.

Effect of Solvent Composition on Gelation

Although solvent composition had a relatively minor effect on
grafting efficiency, it had a major influence on gel content of
graft copolymers. As shown in Figure 7, polymerizations using
neat methylene chloride resulted in significant gelation (about
9% graft copolymer yield). Gelation was sharply reduced in experiments incorporating n-pentane cosolvent. Note that in
Figure 7 as in Table I, isobutylene is an integral component of
the reaction medium. Thus, in an experiment using "1.2 M 1-C$_4$H$_8$,
80% CH$_2$Cl$_2$", total composition of the medium is 80 vol. % methylene chloride, 10% n-pentane, and 10% isobutylene.

Even 10% hydrocarbon cosolvent significantly reduced gelation.
Since the entire solvent composition range is highly polar, it
appears that selective solvation rather than a general polarity
is responsible for the observed variation. Gelation is probably
due to the fact that while polychloroprene is highly soluble in
methylene chloride, polyisobutylene is insoluble, so graft copolymer tends to precipitate from solution as it is formed.
Propagating cations of growing branches are thus brought in close
proximity to double bonds of the backbone and crosslinking (grafting onto) may occur. The hydrocarbon cosolvent apparently increases solubility of growing branches to reduce gelation. A similar
effect has been reported in synthesis of isobutylene/isoprene

copolymers (22). Methyl chloride medium for this polymerization
is a non solvent for the polymers, leading to gelation, however,
incorporation of n-heptane cosolvent reduces gelation.

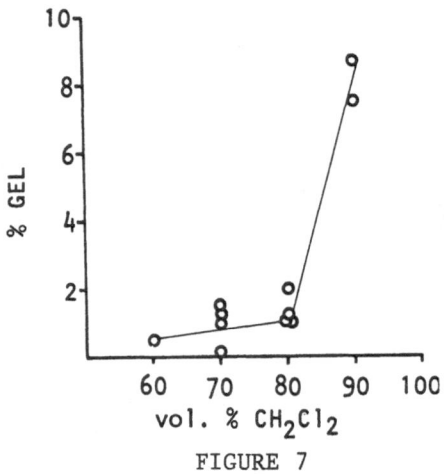

FIGURE 7

Effect of Solvent Composition on Gelation. T = -70°C.
[PCR] = 10 g/l; [Et$_2$AlCl] = 0.03 M.

Effect of Grafting on Backbone Molecular Weight

In the absence of side reactions, graft molecular weight can
be calculated from graft composition (X_b, %) and number-average
molecular weight of the backbone (M_B):

$$M_g = M_B/(1 - X_b/100) \tag{3}$$

Deviation of actual graft molecular weight M_g from calculated
M_g^* indicates crosslinking if $M_g > M_g^*$. Thus, the ratio M_g/M_g^* is
a sensitive diagnostic tool for the identification of side
reactions which involve the backbone.

Graft composition and molecular weight data are summarized
in Table II, accompanied by calculated M_g^* and M_g/M_g^* In nearly
all cases, $M_g/M_g^* < 1$, indicating backbone scission. Molecular
weight decrease was more severe at higher polymerization temperatures and as with high levels of grafting ($X_b > 60$%).

TABLE II

GRAFT COPOLYMER MOLECULAR WEIGHTS

Expt.	T ($^\circ$C)	X_b (%)	$M_g^* \times 10^{-3}$	$M_g \times 10^{-3}$	M_g^*/M_g
N14	-10	55	400	85	.22
N15	-10	43	310	160	.52
N16	-10	57	230	63	.27
N11	-30	69	580	270	.46
N12	-30	37	280	190	.67
N13	-30	41	310	150	.48
N8	-50	54	390	230	.60
N9	-50	0	180	150	.83
N18	-70	75	720	400	.56
N20	-70	50	360	350	.99
N27	-70	48	340	320	.92
N28	-70	78	830	610	.73
N29	-70	82	1000	560	.56
N30	-70	70	590	300	.51
N31	-70	46	330	200	.60
N32	-70	15	210	130	.61
N33	-70	74	700	550	.79
N36	-70	11	200	190	.94

These observations were supplemented by experiments in which polychloroprene was contacted with Et_2AlCl under simulated grafting conditions in the absence of monomer. Backbone scission was observed (Figure 8), followed at $-30^{\circ}C$ by eventual crosslinking. At -70° molecular weights did not change after 20 min. These data suggest scission at a relatively small number of "weak links" in the backbone, rather than by a random process. Although the mechanism of chain scission was not investigated in detail, a mechanism consistent with the observations may be suggested. Chain scission may occur at the graft site after carbenium ion formation prior to initiation.

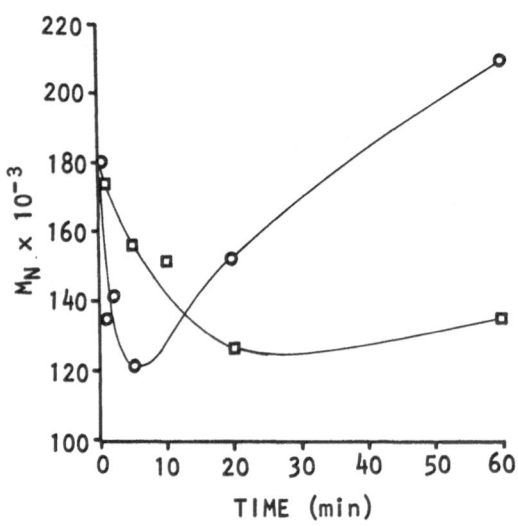

FIGURE 8

Effect of Et_2AlCl on Polychloroprene Molecular Weight. [PCR] = 10 g/l; [Et_2AlCl] = 0.03 M; solvent CH_2Cl_2.

This process is analogous to the β-scission mechanism in degradation of butyl rubber. In addition to a small entropy gain from division of one chain into two, thermodynamic driving force is provided by formation of a conjugated double bond in one fragment and formation of a highly resonance-stabilized ion in the other:

This cation may or may not subsequently initiate grafting.

Characterization of Graft Structure by Backbone Degradation

Two key problems in graft copolymer characterization are determination of the average number of branches per backbone chain (b/B) and average molecular weight per branch (M_b). If backbone molecular weight is known and graft molecular weight or composition can be determined, b/B and M_b are not independent:

$$b/B = \frac{M_g - M_B}{M_b} \tag{4}$$

$$= \frac{M_B \, X_b/100}{M_b \, (1-X_b/100} \tag{5}$$

b/B is thus readily calculable if M_b can be obtained.

Unfortunately, in many grafting systems M_b is inaccessible since branches cannot be isolated from the backbone and characterized independently. One means of circumventing this problem is to assume that M_b is equal to M_h, the molecular weight of ungrafted homopolymer formed in the same reaction. This approach has been employed in several cationic grafting studies (1, 4, 6). However, in the case of PCR-g-PIB, it has been possible to exploit the unsaturation of the backbone and to carry out selective PCR backbone degradation with subsequent direct characterization of PIB branches.

TABLE III

MOLECULAR WEIGHTS OF GRAFTED AND HOMO—POLYISOBUTYLENES

Expt.	grafted PIB			homo-PIB			b/B*	b/B
	$\overline{M}_w$	$\overline{M}_n$	$\overline{M}_w/\overline{M}_n$	$\overline{M}_w$	$\overline{M}_n$	$\overline{M}_w/\overline{M}_n$		
N21	240	100	2.4	240	85	2.8	7.7	–
N22	170	78	2.2	200	73	2.5	2.6	–
N26	270	130	2.1	230	89	2.6	6.8	–
N27	200	74	2.6	220	81	2.8	2.6	2.4
N28	330	120	2.4	460	150	3.1	7.1	5.2
N29	250	110	2.3	300	130	2.4	7.2	4.0
N30	200	96	2.1	210	87	2.4	3.6	1.8
N33	240	110	2.2	250	100	2.5	3.6	2.8
N34	300	130	2.4	260	100	2.5	2.7	–

Degradation of graft copolymer backbones was carried out by epoxidation with m-chloroperbenzoic acid, hydrolysis of epoxides, and cleavage of resulting keto-alcohols with periodic acid:

IR spectra of recovered PIB showed only faint traces of character-istic polychloroprene absorption bands. In control experiments, PIB molecular weight was unaffected by treatment with periodic acid. Treatment with m-chloroperbenzoic acid produced significant degradation, but degradation was largely eliminated in treatment of mixtures of polychloroprene and polyisobutylene, simulating graft degradation experiments. For branches in the molecular weight range of interest (M_n = 100,000) degradation was estimated to be $\sim$ 6%, within experimental error of molecular weight deter-mination. Thus complete backbone degradation was demonstrated, accompanied by very slight degradation of PIB branches.

The degradation was carried out on samples of all graft copolymers prepared at -70°C containing 50% or more PIB. Re-covered branches were characterized by GPC, along with samples of PIB isolated from crude products by n-pentane extraction. Molecular weight data are presented in Table III. With one ex-ception (experiment N28), molecular weights of grafted and homo-PIB's were virtually indistinguishable. The mean deviation of corresponding molecular weights ($1M_g - M_n1/M_n$) for experiments in Table III (excluding N28) was 10% for M_w's and 20% for M_n's. Thus we conclude that in this system, M_n generally provides a good estimate of M_b. (A possible explanation for the deviation of N28 is discussed below). This conclusion supports assumptions based on purely statistical considerations, in previous studies (1, 4, 6).

<u>Effects of Polymerization Variables on</u>
<u>Polyisobutylene Molecular Weight</u>

Effects of initial monomer concentration, solvent composition, and conversion on PIB molecular weight are shown in Figures 9-11,

for both grafted and homopolymer samples. As expected, molecular
weight increased with monomer concentration over the range examined (Figure 9). Molecular weight also increased slightly with
solvent polarity over the range 70%-90% CH_2Cl_2 (Figure 10). Both
trends represent parameter changes favoring chain propagation
over termination, leading to higher molecular weights. Unusually
high molecular weights were obtained at 60% CH_2Cl_2, possible reflecting increased solubility of growing PIB chains in the least
polar solvent composition. This assertion is supported by the fact
that homopolymer molecular weight was substantially higher than
that of grafted PIB at 60% CH_2Cl_2, while at 90% the molecular
weight of grafted PIB was greater. In highly polar media, the
backbone should increase solubility of growing branches, while at
lower polarity the backbone is relatively insoluble and should
decrease solubility of branches. This result leads to the observation that equality of M_g and M_h is less probable when branch
and backbone solubilities in the reaction medium are dissimilar.
It is significant that the experiment with the largest variation
between M_g and M_h (N28) was that conducted in 60% CH_2Cl_2.

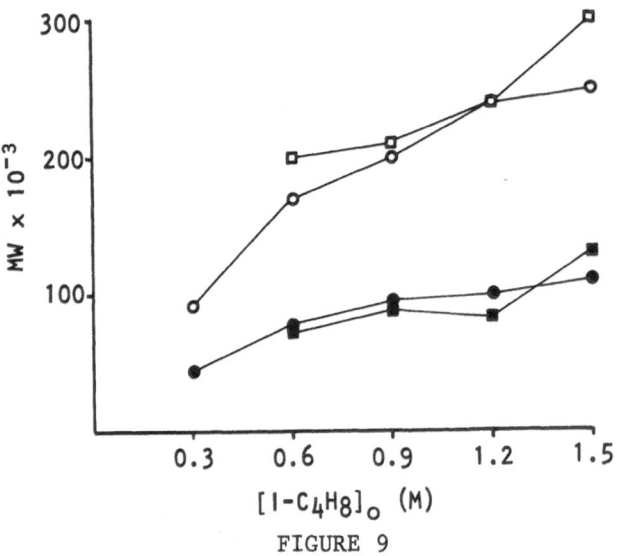

FIGURE 9

Effect of Initial Monomer Concentration on PIB Molecular Weight.
O ●Grafted PIB; □ ■homo-PIB;O□ M_w;● ■ M_n;   T = -70°C;
[PCR] = 10 g/l; [Et$_2$AlCl] = 0.03 M; solvent 80% CH_2Cl_2.

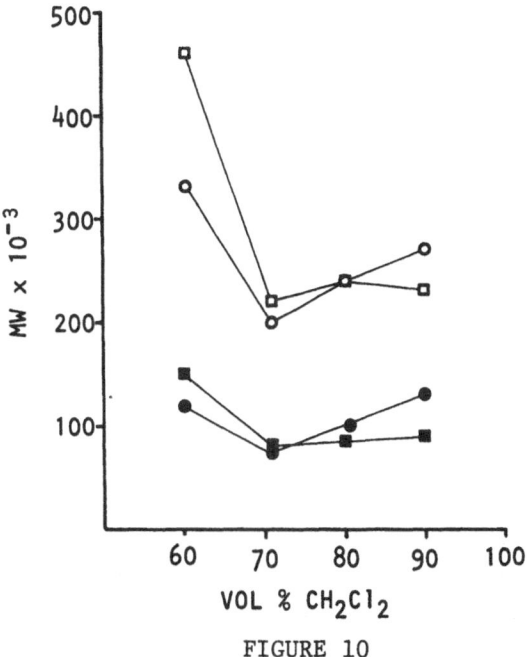

FIGURE 10

Effect of Solvent Composition on PIB Molecular Weight. ○●Grafted PIB; □ ■homo-PIB;○ □M_w;●■M_n. T = -70°C; PCR = 10 g/l; $[i-C_4H_8]$ = 1.2 M; $[Et_2AlCl]$ = 0.03 M.

Conversion has very little, if any, effect on PIB molecular weight. Results of three experiments carried out under conditions are shown in Figure 11. The data suggest a possible slight increase in molecular weight with conversion, but the number of points is insufficient to lend confidence to this conclusion. A similar plot of data from all experiments gave only random scatter.

Effects of Polymerization Variables on Branch/Backbone Ratio:

Also shown in Table III are calculated branch/backbone ratios. Since grafting was accompanied by some backbone scission, ratios were calculated in two ways, reflecting 1) branches per initial backbone chain (b/B*), and 2) branches per backbone in the final product (b/B). Values of b/B* were calculated using equation 5, with M_B = 180,000, as determined by osmometry of the starting material. Values of b/B were obtained by using molecular weight data to adjust b/B^* for backbone scission:

$$b/B = (b/B^*) \times (M_g/M_g^*) \tag{6}$$

Although b/B is important in product characterization, b/B^* is
more informative with regard to the nature of the grafting process,
since it excludes effects of backbone scission.

   Figures 12 and 13 illustrate b/B^* as a function of monomer
conversion and graft copolymer composition. As shown in Figure
12, b/B^* increased approximately linearly with conversion, although substantial scatter is observed due to effects of other
variables, particularly grafting efficiency. As shown in Figure
13, b/B^* increased more systematically with graft composition,
with effects of grafting efficiency variations excluded. The
dashed line in the figure is the calculated relation between
b/B^* and composition for $M_B = 180,000$ and $M_b = 100,000$. Close
conformity of data to the line is a second indication that M_b is
relatively independent of conversion in this system. This result
implies that grafting occurs by sequential growth of individual
chains from each backbone rather than by simultaneous growth of
many branches, since in the latter case, b/B^* would be independent
of composition and M_b would be directly proportional to composition.

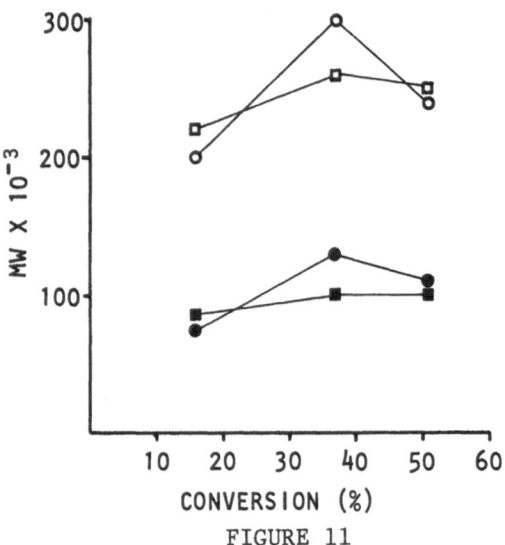

FIGURE 11

Effect of Conversion on PIB Molecular Weight. ○ ●Grafted PIB;
□ ■homo-PIB; ○ □M_w; ● ■M_n.   T = -70°C;   PCR = 10 g/l;
$[i-C_4H_8]$ = 1.2 M; $[Et_2AlCl]$ = 0.03 M; solvent 70% CH_2Cl_2.

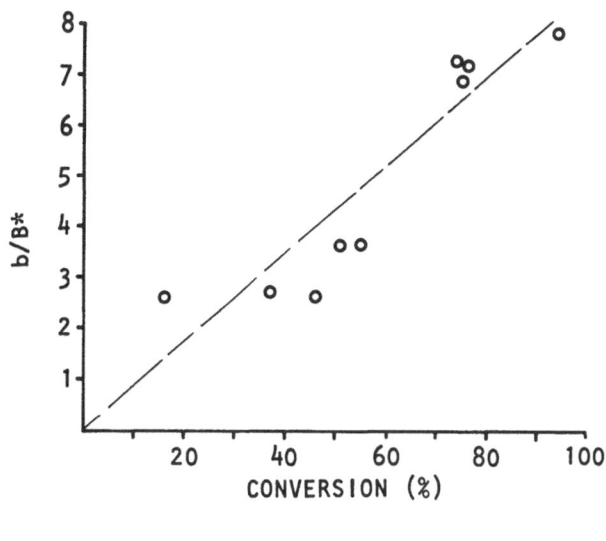

FIGURE 12

Effect of Conversion on Branch/Backbone Ratio

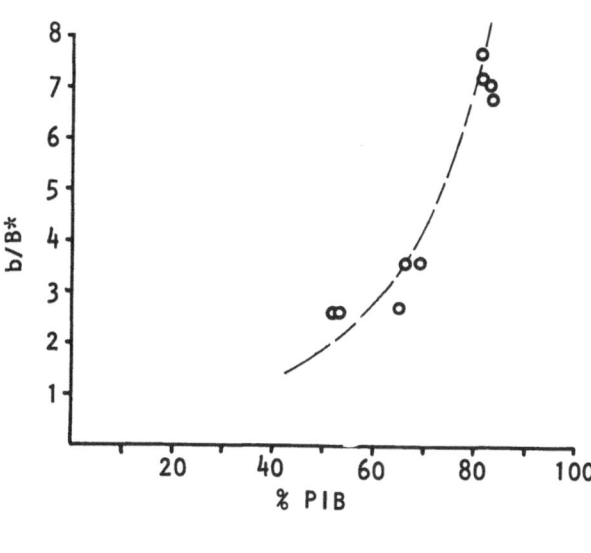

FIGURE 13

Effect of Graft Composition on Branch/Backbone Ratio

CONCLUSIONS

Mechanism of Grafting

Formation of poly(chloroprene-g-isobutylene) occurs by "grafting from" polychloroprene, probably initiated by labile chlorines of 1,2 and isomerized 1,2 structures inherent in the backbone. Propagation of growing branches generally occurs under kinetically similar conditions, leading to similar molecular weights of homo-PIB and grafted branches. Grafting proceeds by sequential rather than simultaneous growth of multiple branches on each backbone.

Polymerization Conditions and Side Reactions

Grafting can be carried out effectively at $-70^{\circ}C$ in methylene chloride/n-pentane mixed solvents, using diethylaluminum chloride as coinitiator. Three important side reactions accompany grafting: Homopolymer formation, backbone scission, and gellation. Homopolymer formation and backbone scission are greatly reduced at low temperatures. Smaller effects of solvent polarity and conversion on homopolymer formation were also observed. Gelation was effectively avoided by use of a hydrocarbon solvent.

Variation of Graft Structure

PIB branch molecular weight can be varied over the range $M_n = 50,000$ to $150,000$ by variation of monomer concentration and solvent polarity. Branch/backbone ratio can be varied by control of conversion.

ACKNOWLEDGEMENTS

This work was supported in part by the National Science Foundation (Grant DMR-75-D4366) and by the Phillips Petroleum Company, through a fellowship to D.K. Metzler.

REFERENCES

1. J. P. Kennedy and R. R. Smith, in Recent Advances in Polymer Blocks, Grafts, and Blends, L. H. Sperling ed., Plenum Press, New York, 1974, p. 303.

2. J. P. Kennedy and A. Vidal, J. Polymer Sci., Polymer Chem. Ed., 13, 1765 (1975); 13, 2269 (1975).

3. J. P. Kennedy and J. Oziomek, J. Appl. Poly. Sci., Symposium Issue, in press.

4. J. P. Kennedy and J. J. Charles Polymer Preprints 16 (1), 393 (1975)

5. J. P. Kennedy and D. K. Metzler, Polymer Preprints 16 (2), 512 (1975)

6. J. P. Kennedy and D. L. Davidson, J. Polymer Sci., Polymer Chem. Ed., in press.

7. J. P. Kennedy and M. Nakao, J. Macromol. Sci., Chem., in press.

8. J. P. Kennedy, J. Macromol. Sci., Chem., A3, 861 (1969).

9. J. P. Kennedy, in Recent Advances in Polymer Blocks, Grafts, and Blends, L. H. Sperling ed., Plenum Press, New York, 1974, p. 5.

10. P. S. Bauschwitz, J. B. Finlay, and C. A. Stewart, in Vinyl and Diene Monomers (Part 2), E. C. Leonard ed., Interscience, New York, 1971, pp. 1149–83.

11. C. A. Hargreave, in Polymer Chemistry of Synthetic Elastomers, J. P. Kennedy and E. G. M. Tornqvist eds., Interscience, New York, 1968, pp.. 227–52.

12. J. T. Maynard and W. E. Mochel, J. Polymer Sci., 13, 251 (1954).

13. M. M. Coleman D. L. Tabb and E. G. Brame, J. Polymer Sci., Polymer Phys. Ed., in press.

14. B. Mandel, J. P. Kennedy and R. F. Kiesel, J. Polymer Sci., Polymer Chem. Ed., in press.

15. J. P. Kennedy, S. C. Feinberg and S. Y. Huang, J. Polymer Sci., Polymer Chem. Ed., in press.

16. C. A. Vernon, J. Chem. Soc., 1954, 423.

17.   J. P. Kennedy and S. Rengachary, Adv. Polymer Sci., 14, 1 (1974).

18.   J. P. Kennedy and R. M. Thomas, Adv. Chem. Ser., 34, 111 (1962).

19.   P. Dreyfuss and J. P. Kennedy, Anal. Chem. 47, 771 (1975).

20.   R. T. Morrison and R. N. Boyd, Organic Chemistry, Allyn and
      Bacon, Boston, 1966, p. 879.

21.   J. P. Kennedy and M. J. le Garlantezec, unpublished results.

22.   W. Thaler, D. J. Buckley, and J. P. Kennedy, U. S. Patent
      3,808,177 (1974).

KINETICS AND MECHANISM OF THE EARLY STAGES OF POLYMERIZATION IN

THE STYRENE-POLYBUTADIENE SYSTEM

W.A. Ludwico* and S.L. Rosen

Carnegie-Mellon University

Pittsburgh, Pennsylvania 15213

INTRODUCTION

The polymerization of an unsaturated rubber dissolved in a monomer has been practiced commercially for over two decades. A common example is the polymerization of styrene in the presence of 5-20% dissolved polybutadiene rubber to produce high-impact polystyrene. The commercial product consists of a dispersion of rubber particles in a continuous glassy phase. These rubber particles enhance the toughness of the otherwise brittle glassy phase. By carrying out the polymerization of the monomer in the presence of the rubber, it is generally agreed that some of the polymerized monomer is grafted to the rubber. The compatibilizing action of this graft copolymer is felt to be important in obtaining the desired mechanical properties. These two-phase polymer systems have been extensively reviewed(1-3).

This type of reaction can be followed on the phase diagram in Figure 1. The diagram is representative of nearly all polymer-polymer-common solvent systems. Despite the fact that each polymer alone is infinitely soluble in the solvent (in this case, the monomer of the glassy polymer), it takes very little of a second polymer to form a heterogeneous system, each layer containing nearly pure polymer. This incompatibility is a direct result of the extremely small entropies of solution for high-polymer pairs. The extent of the homogeneous region depends on the particular system, the temperature, and the molecular weight of the polymers, lower molecular weights and higher temperatures enhancing compatibility. Nevertheless, for polymers with molecular weights high enough to

*Present Address: Mobay Chemical Co., Pittsburgh, Pa. 15205

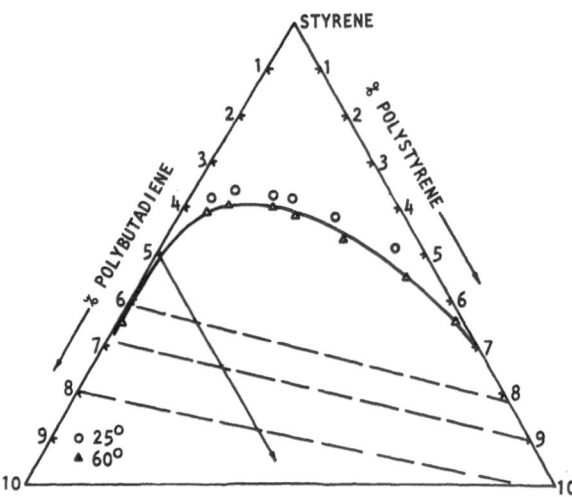

Figure 1. Phase diagram for a styrene-polybutadiene ($M_v = 320,000$)-
polystyrene ($M_v = 360,000$) system. Also shown is the
reaction path for a system containing 5% PBD.

be of commercial interest, the initially homogeneous reaction mass
becomes a heterogeneous system at a very low conversion as the reaction vector reaches the phase boundary on its path parallel to
the monomer-glassy polymer axis.

    If the reaction vector passes to the left of the plait point
(as is the case in most systems of commercial interest), the rubber phase will initially be continuous. As the reaction proceeds,
however, more and more glassy polymer is produced, and a phase inversion occurs, leaving the rubber phase dispersed in a continuous
glassy phase (with small, glassy inclusions trapped in the rubber
particles). The reaction is carried essentially to completion,
giving the final product: 5-20% of micron-sized rubber particles
dispersed in a continuous glassy matrix. It is possible to avoid
phase inversion in quiescent systems. The product then consists
of relatively large glassy globules with rubber in the interstices.
This morphology does not give the desired mechanical properties in
the product and is easily and routinely avoided by agitation. With·
even minimal agitation, the conversion at which phase inversion
occurs is reproducible(4).

PREVIOUS KINETIC STUDIES

    A number of kinetic studies of the initial, homogeneous regime
(which only accounts for a small fraction of the total conversion
in a commercial reaction) have been conducted. Scanlon(5) investigated the effects of dihydromyrcene (DHM; an isoprene dimer) on the
polymerization of styrene, methyl methacrylate, methyl acrylate,
and vinyl acetate. Although the molecular weights of the resulting

polymers could be explained by a simple chain transfer mechanism, he observed a depression in polymerization rate caused by the DHM which could not be explained by a simple chain transfer mechanism. It should be noted that DHM may not be a good model for high molecular weight polyisoprene because of its much higher entropy of solution, i.e., solutions of DHM will more nearly approach true homogeneity than solutions of high molecular weight polyisoprene.

Merrett(6) investigated the polymerization of styrene and methyl methacrylate in the presence of a high molecular weight polyisoprene using benzoyl peroxide (BPO) and azobisisobutyronitrile (AIBN) initiators. He attempted separation of the product into rubber homopolymer, graft copolymer, and glassy homopolymer fractions by solution techniques. He found that a graft fraction was obtained with BPO but not with AIBN, but that the grafted chains were far shorter than the homopolymer chains, contrary to what would be expected from a simple chain transfer mechanism. This led him to postulate direct attack of the rubber chains by BPO radicals, but the number of radicals required to produce the experimentally determined number of grafting sites far exceeded the number of radicals available from the BPO.

Later studies by Allen, Merrett, and Scanlon(7) refuted many of the earlier results. Polymerization of methyl methacrylate in the presence of DHM using BPO and AIBN initiators fit a simple chain transfer mechanism, and no rate reductions were observed. Earlier observations of rate reduction were attriubted to inhibitors in the DHM. With vinyl acetate monomer, they found that the addition of 0.5 mole/l. DHM cut the rate by two orders of magnitude, regardless of initiator. To explain this, they proposed a "degradative chain transfer" mechanism, i.e., radicals produced by the attack of growing vinyl acetate chains on the chain transfer agent (DHM) are resonance stabilized to the point where they are slow to reinitiate polymerization of this monomer.

Allen and Merrett(8) repeated earlier experiments, polymerizing methyl methacrylate in the presence of high molecular weight polyisoprene with AIBN and BPO initiators, and again reached the conclusion that no graft was formed with AIBN initiator. A different separation procedure revealed that the grafted and free poly(methyl methacrylate) had the same chain lengths, in contrast to previous results. Rate reductions were observed with each initiator, eliminating simple chain transfer from consideration and raising the question as to why rate reduction should be observed with AIBN even in the absence of detectable grafting. Note that the Trommsdorff or gel effect would increase the rate of reaction, not reduce it.

Ayrey and Moore(9) used radioactively tagged initiators folflowed by separation of the reaction product to gain further insight into the reaction mechanism. Again, no simple mechanism

could explain their results. Ide, Sasaki, and Deguchi(10) polymerized methyl methacrylate, styrene, and vinyl acetate in the
presence of acrylic rubbers with BPO and AIBN. They separated
their reaction products by solution techniques and claimed grafting for both BPO and AIBN initiators, although more with the former. They also observed rate reductions caused by the rubber,
eliminating a simple chain transfer mechanism from consideration
even though measured chain lengths agreed with the predictions of
a chain transfer mechanism. Interestingly, this investigation
may have carried kinetic measurements into the heterogeneous regime, although it is impossible to tell from the data in the paper. The authors present conversion-versus-time curves carried
to about 96% conversion and label them "bulk" polymerization.
However, their experimental procedure calls for the use of benzene,
an inert solvent, in unspecified amounts. An inert solvent would
delay phase separation, or even prevent it, if the total polymer
concentration remained low enough. Such solvents are not generally used commercially, as they waste valuable reactor space.
The fact that they "poured" their reaction products implies a
rather large percentage of inert solvent, because a mixture of
90% methyl methacrylate and 10% rubber carried to 96% conversion
will not pour, even at their reaction temperature of 75°C. In any
case, further kinetic analyses were based only on rates at 10%
conversion.

Kumar(11) polymerized styrene in the presence of several different molecular weight polybutadienes using AIBN and BPO initiators. He observed rate reductions which increased with the
molecular weight of the polybutadiene and were greater for BPO
than for AIBN. No plausible chemical mechanism could be found
to explain his results adequately.

Kolesnikov and Khamukaeva(12) polymerized styrene in the presence of a styrene-butadiene copolymer rubber with BPO initiator up
to conversions of 40%. Although they never mentioned the undoubtedly heterogeneous nature of their reactions, they observed rate
reductions on the order of 100% at rubber concentrations approaching 10%.

More recently, Riess and co-workers(13-20) have studied the
ABS system: polybutadiene dissolved in styrene and acrylonitrile
comonomers, to which in most cases was added an inert solvent,
benzene.

While there is considerable conflict in prior results, there
appears to be general agreement that the addition of rubber reduces
the rate of polymerization (by an amount greater than what would be
expected from a simple reduction in monomer concentration caused by
the addition of the rubber); systems initiated with benzoyl peroxide (BPO) exhibit greater rate reductions than those initiated
with azobisisobutyronitrile (AIBN); benzoyl peroxide is effective

in producing graft copolymer, while AIBN results in little or no grafted material.

Most previous attempts at explaining the observed results have centered on the chemical role played by the rubber in the reaction, and have been conflicting and unsatisfactory. One of the major difficulties encountered is the uncertainty involved in separating and analyzing grafted and ungrafted components of the reaction product(21). In any event, most of them are not in accord with certain results obtained in this laboratory, initially by Kumar(11), and extended in the work presented here.

EXPERIMENTAL

Phase equilibrium measurements were made on simulated reaction systems of 5%, 10%, 15% and 25% conversion prepared from Firestone Diene 35-NFA polybutadiene ($\overline{M}_v$ = 320,000), Dow Styron 666-K27 polystyrene ($\overline{M}_v$ = 360,000), AIBN and/or BPO initiators, all used as received. After equilibration at 0° or 25°C, the phases were separated at 0° or 25° in a Sorvall RC-2B centrifuge. The phase volumes were measured in calibrated centrifuge tubes and samples of each phase were removed for gravimetric analysis of monomer content, and the initiator concentration in each phase determined by the polarographic technique of Dmitrieva and Bezuglyi(22,23). There were no impurities in either the polymers or the monomer which were detectable by the polarographic measurements. Material balances on initiator always closed to within 2%.

Kinetic measurements were made in simple glass dilatometers, consisting of a spherical bulb of approximately 13 cm^3 volume connected to a 30-cm length of 1-mm precision-bore capillary tubing with a 7/25 ground-glass joint. The geometry and density changes were such that a 10-cm meniscus drop corresponded to a conversion of about 3.6%. All rate data here are initial rates, with conversions generally limited to about 1.5%, and were obtained at $60 \pm .01^\circ$C. Menisci were read with a cathetometer to $\pm$ 0.005 cm at five-minute intervals.

The following chemicals were used in the kinetic measurements.

2,2-Azobisisobutyronitrile (AIBN) from Aldrich Chemical Co. was purified by three recrystallizations from chloroform solution with reagent-grade petroleum ether followed by drying in vacuo.

Benzoyl peroxide (BPO) from K&K Laboratories was purified by three recrystallizations from chloroform solution with methanol, followed by drying in vacuo.

Kinetic runs in styrene for identical concentrations of purified and unpurified initiators gave indistinguishable results, therefore the initiators were subsequently used without further purification.

Styrene monomer (obtained from several sources due to short-ages) was dehydrated with $CaCl_2$ prior to vacuum distillation at 28 in. Hg vacuum. The vacuum was broken with dry nitrogen.

High molecular weight polystyrene was prepared from vacuum-distilled styrene by bulk polymerization at $80^{\circ}C$ with AIBN for 24 hr, which was at least 20 half-lives of the AIBN. The polymer was then dissolved three times in reagent-grade benzene and reprecip-itated dropwise into absolute methanol, dried in a vacuum oven, ground to fine powder, and redried in the vacuum oven. The result-ing polymer had the following properties: $\overline{M}_v$ = 325,000; GPC peak molecular weight = 190,000.

Two low molecular weight, monodisperse, anionically polymerized polystyrene samples were obtained from Pressure Chemicals, Inc., Pittsburgh, Pa., with the following specifications: M = 4000, $\overline{M}_w/\overline{M}_n$ < 1.10; M = 2200, $\overline{M}_w/\overline{M}_n$ < 1.10. These low molecular weight samples were used in kinetic studies without further purification.

Solid Diene 35 NFA polybutadiene was obtained from Firestone. The manufacturer specifies(24) the steric structure as 32-35% cis-1,4 and 7.5% 1,2. An initial batch was purified for kinetic studies by dissolving three times in reagent grade benzene, and dropwise reprecipitating into absolute methanol, followed by drying in vacuo for one month. A subsequent batch was subjected to only one dissolution and reprecipitation. The former had an intrinsic viscosity in toluene at $30^{\circ}C$ of 2.9 dl/g and the latter 2.0 dl/g, indicating that an appreciable portion of low molecular weight material had been fractionated from the first batch during puri-fication. Using the Mark-Houwink constants listed for a poly-butadiene of the most similar steric structure (K = 39 x 10^{-5} dl/g, a = 0.713)(25) gave viscosity-average molecular weights of 270,000 and 160,000 for the two batches.

Liquid polybutadienes from the Lithium Corporation of America were used in kinetic studies without further purificiation: Lithene PM resin, molecular weight = 1300 (manufacturer's specification); Lithene PH resin, molecular weight = 2200 (manufacturer's specification).

2,2-Diphenyl-1-picrylhydrazyl (DPPH) from Eastman Organic Chemicals was used directly out of the bottle.

Viscosity measurements were made with standard glass capil-lary viscometers.

Additional experimental details are available(26-28).

RESULTS AND DISCUSSION

The object of this work is to elucidate the role of dissolved rubber in the polymerization reaction. To this end, kinetic re-sults will be presented in the form of a fractional rate reduction y, which compares the rate of reaction in the system containing rubber to that in a rubber-free system at equivalent monomer and initiator concentrations:

$$y = (r_p/r_{p0})([M]_0/[M])([I]_0/[I])^n \tag{1}$$

where n is the experimentally determined order with respect to initiator concentration, and the subscript zero refers to the system without rubber. In all kinetic experiments for which initiator concentration was not the independent variable, [I] was maintained as nearly constant at 17.5 mmole/l. as individual weighings allowed, so the $([I]_0/[I])^n$ correction was negligible. In all cases, first-order dependence on monomer concentration was assumed. According to this definition, if the rubber acted simply as an inert diluent, y would always equal unity.

Effect of Rubber Level and Initiator on Rate Reduction

Figure 2 shows the effect of rubber level on rate reduction with each initiator, using the high molecular weight polybutadiene which is sold commercially for toughening plastics. The data are limited to about 10% rubber by extremely high solution viscosities. These results are in excellent agreement with those obtained earlier by Kumar for the same system(11) and serve to illustrate the phenomenon of rate reduction and its dependence on the particular initiator used.

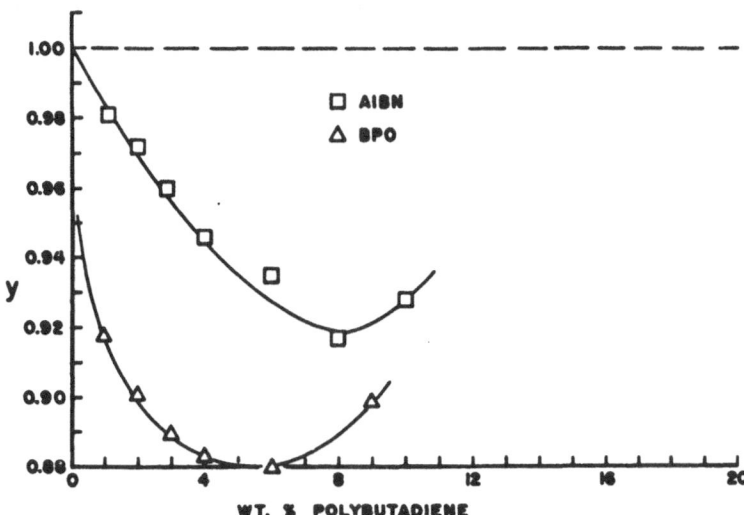

Figure 2. Rate reduction vs weight percent polybutadiene
$(M_v = 270,000)$. [I] = 17.5 mmoles/liter.

Influence of Rubber Molecular Weight

In Figure 3 rate reductions are plotted as a function of rubber level for three different molecular weights of polybutadiene. Again, these results are in excellent agreement with those obtained earlier by Kumar and begin to case doubt on explanations of rate reduction which are based on the chemical reactivity of the rubber. At a given weight percent, the number of repeating units is independent of the molecular weight of the rubber (neglecting end effects, which should be quite small). Kumar (and others) suggested that rate reduction resulted from attack of the rubber by primary free radicals from the decomposed initiator (BPO radicals being more active than AIBN radicals) and that the resulting macroradicals were sluggish in reinitiating polymerization because of resonance stabilization. It is hard to believe, however, that any such resonance effects could be influenced by portions of the rubber molecule more than a few repeating units from the attack site, and therefore, that the results should depend so markedly on molecular weight.

Before proceeding, it must be pointed out that the possibility exists that there could have been differences other than molecular weight in the three polybutadienes employed. Riess has demonstrated an effect of steric structure on rate reduction(17). While no measurements were made of the proportions of cis-1,4, trans-1,4, and 1,2 units in the polymers used here, the two lower molecular weight Lithene resins should have nearly identical steric structures. Furthermore, the manufacturer's representative provided assurances that there were no stabilizers or other species in the Lithene resins which would influence free-radical polymerization. The high molecular weight rubber is specifically formulated by the manufacturer so as not to contain any impurities which would affect

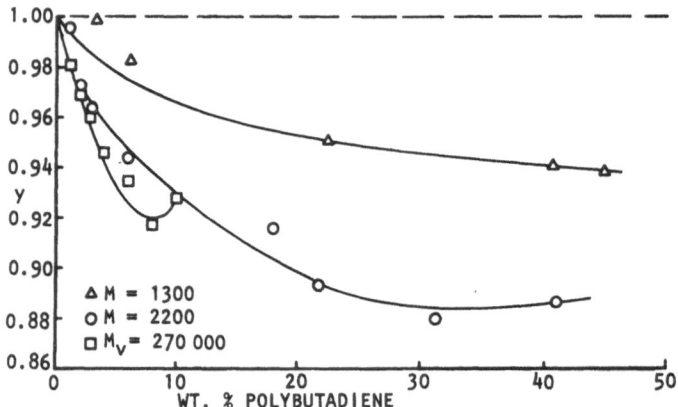

Figure 3. Rate reduction vs weight per cent polybutadiene,
[AIBN] = 17.5 mmoles/liter.

free-radical polymerizations, and in any case, was carefully purified before use.

The Effect of Unequal Initiator Distribution

At this point in the investigation, it was decided to consider the possibility that the observed rate reductions were physical, rather than chemical in origin. In the heterogeneous regime, the half-order dependence of rate on initiator concentration in each phase _must_ give rise to a rate reduction if the initiator is partitioned unequally between the phases. The reason for this is illustrated qualitatively in Figure 4. Quantitatively, each phase may be considered an independent reactor for the conversion of monomer to glassy polymer. Glassy homopolymer formed in the rubber phase diffuses to the glassy polymer phase, and any graft formed diffuses to the interface to stabilize the system. The overall rate of conversion of glassy monomer to polymer in the heterogeneous regime is given by

$$r_p = r_p' (1 - v'') + r_p'' v''$$ (2)

where a single prime denotes the glassy polymer phase, a double prime the rubber phase, and the v's are volume fractions. Under isothermal conditions,

$$r_p' = r_p' ([I]', [M]')$$ (3a)

$$r_p'' = r_p'' ([I]'', [M]'')$$ (3b)

where the overall initiator [I] and monomer [M] concentrations are related to those in each phase by

$$[I] = [I]' (1 - v'') + [I]'' v''$$ (4a)

and

$$[M] = [M]' (1 - v'') + [M]'' v'' \quad .$$ (4b)

The partitioning of monomer and initiator may be represented by distribution coefficients

$$K_I = [I]''/[I]'$$ (5a)

$$K_M = [M]''/[M]'$$ (5b)

which, in general, will be functions of conversion. It might be argued that Eqs. (5a) and (5b) are not necessarily applicable because of the possibility of diffusion limitations in a reacting system. Simple calculations show, however, that because of the

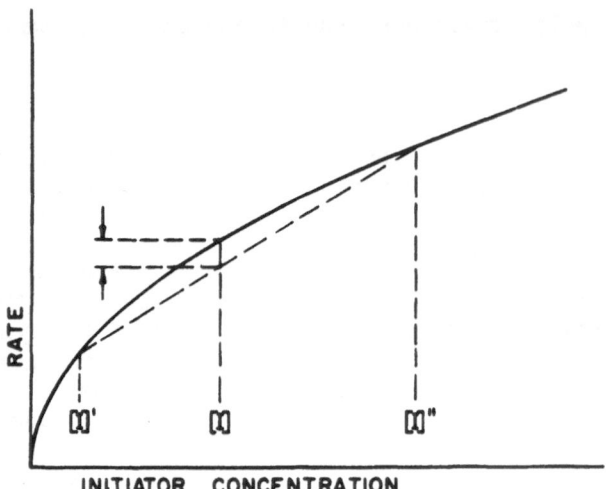

Figure 4. The dependence of rate of polymerization on initiator
distribution. Double prime refers to the polybutadiene
phase; single prime to the polystyrene phase.

very small dimensions of the discontinuities (microns) and the
long initiator half-lives and times of monomer conversion (tens of
hours) in typical systems, these reactions are not limited by diffusion between phases. For the same reason, variations in agitation (beyond that necessary to produce and maintain dispersion)
and the occurrence of phase inversion should not influence the
reaction rate.

If the classical, homogeneous rate expression

$$r_p = k_p \left(\frac{fk_d}{k_t} \right)^{\frac{1}{2}} [I]^{\frac{1}{2}} [M] \quad . \tag{6}$$

is followed in each phase, combining Eqs. (1-6) gives

$$y = \frac{(1 - v'') + K_I^{\frac{1}{2}} K_M v''}{[(1 - v'') + K_M v''] [(1 - v'') + K_I v'']^{\frac{1}{2}}} \quad . \tag{7}$$

If $K_I = 1$, Eq. (7) reveals that $y = 1$ for all v' and K_M, i.e.,
the presence of the rubber producing a second phase does not
alter the overall rate at all. This merely reflects the first-
order dependence of rate on monomer concentration in each phase.
Table I gives y versus v'' for various values of K_I for $K_M = 1$.
A reduction in rate is always obtained. Interestingly enough, the
reductions are similar in nature to those observed in kinetic
investigations.

TABLE I

y Versus v" for Various K_I Values ($K_M = 1$)

v"	K_I = 1	0.6	0.4	0.2	2	4	6	8
1.0	1.0	1.0	1.0	1.0	1.0	1.0	1.0	1.0
0.8	1.0	0.994	0.979	0.930	0.992	0.976	0.983	0.959
0.6	1.0	0.992	0.974	0.927	0.987	0.956	0.935	0.920
0.4	1.0	0.993	0.978	0.945	0.985	0.944	0.912	0.888
0.2	1.0	0.996	0.983	0.971	0.989	0.949	0.912	0.882
0.1	1.0	0.998	0.994	0.985	0.993	0.965	0.935	0.907

On further reflection, the rate reduction described above for the heterogeneous reaction regime due to unequal partitioning of initiator could turn out also to be the explanation for the rate reductions observed in the initial, homogeneous reaction regime. It is doubtful that any solution of a high molecular weight polymer in a low molecular weight solvent is truly homogeneous (whatever that means). In fact, theories of dilute polymer solutions model them as pseudoheterogeneous systems -- polymer molecules with associated "bound" solvent floating around in "free" solvent(29). The size of these polymer-bound solvent domains is a couple of orders of magnitude smaller than the particles in the truly heterogeneous regime -- several hundred angstroms versus a micron, or so. The usual way of defining the phase boundary is to observe the light scattered by the system as the reaction progresses. It is known that turbidity is a function of the difference in refractive index between phases and the particle size of the dispersed phase (increasing with each in the range of interest here)(30). At low conversions, the difference in refractive index can never be very great, since both phases consist largely of the same unreacted monomer. It is entirely possible that the abrupt increase in turbidity associated with phase separation is a result of a rather sudden coalescence of the polymer-bound solvent domains into much larger (micron-sized) particles as more glassy polymer is generated, but without any abrupt change in composition, and therefore with the same sort of initiator partitioning existing before and after visible phase separation. This would imply that there is no discontinuity in rate as the reaction vector crosses the visible phase boundary. None has ever been observed in the course of this work. In summary, the kinetic observations to this point can be explained, at least qualitatively, on the basis of unequal initiator partitioning.

Not only will the distributions of initiator and monomer between phases in these heterogeneous reactions influence the rate of reaction, but they also affect the maximum extent of grafting which can be accomplished. Only that monomer which is converted to polymer within the rubber phase can possibly graft to the rubber.

Monomer which polymerizes in the glassy polymer phase is never in contact with rubber, and so cannot graft to it. This establishes a physical upper limit to the extent of grafting, regardless of the chemical nature of the grafting reactions. This phenomenon has been discussed in detail elsewhere(26,31).

Unfortunately, the measured distributions of initiator and monomer could not account for the kinetic results. Each initiator slightly favored the polystyrene phase ($K_I < 1$), but the K_I's for both were between 0.9 and 1.0 over the (simulated) conversion range of 5%-25%. The monomer favored the rubber phase, with K_M varying from 1.02 at 5% conversion, to 1.05 at 25%. No significant difference in either K was noted between 0° and 25°, so the values obtained at those temperatures should be representative of those at the actual reaction temperature of 60°C. In no case would the experimentally determined K_I, K_M and phase volumes give rise to rate reductions of greater than 1%, according to Eq. (7).

The Effect of Dissolved Polystyrene

As much to check experimental techniques as to document the kinetics in the styrene-polystyrene phase, it was decided to duplicate the previous studies, substituting polystyrene for polybutadiene. The results, for this presumably well understood reaction, the homogeneous polymerization of styrene, were surprising, to say the least.

Figure 5 shows that rate reductions are also observed when the dissolved polymer is polystyrene. At equal concentrations, they were greater for BPO initiator than AIBN, as was the case with dissolved polybutadiene. Furthermore, the rate reductions increase with molecular weight of the dissolved polystyrene, just as they did with dissolved polybutadiene (similar results were obtained with AIBN initiator). Qualitatively, the only difference between dissolved polybutadiene and polystyrene is that ultimately, the dissolved polystyrene accelerated the reaction rate. This is undoubtedly a manifestation of the well-known autoacceleration or Trommsdorff effect, universally attributed to a diffusion-controlled reduction in the termination rate. Acceleration was not observed with polybutadiene, although it appears that it might have been, if data could have been obtained at higher levels of polymer (Figure 2).

While observation of autoacceleration in the polystyrene-styrene system was not surprising, the rate reductions certainly were. Such deviations from classical free-radical kinetics were totally unexpected, and precipitated an intensive review of experimental procedures. It was suggested that perhaps oxidation of monomer and/or polymer in the reaction mass prior to the runs give rise to some inhibiting species. However, bubbling air through the reaction mass for 24 hr (prior to the usual nitrogen purging) did

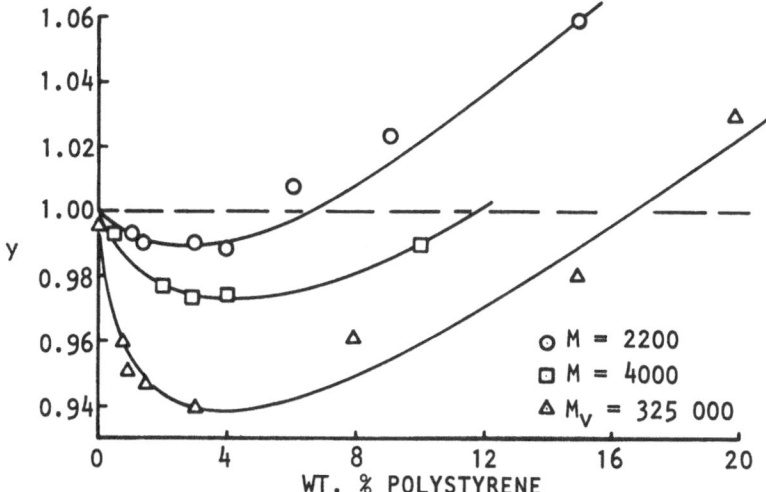

Figure 5. Rate reduction vs weight percent polystyrene,
[BPO] = 17.5 mmoles/liter.

not change the results. In any case, this would not explain the
molecular weight effect.

If reductions in the initial rate of polymerization are produced by dissolved polystyrene, similar rate reductions should be
observed over the course of a reaction carried to high conversion,
caused by the polymer formed in situ earlier in the reaction. To
verify this, several runs were carried up to about 10% conversion,
taking data at 2-min intervals. These data were differentiated by
using a computer technique to provide rate as a function of polystyrene level (conversion). Rate reductions similar to those for
the high molecular weight polystyrene in Figure 5 were again
observed(27,28).

The question arises as to why these rate reductions have not
been noted before in a system which has been studied as exhaustively as the homogeneous, free-radical polymerization of styrene.
For one thing, the effect is not large. The maximum reduction in
rate observed was about 6%. Much of the previous data are not
precise enough to distinguish them from experimental scatter.
Where the data were good enough, reductions were attributed to the
decrease in monomer and initiator concentrations with conversion.
Also, most earlier work was concerned with confirming classical
kinetics, and slight deviations were ignored.

Other Possibilities

At this point, the reasons for rate reduction were not at all
clear, but two other possibilities were investigated. Classical
inhibition tests with 2,2 diphenyl-1-picrylhydrazyl showed that

none of the dissolved polymers used had any significant effect on the efficiency or decomposition rate of either initiator. Furthermore, the rate reductions showed no correlation with the macroscopic viscosity of the reaction mass.

Thus, it appears that the rate reductions observed for some twenty years for systems in which polymerization is conducted in the presence of various unsaturated rubbers are certainly not unique to these types of polymers. In fact, the reactivity of the dissolved polymers does not appear to play a major role at all, because it is generally agreed that no significant branching occurs in the polymerization of styrene, i.e., polystyrene is inert in the polymerization reaction.

At this point, a review of the literature on homogeneous, free-radical polymerization revealed that North and Reed had indeed observed(32) rate reductions similar to those reported here for styrene in their polymerization of methyl methacrylate. They attributed these rate reductions to an enhancement in the diffusion-controlled termination rate of the growing chain radicals. First, they present evidence that the termination rate in homogeneous polymerization is always diffusion-controlled, even at zero conversion. They then break termination into two steps: (a) bulk translational diffusion of two macroradicals together, and (b) the segmental diffusion of the radical chain ends through the macroradical coil to effect termination. While they conclude that the latter step is rate-controlling, they show that the rate of either process is enhanced by reducing the dimensions of the macroradical coil. Horie, Mita, and Kambe have recently presented(33) a much more elaborate analysis of diffusion-controlled termination, with the same qualitative result, namely, that a reduction in the dimensions of the terminating coils enhances the rate of termination.

And that, evidently, is just what the dissolved polymer is doing. The polymer in solution makes the reaction medium a poorer (in a thermodynamic sense) solvent for the growing chain radicals, reducing the overall dimensions of the coils, and therefore enables them to diffuse together for termination more rapidly, at least up to the point at which the concentration of dissolved polymer becomes high enough to produce entanglements, which impede the progress of the chain ends toward each other. Note that the entropic repulsion between growing polystyrene chains and dissolved polybutadiene (which ultimately leads to phase separation) makes entanglements less likely than the case where the dissolved polymer is polystyrene. Therefore, autoacceleration should be deferred to higher concentrations of dissolved polymer, which is just what is observed experimentally.

Fixman analyzed theoretically the concentration dependence of polymer chain dimensions(34). If his results are combined with the results of either North and Reed or Horie, Mita, and Kambe relating dimensions to the termination rate constant, reasonably

good qualitative representations of the molecular weight and concentration dependence of experimental rate reductions can be obtained, at least at lower concentrations, before entanglements
(which Fixman does not treat) become important(28). However,
there are enough adjustable parameters involved to prevent significant quantitative conclusions.

If the above explanation is valid, rate reductions should be
observed simply by replacing the dissolved polymer with a poor,
low molecular weight solvent. Unfortunately, the decomposition
rates of free-radical initiators are known to depend on the solvent
in which they are dissolved, and it would be impossible to separate this effect from true rate reduction. A few DPPH induction
time runs with mixtures of styrene and isopropanol showed this to
be the case, and this line of attack was abandoned.

Effect of the Length of the Terminating Chains

The fact that BPO produces greater rate reductions than AIBN
at equal concentration remains to be explained. In polymer-polymer
systems, compatibility is a function of the molecular weights of
both macromolecular species. If, as postulated, higher molecular
weight dissolved polymer makes the reaction medium a poorer solvent
for growing polymer macroradicals, the same effects should be observed if the molecular weight of the dissolved polymer remains
constant but the length of the terminating macroradicals is increased. Now, at the 60°C reaction temperature, AIBN is a more
active initator than BPO (k_d = 1.46 x 10^{-5} sec^{-1} vs. 4.36 x 10^{-6}
sec^{-1} in styrene)(35). At equal concentrations, therefore, it
generates more initiating radicals per unit time and gives rise to
terminating macroradicals with a lower average chain length, which
should be less subject to the influence of the dissolved polymer.
According to this reasoning, for a given amount, type, and molecular
weight of dissolved polymer, the magnitude of the observed rate
reduction will vary with the length of the terminating chains, no
matter how that chain length is varied.

One means of varying the chain length is to vary the initiator
level. The chain length is inversely proportional to the square
root of initiator concentration in classical, free-radical polymerization. Therefore, increasing initiator concentration should
diminish rate reduction, other things being equal. It does, as
shown in Figure 6, for both initiators. These data compare directly reacting systems with and without rubber, at the same initiator concentrations.

While these results are significant in that they tend to confirm a physical basis for rate reductions, they may be of greater
significance in a negative sense, as they are very hard to rec-
oncile with any chemical interpretation of rate reduction. The
frequency of attack of dissolved polymer molecules by either primary

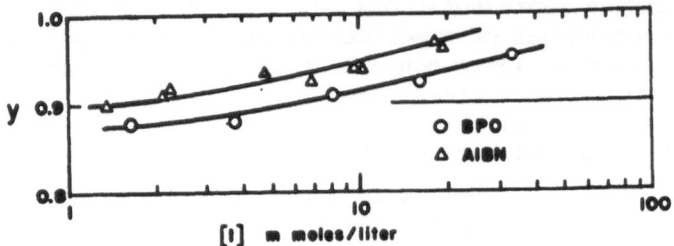

Figure 6.   Rate reduction vs initiator concentration.
            5.7% polybutadiene (M_V = 160,000).

initiator radicals or growing chain radicals should increase with
initiator concentration, whereas, experimentally, the kinetic influence of the dissolved polymer is diminished.

Still another method by which the average chain length may be
reduced is to add a chain-transfer agent to the reaction mass. The
decrease in rate reduction caused by increasing concentrations of
n-dodecyl mercaptan is shown in Figure 7. The data were obtained
at constant rubber level (the same as that used in Figure 6) and
constant AIBN level.

Qualitatively, then, three different methods of varying the
length of the terminating chains (different initiators, varying
initiator level and mercaptan addition) produce the same effect.
But are they really doing the same thing? By using the standard
relations for free-radical polymerization and representative literature values for the necessary constants (Table II) the average
length of the terminating chains may be calculated for all the

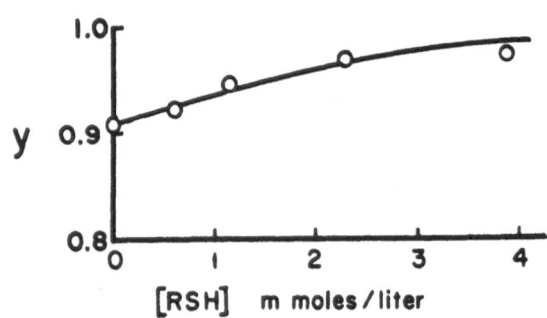

Figure 7.   Rate reduction vs dodecyl mercaptan concentration.   5.7%
            polybutadiene (M_V = 160,000), [AIBN] = 2.13 mmoles/liter.

TABLE II

Rate Constants Used in Calculation of the Number-Average
Length of Terminating Chains

			Reference
k_d (AIBN)	Decomposition rate constant for AIBN	$1.46 \times 10^{-5} \sec^{-1}$	35
k_d (BPO)	Decomposition rate constant for BPO	$4.36 \times 10^{-6} \sec^{-1}$	35
k_p^2/k_t (styrene)	$\dfrac{(\text{propagation constant})^2}{\text{termination constant}}$	1.19×10^{-3}	36
c_S	Chain-transfer constant to dodecyl mercaptan	17	36
c_I (BPO)	Chain-transfer constant to BPO	0.055	37

data in Figures 6 and 7. Rate reduction as a function of this calculated chain length is shown in Figure 8. Considering that the
assumption of a constant k_t is at best an approximation and that
there is considerable variation in literature values for the constants used, and no attempt was made to pick out "optimum" (i.e.,
best-fit) values, the correspondence is remarkably good. Thus, in
systems containing a fixed amount of a particular dissolved polybutadiene, rate reduction is largely, if not exclusively, a function of the molecular weight of the polystyrene being formed in
the reaction.

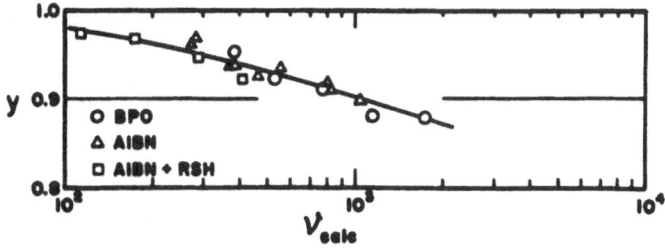

Figure 8. Rate reduction vs calculated number-average length of
terminating chains. 5.7% polybutadiene (M_v = 160,000).

CONCLUSIONS

In the styrene-polystyrene-polybutadiene system studied, rate reductions at low conversions are largely, if not exclusively, physical in origin, resulting from the action of dissolved polymer making the reaction medium a "poorer" solvent for the growing polymer chains. This, in turn, increases the rate at which the chain radicals diffuse together for termination, lowering the overall observed rate of polymerization. Rate reductions increase with the molecular weights of both the initially dissolved polymer and of the polymer being formed in the reaction, as expected on the basis of thermodynamic compatibility.

Acknowledgements

This work was supported by a Fellowship from the Standard Oil Company of Ohio and by National Science Foundation Grant No. GK-40913. The authors wish to thank Miss Sharon Smith for making the viscosity measurements, and Drs. G. Berry, R. Carlin, and T. Fox for their helpful suggestions.

References

1. S.L. Rosen, Trans. N.Y. Acad Sci., 35, 480 (1973).

2. N.A.J. Platzer, Ed., Adv. Chem. Ser., 142, American Chemical Society, Washington, D.C., 1975.

3. P. Bruins, Ed., Polyblends and Composites (Appl. Polym. Symp., 15), Wiley-Interscience, New York, 1970.

4. G.E. Molau and H. Keskkula, J. Polym. Sci. A-1, 4, 1602 (1966).

5. J. Scanlon, Trans. Faraday Soc., 50, 756 (1954).

6. F.M. Merrett, Trans. Faraday Soc., 50, 759 (1954).

7. P.W. Allen, F.M. Merrett, and J. Scanlon, Trans. Faraday Soc., 51, 95 (1955).

8. P.W. Allen and F.M. Merrett, J. Polym. Sci., 22, 193 (1956).

9. G. Ayrey and G.C. Moore, J. Polym. Sci., 36, 71 (1959).

10. F. Ide, I. Sasake, and S. Deguchi, J. Appl. Polym. Sci., 15, 1791 (1971).

11. A. Kumar, Ph.D. Thesis, Carnegie-Mellon University, 1972.

12. G.S. Kolesnikov and I.A. Khanukaeva, Int. Chem. Eng., 8 (4), 698 (1968).

13. J.L. Locatelli and G. Riess, Angew, Makromol. Chem., 27, 201 (1972).

14. J.L. Locatelli and G. Riess, Angew. Makromol. Chem., 28, 161 (1973).

15. J.L. Locatelli and G. Riess, Angew. Makromol. Chem., 32, 101 (1973).

16. J.L. Locatelli and G. Riess, Angew. Makromol. Chem., 32, 117 (1973).

17. J.L. Locatelli and G. Riess, Angew. Makromol. Chem., 35, 57 (1974).

18. J.L. Locatelli and G. Riess, J. Polym. Sci. Polym. Chem. Ed., 11, 3309 (1973).

19. J.L. Locatelli and G. Riess, J. Polym. Sci. Polym. Letters Ed., 11, 257 (1973).

20. J.L. Refregier, J.L. Locatelli, and G. Riess, Eur. Polym. J., 10, 139 (1974).

21. E.R. Wagner and R.J. Cotter, J. Appl. Polym. Sci., 15, 3043 (1971).

22. V.N. Dmitrieva and V.D. Bezuglyi, Vysokomoll. Soedin., 4, 1672 (1962).

23. V.D. Bezuglyi and V.N. Dmitrieva, Zavodskaya Lab., 24, 941 (1958).

24. G. Alliger, B.L. Johnson, and L.E. Forman, Synthetic Rubber Facts, Firestone Tire and Rubber Co., Akron, Ohio, 1964.

25. J. Brandrup and E.H. Immergut, Eds., Polymer Handbook, 2nd ed., Wiley, New York, 1975.

26. W.A. Ludwico and S.L. Rosen, J. Appl. Polym. Sci., 19, 757 (1975).

27. W.A. Ludwico and S.L. Rosen, J. Polym. Sci.:Polym. Chem. Ed., 14, 2121 (1976).

28. W.A. Ludwico, Ph.D. Thesis, Carnegie-Mellon University, 1974.

29. P.J. Flory and W.R. Krigbaum, J. Chem. Phys., 18, 1086 (1950).

30. B. Conaghan and S.L. Rosen, _Polym. Eng. Sci._, _12_ (2),
     134 (1972).

31. S.L. Rosen, _J. Appl. Polym. Sci._, _17_, 1805 (1973).

32. A.M. North and G.A. Reed, _Trans. Faraday Soc._, _57_, 859 (1961).

33. K. Horie, I. Mita, and H. Kambe, _Polym J._, _4_, 341 (1973).

34. M. Fixman, _Ann. N.Y. Acad. Sci._, 89, 657 (1961).

35. National Polychemicals, Inc., Technical Bulletin XIN-01-0264.

36. P.J. Flory, Principles of Polymer Chemistry, Cornell Univ.
     Press, Ithaca, N.Y., 1953.

37. R.W. Lenz, Organic Chemistry of Synthetic High Polymers,
     Wiley, New York, 1967.

GRAFT COPOLYMERIZATION OF VINYL MONOMERS WITH SYNTHETIC RUBBERS

IN THE PRESENCE OF COMPLEXING AGENTS

R. V. Kucher, Ju. S. Zaitsev, A. A. Kuznetsov,
V. V. Zaitseva
Institute of Physical Organic Chemistry and Coal
Chemistry, Academy of Sciences of the Ukrainian SSR
340048, Donetsk-48, R. Luxemburg Str., 70

An increase in the process rate and the formation of vinyl copolymer alternating structures in the presence of complexing agents (CA) presupposes that graft copolymerization may involve a specific grafting reaction on elastoplastic macromolecules and a change of the morphology of the systems under investigation. Taking into account that copolymerization of vinyl monomers with synthetic rubber is a heterophase process, one should expect heterogeneity to increase upon introducing a CA, and it will be ambiguous in different stages of conversion. In this paper we have presented the experimental results of the influence of zinc halides on the reaction rate of graft copolymerization, the morphology and the composition of the resultant copolymers. The bulk copolymerization of styrene(S) with acrylonitrile(AN) and rubber(R), of styrene with methyl methacrylate(MMA) and rubber at initial and high conversion levels was carried out at 75° and 100°C using 0.001 mole of benzoyl peroxide(BP) as initiator per mole of a mixture of the monomers. Butadiene rubber "Inten-55NFA" ($M=1.7 \times 10^5$) containing 35% of 1.4 cis-, 54% of 1.4 trans-chains and 11% of 1.2 groups and butadiene styrene thermoplastic (DST-30,50 and 80) containing 31,50, and 80 mole % styrene chains were used as elastoplastics. The rubbers were purified by precipitating twice with methanol from benzene solutions. The zinc halides were purified by the procedure of Golubjev et (1) and used at levels of from 0 to 0.05 mole per mole of AN or MMA. The ratio of vinyl monomers was either the azeotropic composition or equimolar. The copolymerization conversion was determined by the gravimetric method. The distribution coefficient was calculated from the formula

$$K_D = c_{R.ph.}/c_{C.ph.} \quad c_{R.ph.} \text{ and } c_{C.ph.}$$

Where $C_{R.ph.}$ and $C_{C.ph.}$ are the concentrations of the component being determined in the rubber and copolymer phases, respectively. The monomer concentrations in both phases were measured using a "Vyruchrom" chromatograph employing 15% Apiezon-L on chromosorb W 60 as the stationary phase, the temperature of the evaporator and the column being 180 and 140°C, respectively. The concentrations of BP and zinc halides were measured by iodometry (2) and complexonometry (3), respectively. The model system was separated into the rubber and copolymer phases using a centrifuge with g = 6500.

A study of the graft copolymerization in the initial stages allows us to conclude that the relative influence of zinc halides on the reaction rates is (4): $ZnCl_2 \rangle ZnBr_2 \rangle ZnJ_2 \rangle ZnHaL_2 = 0$.

The most interesting results have been obtained by studying the high conversions. Figure 1 shows, for example, the variation of the process rate with conversion for various types of rubbers in the presence of zinc chloride. It is significant that the self acceleration of the process (gel-effect) observed in the copolymerization of styrene with acrylonitrile and rubber shows up at higher levels of complexing agent and finally disappears. Thus the conditions for carrying out graft copolymerization at a constant rate practically as far as 100% conversion have been found.

The influence of the concentration of zinc halides for instance, zinc chloride, is clearly seen from Fig. 2. The rate is proportional to the zinc chloride concentration at low conversions (~10%). An increase in conversion results in the gel effect over the zinc chloride concentration range 0 to 0.01 mole/mole of AN. With further increase of the CA, the process rate acquires a linear nature and the values fall on a straight line. The observed change in the rate may be explained by the influence of the CA on all elementary graft copolymerization reactions.

Upon replacing butadiene rubber by butadiene-styrene rubber, the rate increases (Fig. 1, curves 1,2) at the cost of a decrease in the formation of macroradicals of the allyl type in a chain transfer reaction to the elastoplastic macromolecules.

The rate of graft copolymerization changes with extremes in the S-MMA-R system. This is due to formation of stable complexes of Lewis acids with MMA and an initiator containing a carbonyl group (5). An increase in the rate constant of the initiator decomposition takes place together with an increase in the reactivity of the complexed MMA. This favors a more intensive rise of the concentration of the active centers in the system studied. The major portion of the active centers is destroyed before a conversion level of 50-60% is reached, the process rate falls sharply, approaching thermal copolymerization (Fig. 3, curves 3,4).

The change in the composition of the grafted and free copolymers of styrene with acrylonitrile (SAN) and methyl methacrylate (MS) with conversion level is presented in Table I.

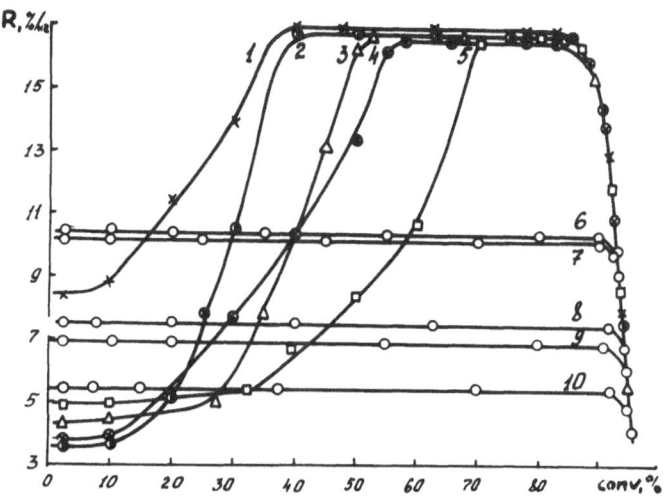

Fig. 1. Relationship between the overall rate of copolymerization and the conversion of the S-AN-R system at 75°C: BP
concentration is 0.001 mole/mole of mixture; the monomer ratio is
azeotropic (2-5, 7-10) and 50:50 mole% (1,6) with various moles of
$ZnCl_2$/AN; Inten-55NFA: 2-0, 3-0.004, 5-0.006, 7-0.040, 9-0.020, and
10-0.010; DST: 1-0, 4-0, 6-0.030 and 8-0.030.

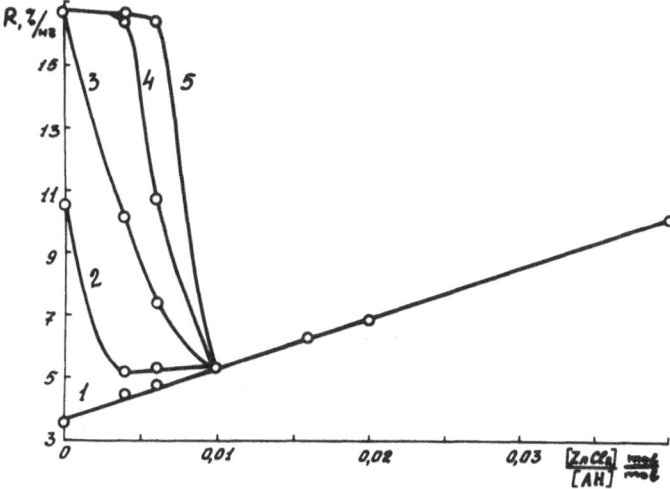

Fig. 2. Dependence of the overall rate of copolymerization of
styrene with acrylonitrile(azeotropic ratio) and Inten-55NFA(5wt%)
on the zinc chloride concentration with various conversions (%):
1-10, 2-30, 3-40, 4-60 and 5-80.

Table I.
The composition of grafted and free SN or MS - copolymers (rubber concentration 5 weight %, PB = 0.001 mole/mole of mixture; copolymerization temperature 75°C)

Type of rubber	ZnHaL$_2$	mole ZnHaL$_2$ / mole AN(MMA)	Conversion,%	Contents (mole%) of AN- or MMA-chains in	
				graft copolymer	free SN or MMA copolymer
*Inten-55NFA	–	0	2.3	35.3	37.2
	–	0	3.4	34.6	38.1
	–	0	5.3	35.1	38.8
*Inten-55NFA	ZnCl$_2$	0.01	3.2	49.9	39.2
		0,01	9,3	49.4	38.6
		0,01	21.8	–	36.5
		0.01	46.8	47.5	37.0
**DST-30	–	0	15.2	46.1	43.8
	–	0	42.3	–	40.7
	–	0	59.4	37.4	39.7
**DST-30	ZnCl$_2$	0.03	25.7	49.9	49.8
		0.03	45.1	49.8	49.9
		0.03	63.4	49.7	49.8
***Inten-55NFA	ZnCl$_2$	0	60.7	49.0	61.0
		0.03	60.7	50.0	50.0

Monomer ratio: *S-AN - azeotropic; **S-AN - 50:50mole%
          ***S-MMA - 50:50 mole %, the initiator
          is dicumyl peroxide 0.001 mole/mole
          of mixture, temperature 110°C.

The data of Table I show that the composition of graft copolymers tends to be an equimolar one irrespective of the composition of the initial mixture. In addition, a free SAN is enriched in styrene when the monomer ratio of S and AN is azeotropic, and in the absence of the CA it contains more acrylonitrile chains than a graft one. It may be explained by the heterogeneity of the process whose consequences are a non-uniform distribution of monomers, the initiator and the complexing agent being partitioned between the rubber and copolymer phases. The values of distribution coefficients are given in Table II, which implies that BP and S are more strongly solvated by the rubber phase.

This distribution becomes more uniform with increasing content of styrene chains in the rubber macromolecule (DST-30, 50 and 80).

Upon raising the process level, K_D varies ambiguously (Fig.4). For example, in the presence of the rubber Inten-55NFA, a directly proportional change of the distribution coefficients is observed before a conversion level of 10% is reached, with K_{ZnCl_2} and K_{AN} decreasing by a factor of about 2.1 and 1.1, respectively, and with K_{BP} and K_S rising only slightly. Our K_D - values obtained with model systems have been extended to cover the real values (graft polymerization),   as was demonstrated by Ludwico and Rosen (6) on the styrene-rubber system. The errors can be ignored in this case. In studying the morphology of ABS - plastics it was detected that by increasing the concentration of zinc chloride from 0 to 0.02 mole/mole of AN, the size of rubber particles increases about 10 times, and the major fraction (with a particle size of 3.2) constitutes about 70%.

The thermo stability of the copolymers obtained has been studied by the method of thermogravimetry and is characterized by a number of parameters whose values are presented in Table III.

It is clear from the data of this table that copolymers synthesized in the presence of zinc halides have high thermo stability, and the principal contribution to the value of the E_D - parameter for ABS plastics is made by the gel-fraction.

Thus introduction of complexing agents into a polymerization system allows the process to be carried out at high rates, obtaining of graft polymers of alternating composition, control of the size of particles in the rubber phase, and improvement in the thermo stability of ABS - plastics.

Table II.

Distribution coefficients for S-AN-R systems with modelling conversion of 5% (S-AN ratio is azeotropic, concentration of rubber
and DP is 5wt.% and 0.001 mole/mole of mixture, respectively, temperature of measurement is 20°C).

K_D	:mole $ZnCl_2$:mole AN	Type of rubber			
		: Inten-55NFA:	DST-30:	DST-50:	DST-80
K_S	0	1.21	1.11	1.09	0.96
	0,03	1.13	1.06	1.04	0.98
K_{AN}	0	0.76	0.80	0.86	0.97
	0.03	0.82	0.89	0.93	0.99
K_{PB}	0	1.19	1.14	1.06	1.03
	0,03	1.05	1.02	1.00	0.99
K_{ZnCl_2}	0.03	0.32	0.46	0.60	-

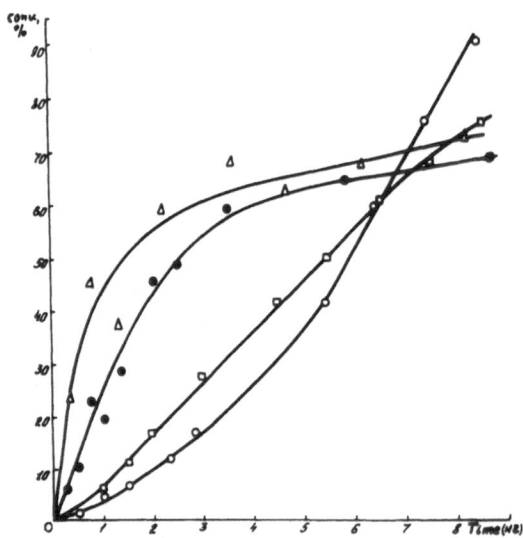

Fig. 3. The conversion time curves for the S-MMA-Inten-55NFA
system(5wt%) with equimolar monomer ratio in the presence of 0.001
mole dicumyl peroxide per mole of monomer mixture at 110°C and at
various concentrations of zinc chloride(mole/mole of MMA): 1-0,
2-0.006, 3-0.050 and 4-0.030.

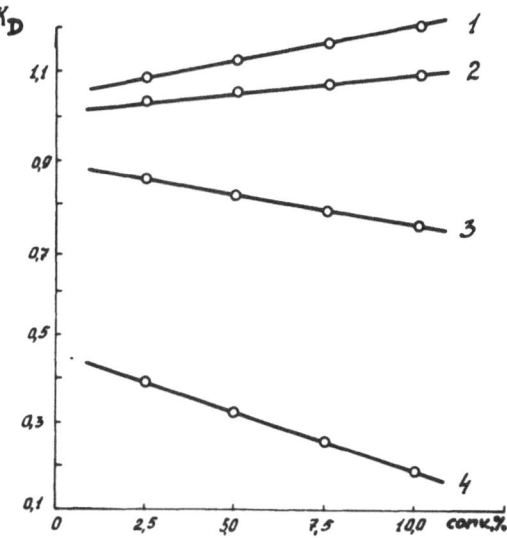

Fig. 4. Dependence of K_D on conversion for the S-AN-Inten-55 NFA model system(5wt.%)with azeotropic monomer ratio at 20°C: 1-K_S, 2-K_{PB}, 3-K_{AN} and 4-K_{ZnCl2}.

Table III.
Destruction parameters of ABS - plastics

Polymer Type	:mole ZnHaL$_2$:mole AN	T_N*,°C	T_M*,°C	ZnHaL$_2$	E_D**ccal/mole
	0	319	387	–	43.0
	0.006	329	410	ZnCl$_2$	49.7
ABS	0.040	330	400	ZnCl$_2$	48.1
	0.030	327	390	ZnJ$_2$	67.0
	0	319	387	–	–
Free SAN	0.040	318	390	ZnCl$_2$	–
	0.030	320	395	ZnJ$_2$	–
Gel-	0	290	385	–	22.9
fraction	0.040	290	395	ZnCl$_2$	41.9
	0.030	323	400	ZnJ$_2$	44.5

*T_N and T_M are the starting breakdown temperature and the temperature of maximum breakdown rate, respectively.
**E_D is the quantity characterizing the relative activation energy of destruction at a heat rate of 6°/min.

SUMMARY

   The influence of Lewis acids on copolymerization reaction was
studied in relation to the following systems: styrene-acrylonitrile-
rubber and styrene-methyl methacrylate-rubber. The findings show
the rate of the process to be dependent on the nature and concentration of complexing agents. The self-acceleration of the process
with increasing complexing agent concentration observed in the copolymerization of styrene with acrylonitrile and rubber manifests
itself at higher conversions and finally degenerates. This allows
the polymerization process to be carried out to high conversions
(85-90%) at a constant rate. For the styrene-methyl methacrylate
system in the presence of zinc chloride, a sharp decrease in rate
takes place with the conversion level of about 40-50%. The graft
and free copolymers differ in compositions due to different distributions of monomers, initiators and complexing agents in the
copolymeric and rubber phases. This distribution becomes more
uniform with increase in the contents of styrene chains in the
rubber macromolecule. The copolymers obtained show high thermo
stability.

REFERENCES

1.   N. Golubjev, V.P. Subov, L. T. Valujev, G. M. Naumov, V. A.
     Kabanov, V. A. Kargin, Vysokomol. sojid., All, 2686, 1969.

2.   T. A. Rasuvajiv, N. A. Kartashova, L. S. Boguslavskaja, Jh.
     O. Ch. 1, 1926, 1965.

3.   R. Prshibbie, Complexons in Chemical Analysis, 1, L., M.,
     1955, s.57.

4.   A. A. Kuznetsov, Ju. S. Zaitsev, Ja. V. Demko, R. V. Kucher,
     B. R. Subov, Doklady Akademii Nauk Ukr. SSR, ser. B, 5, 441
     1975.

5.   A. A. Kuznetsov, Ju. S. Zaitsev, Doklady Akademii Nauk Ukr.
     SSR, ser. B, 5, 433, 1976.

6.   N. A. Ludwico, S. L. Rosen, J. Appl. Polym. Sci., v. 19, 3,
     s. 757, 1975.

VISCOELASTIC PROPERTIES OF POLY(ETHYLENE-G-STYRENE)

J. Diamant, D. R. Hansen* and M. Shen

* Shell Development Company, Houston, Texas 77001
Department of Chemical Engineering
University of California
Berkeley, California 94720

ABSTRACT

The Time Temperature Superposition Principle was applied to
the stress relaxation isotherms of two samples of poly(ethylene-g-
styrene). The graft copolymers were prepared by mutual irradiation
of polyethylene powder and styrene monomer in a nuclear reactor
source, one containing 26% (PEGS/26) and the other 58% styrene
(PEGS/58). The shift factor data when plotted versus the reciprocal
of temperature yielded a straight line in the region of -20°C to 70°C
with an activation energy of 32 kcal/mole. From the viscoelastic
master curves the relaxation spectra were determined by the second
approximation method. The dynamic moduli computed by the intercon-
version technique can then be shifted from the logarithmic time axis
to temperature axis by using the shift factor data obtained in
stress relaxation isotherms. The calculated curves were compared
with the experimental dynamic mechanical data. Agreement was semi-
quantitative for the storage moduli, but only qualitative for the
loss moduli.

INTRODUCTION

Application of the Time Temperature Superposition Principle
to the construction of viscoelastic master curves has been eniment-
ly successful for homogeneous amorphous polymers. (1-3) Fro semicrystalline polymers, however, there does not seem to be a complete agreement in the literature on its applicability. Some
workers have found it necessary to employ the vertical shift factors; (4-7) while others showed that simple horizontal shifting
would be sufficient provided that the strain does not exceed the
linear viscoelastic range and that the temperature remains in the
region in which the degree of crystallinity does not change. (8-11)

More recently there has been considerable interest in the
studies of the viscoelastic behavior of heterogeneous polymers,
such as block and graft copolymers. Because of their heterophase
nature, they are generally thermorheologically complex in the sense
that the Time Temperature Superposition Principle is not strictly
valid. It is nevertheless still possible to construct viscoelastic
master curves by using simple horizontal shifting along the logarithmic time axis if the procedure is restricted to the region
where only one relaxation mechanism dominates, (12,13) or if some
method is available for separating the contributing relaxation
mechanisms. (14,15)

In this work, the viscoelastic behavior of a semicrystalline
graft copolymer will be studied. This material is rather complex
in that it is both semicrystalline and heterophase in nature. The
specific polymer chosen is poly(ethylene-g-styrene), primarily
because the respective parent homopolymers have already been extensively investigated. Previous workers have studied various
aspects of the mechanical properties of this graft copolymer. (16-18)
The purpose of the present communication is to scrutinize the applicability of Time Temperature Superposition Principle to semicrystalline multicomponent polymers.

EXPERIMENTAL TECHNIQUES

Polyethylene (PE) pellets were obtained from the Phillips
Petroleum Company as Marlex 6009. According to the manufacturer's
information, this polymer has a density of 0.960 gm/cc, and a melt
index of 0.9. Its M_w and M_n are respectively 133,000 and 12,000.
The pellets were dissolved in boiling xylene. Polyethylene powder
was obtained by cooling the 1% solution to ambient temperature. It
was subsequently washed with benzene and with methanol. The dried
powder was then mixed with distilled styrene monomer and methanol
and sealed in ampules after a nitrogen purge. Methanol was added
to increase the grafting. (19-20) The ratio of styrene monomer to
methanol was 25/75 and 100/0 respectively for samples PEGS/58 and
PEGS/26. Grafting was carried out using radiation from a nuclear
reactor during the reactor cool-down immediately after the control
rods were inserted into the core of the reactor. Dosimeters from
Far West Technology were used to measure the exposure. Samples
PEGS/58 and PEGS/26 received 0.5 and 0.2 M rads of radiation respectively. The distance between the nuclear reactor core and the
samples determined the dosages. After irradiation, the polymers
were washed with methanol and dried in vacuo. Homopolymer styrene
was extracted by repeatedly washing with benzene. Sample PEGS/58
was extracted with benzene in a Soxlet extractor for an additional
24 hours with only a four percent change in composition. After
washing the compositions were determined by weight-gain and by
carbon-hydrogen (C/H) analysis. It was found that the samples

receiving the lower radiation dosage contained 26% styrene by
weight and by C/H analysis and will be designated PEGS/26. The
other sample which received the larger dosage and had the methanol
accelerators contained 58% styrene by weight and 56% by C/H analysis and is designated as PEGS/58.

Samples were molded into sheet forms in a hydraulic press at
150°C and initial pressure of 3,000 psi for 5 minutes and cooled
to room temperature approximately over a 4 hour period. For stress
relaxation experiments, the sample was cut to the approximate dimensions of 0.15 x 0.5 x 5.0 cm^3. An Instron Universal Testing
Machine Model TMS equipped with a Conrad-Missimer Temperature
Chamber was used for these experiments. For moduli above 10^9
dynes/cm^2, the flexural mode was used; and below that value the
tensile mode was employed. The strains were kept to less than
0.5% in order not to exceed the limits of linear viscoelasticity.
Dynamic mechanical measurements were performed on a Rheovibron
Viscoelastometer Model DDV-II. For this purpose thinner (~0.02 cm)
samples were used. They were prepared by compression molding at
the same temperature and pressure. The polymer was subsequently
cooled slowly to 120°C, and then annealed for 48 hours at 120°C.

RESULTS AND DISCUSSION

The stress relaxation isotherms of PEGS/26 and PEGS/58 are
shown in Figures 1 and 2 respectively. Moduli obtained by flexural
and tensile modes are seen to be consistent. Viscoelastic master
curves (Figure 3) for these samples were constructed from isotherms
by simple horizontal shifting along the logarithmic time axis. The
extraordinarily wide span of time (around 30 logarithmic decades)
covered by these master curves is not unusual for semicrystalline
polymers.[6] The more precipitous drop in relaxation modulus for
PEGS/58 is obviously a consequence of the high polystyrene content,
which undergoes glass transition in this region.

In constructing the master curves for amorphous polymers,
the rubber elasticity correction factor (T/T_0) was often introduced:

$$E_r (T, a_T t) = E_r(T_0, t) \ T/T_0 \qquad (1)$$

where T_0 is the reference temperature. The correction is valid in
such instances since most of the superposition was applied to the
temperature region of T_g to $T_g + 100°C$ of amorphous polymers. For
our graft copolymer systems, the utility of this correction is not
clear due to the semicrystalline nature of the material, therefore
the correction was not used. Similar conclusions were drawn by
Faucher[10] and by Nakayasu, Markowitz and Plazek[11] for PE.

The shift factor data (log a_T) used for the superposition

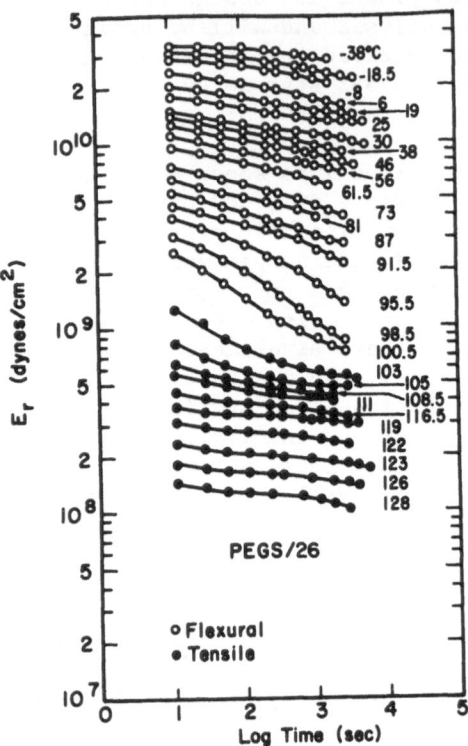

Figure 1:   Stress-relaxation isotherms for poly(ethylene-
g-styrene) containing 26% styrene.

are given in Figure 4 as a function of reciprocals of temperature.
From the slope of this plot, an activation energy of 32 kcal/mole
was obtained for the temperature range of –40°C to 70°C.   This
value is in good agreement with those reported by Faucher for pure
polyethylene (10) and by Locke & Paul for PEGS. (18) Above 70°C the
degree of crystallinity of PE has been found to decrease with increasing temperature. (10)   Thus, it is not surprising to find a
discontinuity at this temperature for the graft copolymer.   In
their dynamic creep measurement of PE, Nakayasu, et al. (11) have
found the activation energy of 28 kcal/mole.   These workers,
however, used a procedure that separated out two contributing
mechanisms to the viscoelastic relaxation which was not adopted
in this paper.

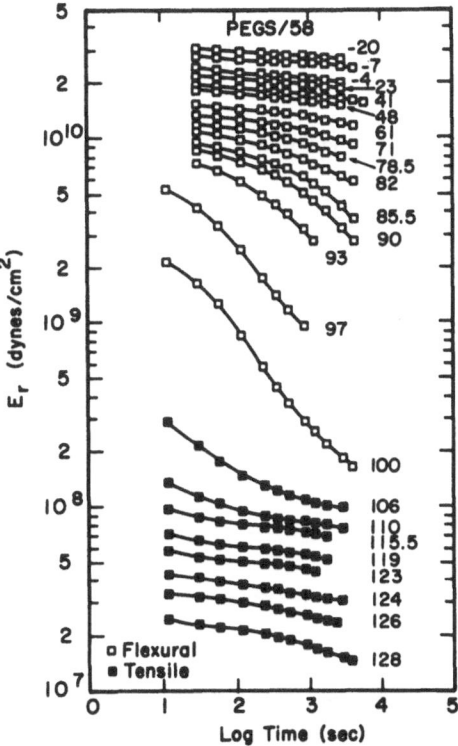

Figure 2: Stress-relaxation isotherms for poly(ethylene-g-styrene) containing 58% styrene.

From the viscoelastic master curves, relaxation spectra were calculated using the Ferry-Williams second approximation method:[2]

$$H_{2B}(\tau) = -2.303 \ M(m) \ E(t) \ dlog \ E(t)/d \ log \ t|_t = \tau \qquad (2)$$

where τ is the relaxation time, m is the slope of the doubly logarithmic plot of the relaxation spectrum, and $M(m) = 1/\Gamma \ (1+m)$ where Γ is the gamma function. Dynamic moduli were then computed by[2]

$$E'(\omega)|_{\omega=1/\tau} = E(t) + H(\tau) \ \psi \ (m)|_{\tau=t} \qquad (3)$$

and

$$E''(\omega)\big|_{\omega=1/\tau} = \left(\frac{H\pi}{2}\right) \; \mathrm{Sec} \; \left(\frac{m\pi}{2}\right)\bigg|_{\tau=1/\omega} \tag{4}$$

where $H = H_{2B}/2.303$ and $\psi(m) = (\pi/2) \; \csc(m\,\pi/2) - \Gamma(m)$.

These computed viscoelastic functions are plotted against logarithmic frequency or reciprocal time scales. Then by using shift factor data given in Figure 4, they are converted to temperature axis. The results are given in Figures 6 and 7 for PEGS/26 and in Figures 8 and 9 for PEGS/58. Also given in the same figures are similar data separately determined on the Rheovibron at corresponding frequencies in the same temperature range. It can be seen that there is reasonable agreement between the calculated and experimental values of dynamic storage moduli for both samples, particularly in the regions below 110°C. Above this temperature

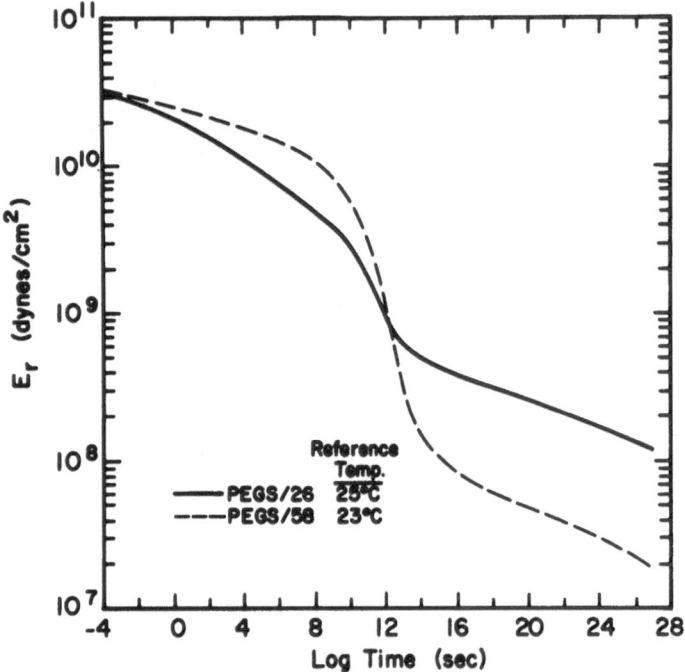

Figure 3: Viscoelastic master curves obtained by simple horinzontal shifting along the logarithmic time axis for poly(ethylene-g-styrene) containing respectively 26 and 58% styrene.

the short plateaus predicted by the computed curves cannot be verified in the data for PEGS/26. However, these plateaus are in evidence for PEGS/58 (Figure 8) particularly for the low frequency data. The expected shifts in the transition temperatures with frequency changes are also clearly discernable.

Table 1 shows numerical data for E' at several selected temperatures. It can be seen that in most instances the discreppancies between calculated and experimental values are less than 20%. Such semi-quantitative agreement is, however, restricted to

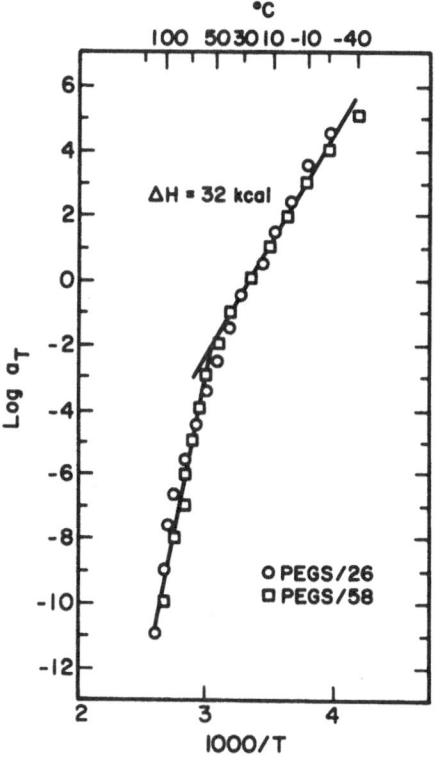

Figure 4:   Logarithmic shift factors as a function of the reciprocals of temperature for poly-(ethylene-g-styrene) containing respectively 26 and 58% styrene.

a limited temperature region. The upper temperature limit is
imposed by the onset of changes in the degree of crystallinity,
in this instance approximately 70°C; and the lower temperature
limit is defined by the region in which the stress-relaxation isotherms become too flat for meaningful superposition.

For the dynamic loss moduli, there seems to be only qualitative resemblance for PEGS/26 shown in Figure 7. The agreement is
especially poor below 20°C. On the other hand, similarity in the
qualitative features between the experimental and calculated results is improved for PEGS/58 (Figure 9), although the predicted
shifts in the transition temperatures are not as great as the observed ones. This poor agreement may partly be due to the fact
that transient mechanical measurements are generally less sensitive
than the dynamic mechanical measurements at low temperatures or
at high frequencies.

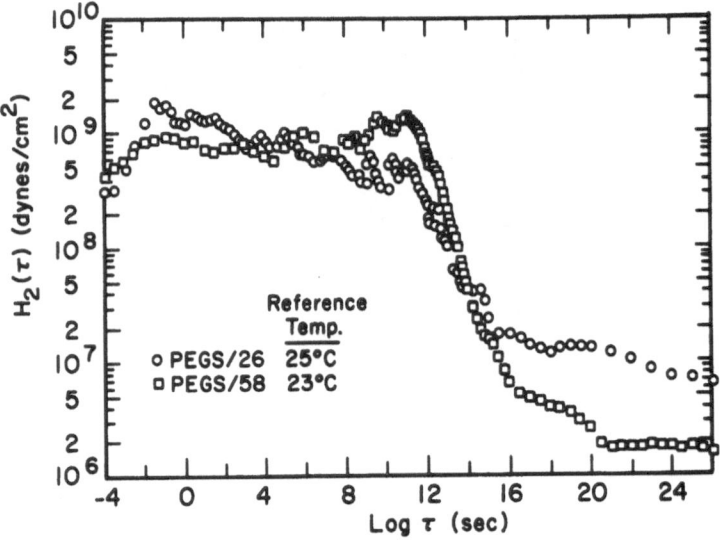

Figure 5: Viscoelastic relaxation spectra obtained from
           the master curves of poly(ethylene-g-styrene)
           by the second approximation method.

The limited applicability of the Time Temperature Superposition Principle to semicrystalline heterophase polymers here
is perhaps attributable to the fact that any errors may have been
cancelled out in using the same set of shift factor data to convert
the computed data back to the temperature axis. Another possible
reason for the partial success may be due to the temperature proximity of the primary relaxations for both polyethylene (~120°C)
and polystyrene (~100°C) in PEGS. Thus with caution simple time-
temperature superposition technique may yield some useful information for the apparently thermorheologically complex semicrystalline multicomponent polymers.

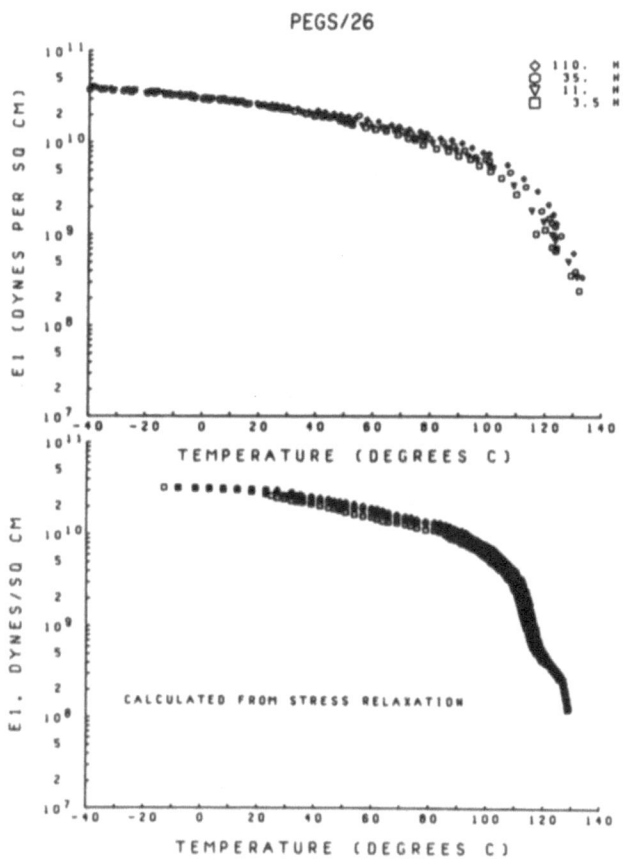

Figure 6: Comparison of the experimental dynamic storage
           moduli for poly(ethylene-g-styrene) containing
           26% styrene (upper curve) with those computed
           from the viscoelastic relaxation spectrum
           (lower curve).

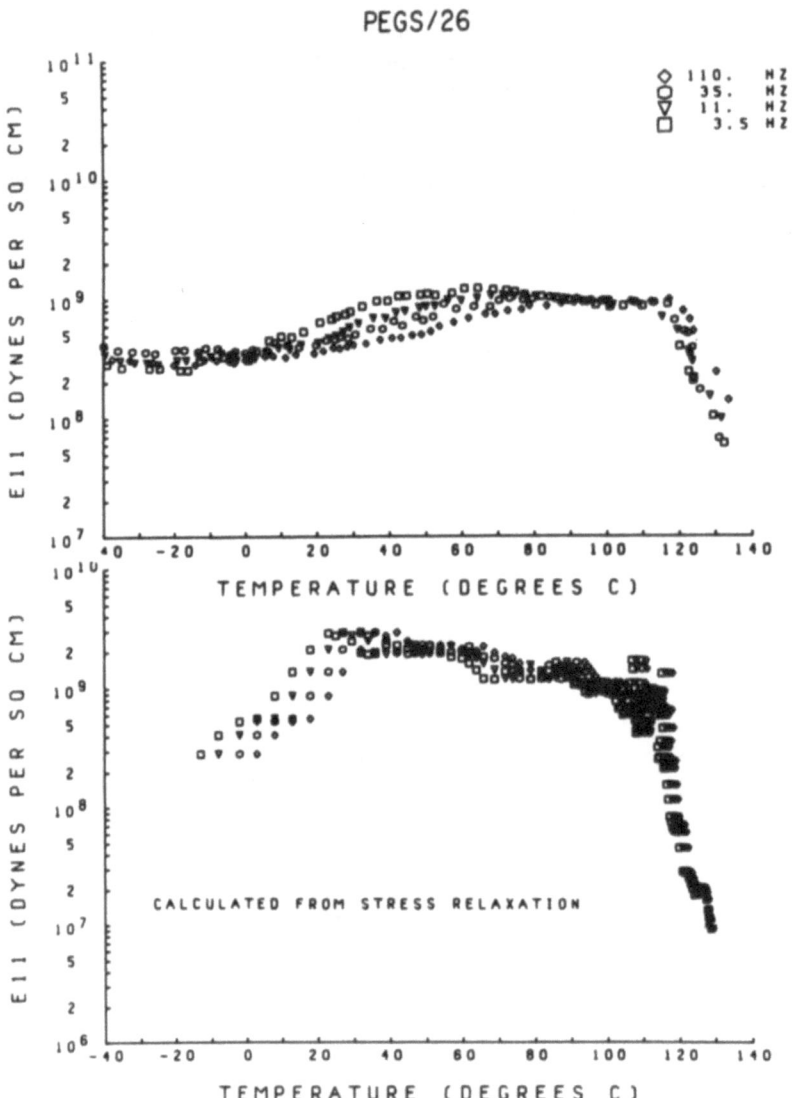

Figure 7:   Comparison of the experimental dynamic loss
            moduli for poly(ethylene-g-styrene) containing
            26% styrene (upper curve) with those computed
            from the viscoelastic relaxation spectrum
            (lower curve).

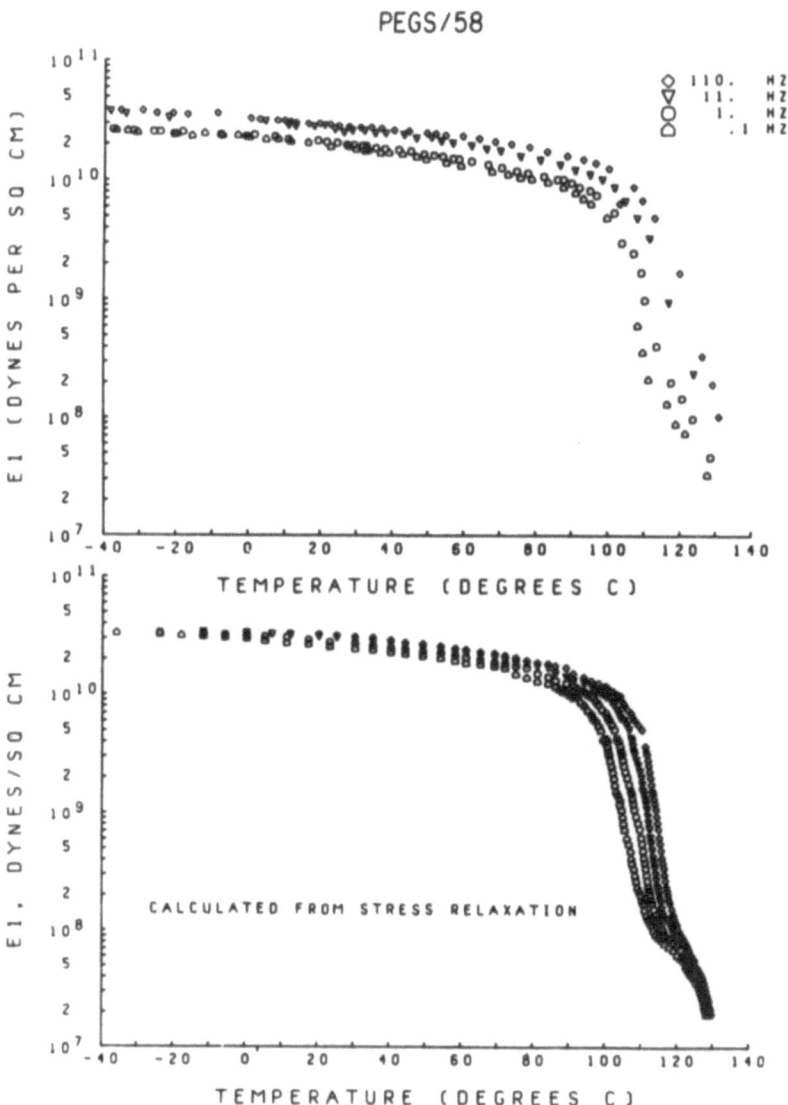

Figure 8:   Comparison of the experimental dynamic
            storage moduli for poly(ethylene-g-styrene)
            containing 58% styrene (upper curve) with
            those computed from the viscoelastic relaxation
            spectrum (lower curve).

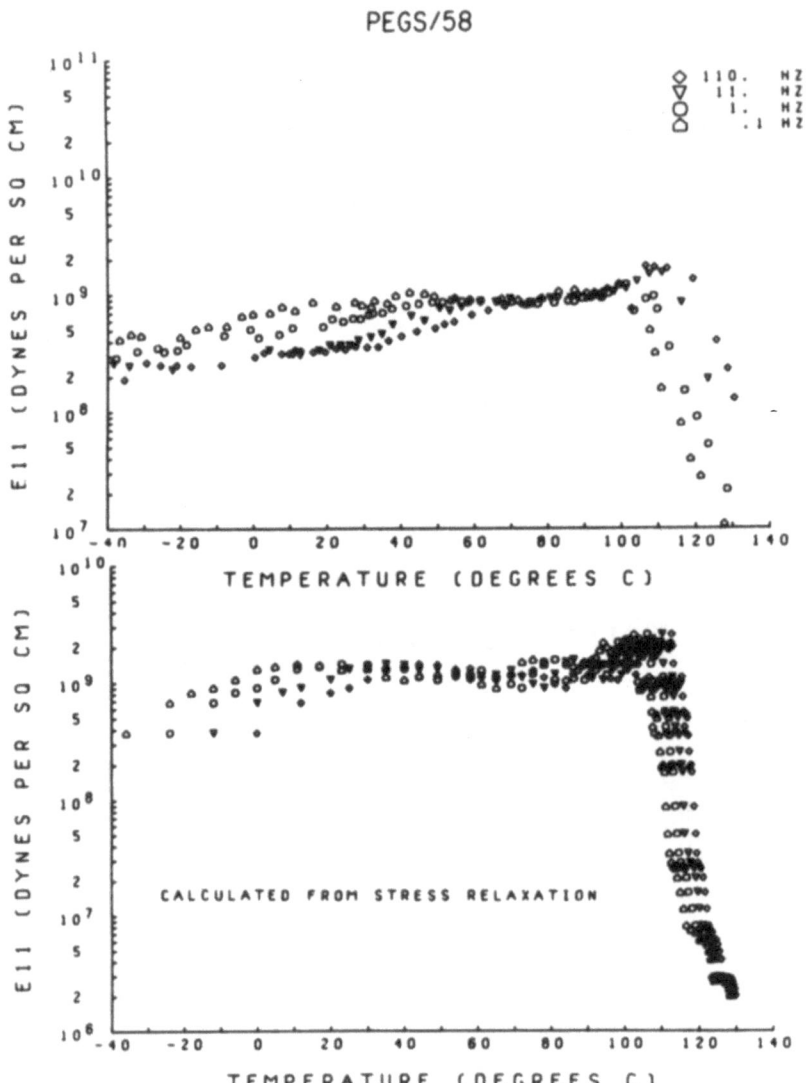

Figure 9:   Comparison of the experimental dynamic
            loss moduli for poly(ethylene-g-styrene)
            containing 58% styrene (upper curve) with those
            computed from the viscoelastic relaxation
            spectrum (lower curve).

Table 1.   Calculated and Experimental Dynamic Storage Moduli for Poly(ethylene-g-styrene)

Temperature (°C)	E' (10^{10} dynes/cm^2) at 110 Hz			E' (10^{10} dynes/cm^2) at 11 Hz		
	Expt'l	Cal'd	% Discrepancy	Expt'l	Cal'd	% Discrepancy
PEGS/26						
0	3.10	3.20	3	2.95	3.20	8
20	2.65	3.09	17	2.55	3.09	21
40	2.20	2.51	14	1.97	2.40	22
60	1.70	1.93	14	1.44	1.81	26
80	1.20	1.38	15	0.99	1.32	33
PEGS/58						
0	3.30	3.33	1	3.0	3.27	5
20	2.97	3.20	8	2.81	3.06	9
40	2.62	2.87	10	2.42	2.68	11
60	2.33	2.38	2	1.93	2.23	16
80	1.82	1.85	2	1.48	1.75	18

ACKNOWLEDGEMENT

This work was supported by the Office of Naval Research.

REFERENCES

1. A. V. Tobolsky, Properties and Structure of Polymers, Wiley, New York, 1960.

2. J. D. Ferry, Viscoelastic Properties of Polymers, 2nd ed., Wiley, New York, 1970.

3. J. J. Aklonis, W. J. MacKnight and M. Shen, Introduction, to Polymer Viscoelasticity, Wiley-Interscience, New York, 1973.

4. K. Nagamatsu, T. Takemura, T. Yoshitomi and T. Takemoto, J. Polymer Sci., 33, 515 (1958).

5. T. Takemura, J. Polymer Sci., 38, 471 (1959).

6. K. Nagamatsu, Kolloid Z., 172, 141 (1960).

7. R. Chujo, J. Phys. Soc. Japan, 18, 124 (1963).

8. T. Yoshitomi, K. Nagamatsu and K. Kosiyama, J. Polymer Sci., 27, 335 (1958).

9. K. Nagamatsu, T. Yoshitomi and T. Takemoto, J. Colloid Sci., 13, 257 (1958).

10. J. A. Faucher, Trans. Soc. Rheo., 3, 81 (1959).

11. H. Nakayasu, H. Markovitz and D. J. Plazek, Trans. Soc. Rheo., 5, 261 (1961).

12. M. Shen, V. A. Kaniskin, K. Biliyar and R. H. Boyd, J. Polymer Sci.- Phys. Ed., 11, 2261 (1973).

13. A. Kaya, G. Choi and M. Shen, in H. H. Kausch, J. A. Hassell and R. I. Jaffe, eds., Deformation and Fracture of High Polymers, Plenum, New York, 1974, p. 273.

14. D. G. Fesko and N. W. Tschoegl, J. Polymer Sci., C35, 51 (1971).

15. D. Kaplan and N. W. Tschoegl, Polymer Eng. Sci., 14, 43 (1974).

16. L. C. Anderson, D. A. Roger and J. K. Rieke, J. Polymer Sci., 43, 423 (1960).

17. S. Machi and J. Silverman, J. Polymer Sci., Part A-1, 8, 3329 (1970).

18. C. E. Locke and D. R. Paul, J. Appl. Polymer Sci., 17, 2597 (1973).

19. G. Odian, A. Rossi and E. N. Trochtenberg, J. Polymer Sci., 42, 575 (1960).

20. G. Odian, T. Acher and M. Sobel, J. Appl. Polymer Sci., 7, 245 (1963).

Synthesis and Thermal Analytical Characterization of

Chlorinated Polyethylene-g-styrene

Carl E. Locke and Larry Watters[*]

University of Oklahoma

Norman, Oklahoma 73019

INTRODUCTION

Chlorinated polyethylene has been reported to be an excellent modifier for blends of polyethylene and poly(vinyl chloride) (1,2). Properties of blends of polyethylene-polystyrene and poly(vinyl chloride)-polystyrene have been improved but not to the extent possible for the polyethylene-poly(vinyl chloride) blends. The improvement in blend properties was attributed to the blocklike structure of the slurry-produced chlorinated polyethylene (3,4).

Graft copolymers have been used for impact modification of plastics with rubbers (5). Polyethylene-g-styrene was prepared and shown to be effective in improving properties of polyethylene-styrene blends (6,7). Conceivably, then, chlorinated polyethylene-g-styrene should be a good modifier for poly(vinyl chloride)-polystyrene blends as well as polyethylene-polystyrene blends.

A few investigators have explored the synthesis of graft copolymers involving chlorinated polyethylene (8-13). However, little has been disclosed in the literature concerning chlorinated polyethylene-g-styrene synthesis or properties.

This paper is a progress report on the initial work on synthesis of chlorinated polyethylene-g-styrene using chemical initiation and an investigation of the properties of this material using thermal analytical techniques. Further work on the synthesis and characterization is anticipated.

---

*Halliburton Services, Duncan, Oklahoma 73533

EXPERIMENTAL

The chlorinated polyethylene-g-styrene was formed by contacting styrene with chemically-initiated active sites formed on the chlorinated polyethylene. The reaction mixture was solvent fractionated and the resulting fractions were characterized utilizing thermoanalysis. The details of the experimental procedure will be described in this section.

Materials

The chlorinated polyethylene was obtained from Dow Chemical Company and characteristics of the two materials used are described in Table I. Both grades of chlorinated polyethylene were prepared by chlorinating high density polyethylene in a slurry. By control of reaction conditions, it is possible to produce a chlorinated polyethylene with residual crystallinity as in the CPE 2552. The styrene used in the study was supplied by Monsanto and contained a polymerization inhibitor. The inhibitor was washed from the styrene with caustic prior to the grafting reactions. The polystyrene used in the blends was Styron 685 manufactured by Dow Chemical Company.

TABLE I

Characteristics of CPE 2552 and CPE 4814
Supplied by Dow Chemical Company

Property	Resin Type	
	CPE 2552	CPE 4814
chlorine content %	25%	48%
*residual crystallinity %x-ray	25	2
*DSC	15	0
specific gravity	1.10	1.25
tensile strength (PSI)	2400	1575

*X-ray data provided by Dow Chemical Company.
Residual crystallinity measured by using DSC in this work.

Synthesis

Strips from melt formed films of CPE were immersed in benzene/methanol mixtures in which the initiator, benzoyl peroxide, was dissolved. Methanol/benzene mixtures were used in order to swell the CPE, but not completely dissolve it. It was necessary to use different concentrations of the solvent/non-solvent mixture for each CPE polymer. A 15/85 (volume) methanol/benzene mixture was used for the CPE 2552 which had some residual crystallinity. A 50/50 (volume) methanol/benzene mixture was used for the CPE 4814 which was essentially amorphous. About 40 grams of CPE film was placed in the initiating solution which contained about 1% by weight benzoyl peroxide based on CPE. The system was deaerated with nitrogen and heated to 70°C. The initiation reaction time was 7 hours and 40 minutes. The solid CPE was removed from the kettle after this initiation reaction and immediately placed in 300 milliliters of styrene monomer which had previously been deareated with nitrogen. The film was left in the styrene at 70°C for 28 hours.

The reaction product, a very viscous liquid, was removed from the kettle at the end of the reaction time. The residual styrene monomer was removed by evaporation and heating below 100°C in a vacuum oven. The reaction product was weighed to determine the total yield of polystyrene formed.

Separation

A two-step solvent fractionation procedure was used to separate the homopolymers and graft copolymer. Five grams of the reaction product were dissolved in 495 grams of benzene at 100°C for CPE 2552 products and 70°C for CPE 4814 products. After the reaction product had dissolved, the mixture was allowed to cool to room temperature. The resulting solid material was separated by centrifuging for 1 hour. The liquid phase was decanted from the solid and the amount of polymer material in each phase was obtained gravimetrically.

The second step of the separation was performed by adding 20 volume percent of methanol to the liquid phase from the first separation. The mixture was heated for one hour at the same temperature used for the initial separation. The solid and liquid phases resulting from this step were separated by centrifuging followed by decantation. Figure 1 is a flow diagram of this separation technique with the polymer species in each phase identified. The same separation technique was conducted on the pure CPE compounds, polystyrene, and melt blends of polystyrene/CPE.

Characterization

All reaction products and parent homopolymers, as well as each fraction of the separations performed on the reactions products and homopolymers, were analyzed using differential scanning calorimetry. The Perkin Elmer Model DSC/1B was utilized for this analysis. Heat capacities for each sample were determined over the temperature range 40-55°C. The crystallinity of those samples involving CPE 2552

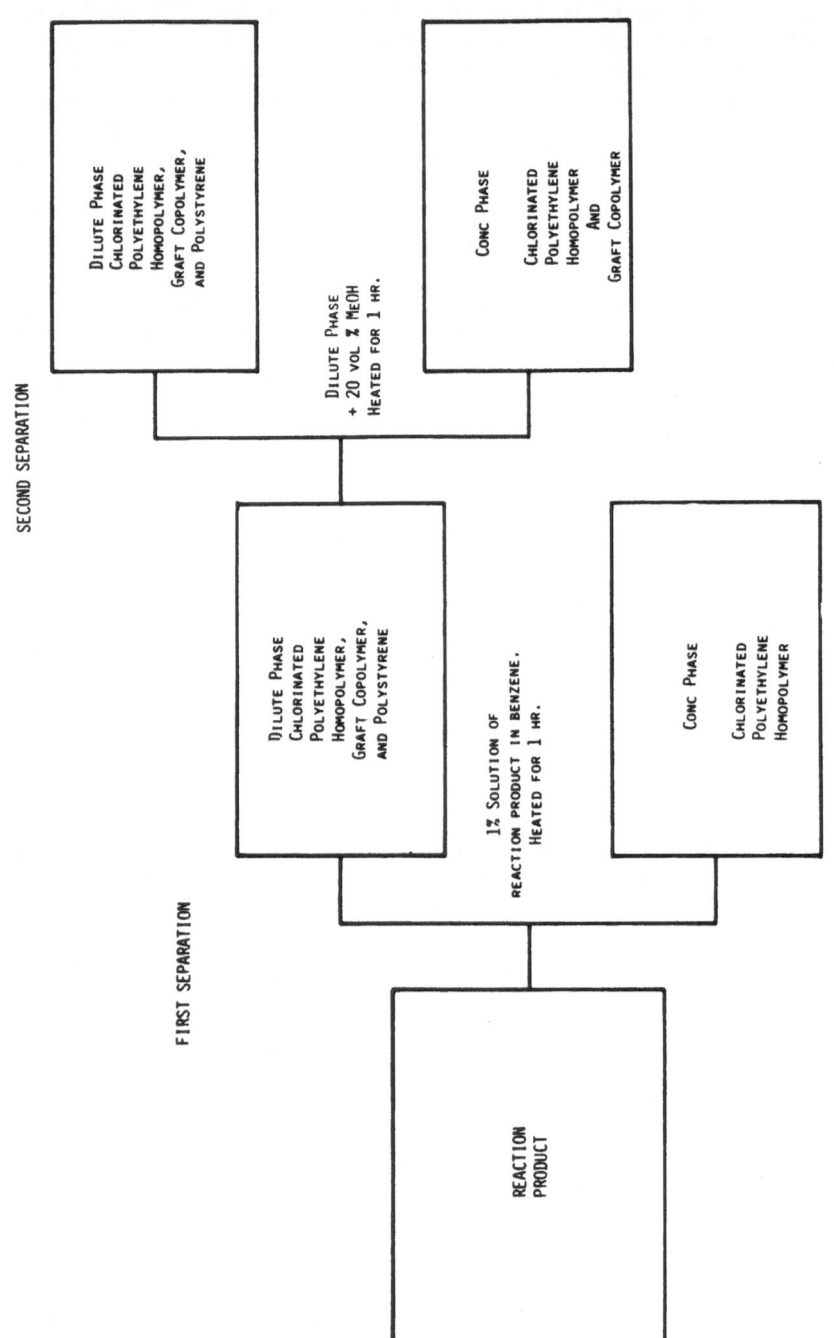

Figure 1.   Schematic Representation of Separation Technique

were determined by heating the material through the melting point and determining the heat of fusion. The heat of fusion was then compared to the heat of fusion for pure polyethylene. Crystallinity was calculated from these data.

Melt Blends

Melt blends were formed of chlorinated polyethylene/polystyrene utilizing the Brabender Plasticorder type PL-V150 with a REE-6 laboratory mixing head. Interstab BC26 and Interstab Plastoflex ESO which were obtained from Cincinnati-Millacron were used as heat stabilizers for these melt blends. Three phr of each stabilizer was used. After the blends were formed at 180°C at a mixer rotation speed of 20 RPM, the resulting mixture was molded at 180°C in a laboratory press. These blends were subjected to the same fractionation and characterization techniques as the graft copolymers.

RESULTS

Synthesis

Table II contains a summary of the yield of the reactions conducted in this study. The weight CPE reacted is the amount of material that remained after the pre-initiation step. The total product weight is the amount of CPE and polystyrene remaining after the grafting reaction had been conducted and the benzene solvent had been evaporated. The weight percent CPE is the amount of chlorinated polyethylene in the reaction product. The yield of polystyrene was calculated as the grams of polystyrene produced per gram of chlorinated polyethylene reacted.

TABLE II

Yield of Grafting Reaction

Reaction	Wt CPE Reacted	Total Product Weight	Wt % CPE	Yield Polystyrene
2552-a	9.3640 gm	46.0025 gm	20.36%	3.91
2552-b	15.8969 gm	73.1796 gm	21.73%	3.60
4814-a	34.0587 gm	110.5917 gm	30.08%	2.28
4814-b	40.9847 gm	94.2723 gm	43.5 %	1.30

Thermal Data

The heat capacity and crystallinity of each of the polymers as received and after the initiation step are tabulated in Table III. The initiation step obviously resulted in some fractionation of the CPE's. The fraction removed from CPE 2552 had lower crystallinity than the fraction that was used in the grafting reaction.

TABLE III

| Polymer | Heat Capacity Cal/gm^{o}C | | % Crystallinity | |
	As received	As initiated	As received	As initiated
CPE 2552	.466	.467	14.6%	17.0
CPE 4814	.366	.360	0	0
Polystyrene	0.366			

The crystallinity of blends containing CPE 2552 and polystyrene were determined using DSC. These were compared to the crystallinity calculated on a weight basis for these blends assuming polystyrene acted as a diluent and had no interaction with the chlorinated polyethylene. The results of these data are presented in Figure 2. There was little difference in the calculated and experimental data.

The heat capacity was determined for the blends and compared to that calculated using the heat capacities of the polystyrene and chlorinated polyethylenes additive on a weight basis. Figure 3 contains data for the experimental and calculated data for the blends of CPE 2552 and polystyrene. Similar data were obtained for blends containing CPE 4814 and styrene. Therefore, experimental heat capacity and crystallinity data were used to calculate composition of blends.

Separation of Reaction Products

Figure 4 is a flow diagram of the separation of a reaction product from grafting styrene onto CPE 2552. The weight, heat capacity, crystallinity, melting point, and glass transition temperatures are given for each fraction. In addition, the CPE concentrations, in weight percent, are listed. These data for the reaction products, blends and pure components are compared below. Similar data were obtained for all reaction products. These will be tabulated below.

Tables IV and V contain the summary of the data for weight of each fraction of each product, pure polymer, and polymer blend. These data indicate there were difficulties in achieving quantitative

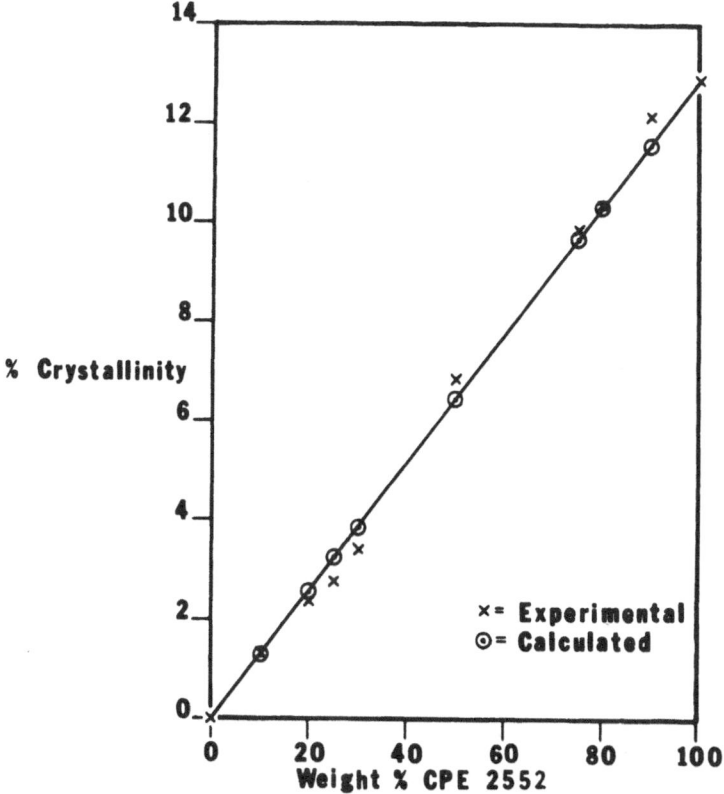

Figure 2.   Experimental vs. Calculated Crystallinity for CPE 2552 –
            Polystyrene Blends

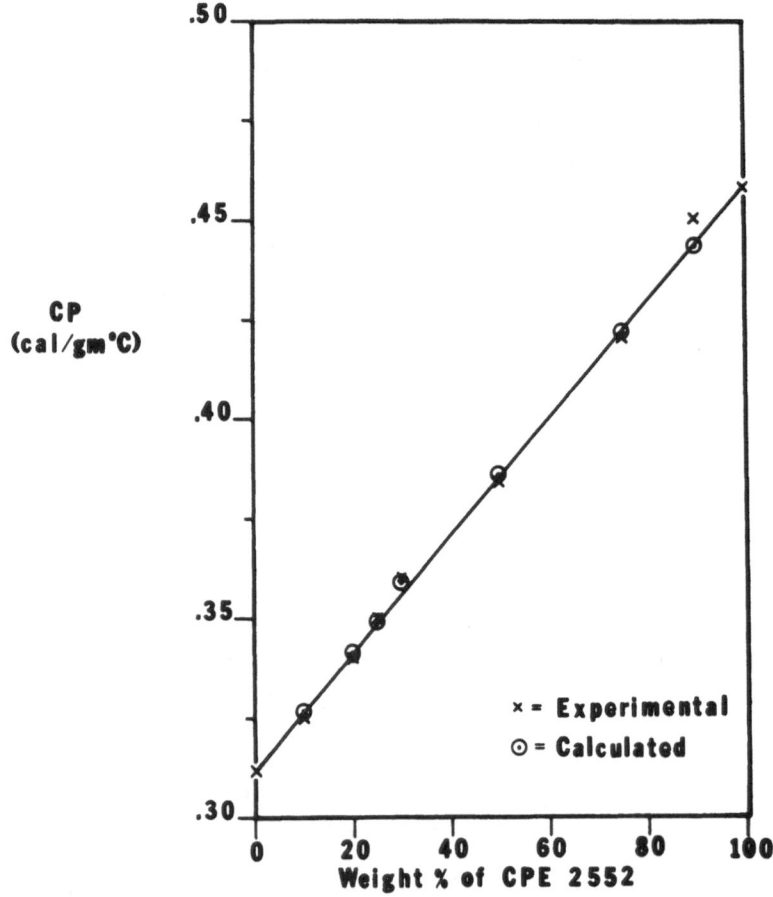

Figure 3.   Experimental vs. Calculated Heat Capacity for CPE 2552 -
            Polystyrene Blends

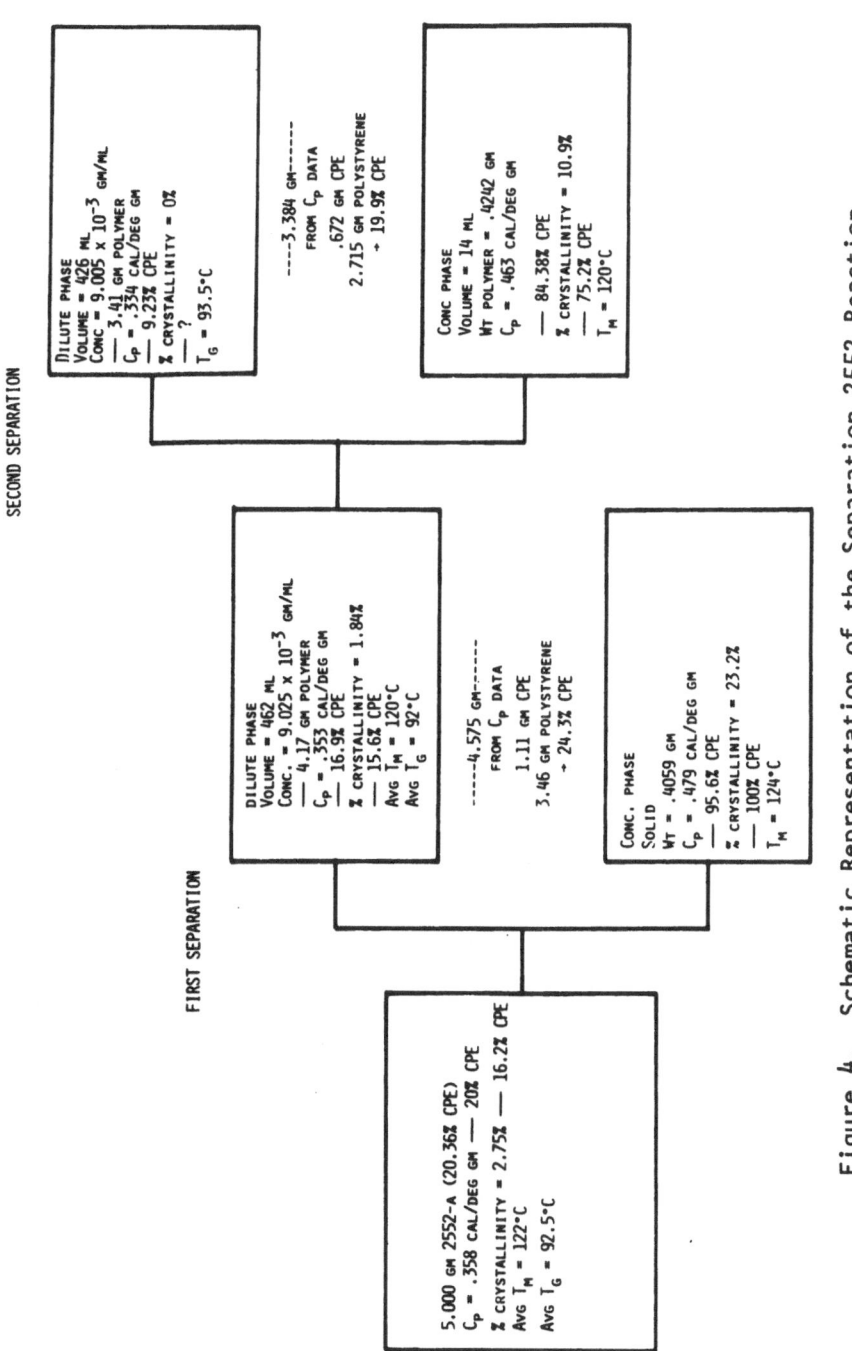

Figure 4. Schematic Representation of the Separation 2552 Reaction Product

separations due to experimental difficulties with the viscous solutions and separation methods. There is no discernable difference
in the separation of the blends and the reaction products when compared on a weight fraction basis.

The heat capacity and crystallinity data indicated that the
first separation of blends and reaction products resulted in a concentrated phase with CPE only.

TABLE IV

Weight of Each Fraction
in Grams - Reaction Products

| | | 1st Separation | | 2nd Separation | |
| | Initial | Conc. | Dilute | Conc. | Dilute |
Sample	Wt.	Phase	Phase	Phase	Phase
2552-a	5.000	0.406	4.17	0.424	3.41
2552-b	5.000	0.490	4.33	0.213	3.00
4814-a	5.000	0.012	4.87	0.398	4.60
4814-b	5.00	0.144	5.04	0.645	3.73

TABLE V

Weight of Each Fraction in Grams -
Pure Polymers and Melt Blends

| | | 1st Separation | | 2nd Separation | |
| | Initial | Conc. | Dilute | Conc. | Dilute |
Sample	Wt.	Phase	Phase	Phase	Phase
2552	1.000	0.301	0.822	0.566	0.139
4814	1.850	0.0703	0.639	0.478	0.332
Polystyrene	2.835	No Separation		No Separation	
Blends					
2552/PS 20/80	5.000	0.355	2.930	0.347	2.437
4814/PS 37/63	5.000	0.292	4.746	0.385	4.051

Table VI is a tabulation of the heat capacity, crystallinity and CPE concentration for the concentrated phase of the second separation. The results for the reaction products and blends are also presented. The heat capacity and crystallinities of these products were compared to the values obtained for the comparable fractions from homopolymer separations. These heat capacities and crystallinities were used as base values from which the compositions of the reaction products were calculated.

TABLE VI

CPE Concentrations in Concentrated Phase -
Second Separation

Sample	Heat Capacity		Crystallinity	
	C_p Cal/gm$^\circ$C	Wt % CPE	% Cryst.	Wt % CPE
Blends				
2552/PS 20/80	0.511	100	12.7	87.6
4814/PS 37/63	0.372	100	--	--
Reaction Prod.				
2552-a	0.463	84.4	10.9	75.2
2552-b	0.453	78.1	10.4	72
4814-a	0.353	58.1	--	--
4814-b	0.367	26.7	--	--

Table VII contains the same data for the dilute phase of the second separation.

The melting points of each fraction from the separations of the reaction products, blends, and homopolymer involving CPE 2552 are given in Table VIII. This data indicates the CPE in the dilute phase of the second separation is amorphous material. The melting points of the concentrated phase of the reaction products are slightly lowered from the blend and the homopolymer. The amount of lowering, however, does not seem to be significant in relationship to the accuracy of determining melting points using DSC.

TABLE VII

CPE Concentrations in Dilute Phase -
Second Separation

| Sample | Heat Capacity | | Crystallinity |
	$Cal/gm^{o}C$	Wt %	%
Blends and Homopolymers			
2552/PS 20/80	0.335	10.8	0
4814/PS 37/63	0.335	14.9	-
2552	0.393		0
4814	0.332		-
Reaction Prod.			
2552-a	0.334	9.2	0
2552-b	0.331	4.6	0
4814-a	0.341	27.7	0
4814-b	0.351	61.7	0

TABLE VIII

Melting Points
for 2552
in ^{o}C

Sample	Initial	Conc. Phase	Dilute Phase	Conc. Phase	Dilute Phase
2552-a	122	124	120	120	---
2552-b	121	126	121	119	---
Blend		123.5	121.5	121	---
Pure	121.5	125.5	121.5	120.5	---

DISCUSSION OF RESULTS

The yields of polystyrene varied substantially dependant upon the CPE material. The products containing 2552 has substantially higher yield than those containing 4814. This possibly can be explained due to the higher freedom of molecular motion with the totally amorphous 4814 than in the partially crystalline 2552. The hindered molecular motion of the polymer chains in the solid state might well hinder the mutual destruction of the trapped free radicals. This relationship between molecular motion and stability of trapped-free radicals was observed by Ballantine et al. (14).

The yield also varied with the concentration of styrene in each reaction. Recall that a fixed amount of styrene was used for each reaction regardless of CPE remaining after initiation. The yield increased with increasing concentration. This same behavior was observed by Kennedy and Smith (15) in grafting styrene to ethylene-propylene rubber.

There was substantial formation of homopolymer polystyrene when the yield values are compared to the weights of the fractions remaining in the solid phase after the second separation. The homopolymer polystyrene may have resulted from thermal initiation of styrene. In addition, some benzoyl peroxide certainly remained in the swollen CPE homopolymer when it was removed from the initiation mixture and introduced into the styrene monomer.

The concentrated phase of the second separation is thought to contain the majority of the graft copolymer produced. In the 2552 graft, notice that the crystallinity of the 2552 is lower than that of a similar blend. The crystallinity of the blend is lower than that of a fraction obtained from the pure 2552, but this difference is due to the presence of the glass transition appearing close to the melting point of the 2552. Also notice that this lowered crystallinity of the graft is thought to be due to the interference by the grafted chains in the formation of the crystallites. In the 4814, notice that there is a substantial amount of polystyrene in this separation product. It is even more strikingly apparent since there was apparently no polystyrene in the blend separation. However, there is possibility of experimental error in the 4814 polystyrene blends since the heat capacities differ by only 10%.

The important consideration in this work revolves about the formation and identification of the graft copolymer. The weight distribution of the fractions obtained with blend compared to those obtained from the reaction products indicate little difference between those. However, examination of the heat capacity and crystallinity data from these fractions do indicate some difference in properties of the fractions. Those differences are significant for the concentrated and dilute phases of the second separation.

The concentrated phase of the second separation should contain no polystyrene as determined from the data from the pure polystyrene. The blends of polystyrene with each CPE type indicate no polystyrene in this phase when calculated from the Cp data. The heat capacity of the like fractions from the graft copolymer products indicate there is substantial amounts of polystyrene present.

The crystallinity data for this phase for the blends containing 2552 indicate the CPE content was 87.6%. There is some disagreement with the Cp data. The concentration calculated from crystallinities for the concentrated phase from graft product was lower than that calculated from the heat capacities. Also it was lower than for the blends. These facts may be interpreted as indicating the presence of the graft in this fraction. The grafted chains of PS would necessarily lower the concentration of CPE. In addition, the branches might be expected to interfere with the crystallization of the CPE. Therefore, the concentrated fraction of the second separation is thought to contain graft copolymers.

The dilute fraction from this separation may contain some graft but with 2552 the evidence is not compelling. There is a much larger concentration of the 4814 in this phase for the graft reaction product than for the blend. Therefore, there may be much more grafted branches of polystyrene on the 4814 which tend to make the CPE more prevalent in the dilute phase.

The crystallinity and heat capacity data allow the conclusion that the concentrated phase does contain graft copolymer. There is some evidence graft copolymer is also present in the dilute phases. Table IX is a listing of the weights of the CPE's in each fraction of the second separation step based on heat capacity measurements, for the graft products and blends. There is a different ratio of separation in each phase for the grafts than for the blends. It seems that more of the CPE may be dissolved in the dilute phase for the grafts than found with the blends. This may be due to the presence of the grafted branches which increase the solubility of the CPE chains. There are some discrepancies in the material balances in these separations. These are due to analytical errors plus experimental difficulties with the concentrated polymer solutions.

Many questions yet need to be answered. It may be possible to increase yield of the graft copolymer by adjustment of initiation and reaction conditions. The fractions should be characterized using other analytical techniques like IR, NMR and possibly Laser Raman spectroscopy. It may be possible to further separate the graft copolymer by an additional separation step. Work in all these areas is planned.

TABLE IX

Calculated Weights of CPE Based on Heat Capacity
in 2nd Separation

| Sample | Fractions in gm | |
| | Phase | |
	Dilute	Conc.
2552-a	0.314	0.357
2552-b	0.175	0.166
4814-a	0.231	1.27
4814-b	0.494	2.3
2552 blend	0.511	0.263
4814 blend	0.372	0.603

CONCLUSIONS

1. A graft copolymer of CPE-g-styrene was formed using chemical initiation.

2. Solvent fractionation separated the graft copolymer from unreacted chlorinated polyethylene. The graft was formed on fractions of the CPE 2552 that had some crystallinity and on the higher molecular weight fractions of the 4552. In addition grafting occurred on the totally amorphous and lower molecular weight portions of the CPE's.

3. This work needs to be followed by further characterization and fractionation steps.

LITERATURE CITED

1. Locke, C.E., Paul, D.R., and Vinson, C.E. Polym. Engr. Sci., 13, 202 (1973).

2. Locke, C.E. and Paul, D.R. Polym. Engr. Sci., 13(4), 308 (1973).

3. Locke, C.E. Ph.D. Dissertation, University of Texas at Austin, 1972.

4. Oswald, H.S. and Kuby, E.T. SPE Trans., 3, 168 (1963).

5. Rose, S.L. Polym. Engr. Sci., 7, 115 (1967).

6. Locke, C.E. and Paul, D.R. J. Appl. Poly. Sci., 17, 2597 (1973).

7. Locke, C.E. and Paul, D.R. J. Appl. Poly. Sci., 17, 2791 (1973).

8. Takahashi, A., Ohusga, J., and Kobayshi, S. Japanese Patent 20, 514 (Sept. 19, 1964). Chem. Abst. 63, 1954a (1965).

9. Takahashi, A., Kojima, H., Ogawa, M., and Osuga, H. German Patent, 1,282,980 (Nov. 14, 1968). Chem. Abst. 70, 29, 729a, (1968).

10. Nishi, Y. and Kodama, T. Japanese Patent 72, 39,390 (Dec. 7, 1972). Chem. Abst. 78, 98305n (1972).

11. Yokoyama, K. and Nose, S. Japanese Patent 73 21,789 (Nov. 18, 1973). Chem. Abst. 79, 674323 (1973).

12. Yoshimoto, H. and Nose, S. Japanese Patent 74 29,391 (March 18, 1974). Chem. Abst. 81, 170465v (1974).

13. Platzer, N. French Patent 1, 450, 458 (Aug. 26, 1966). Chem. Abst. 67, 33145q (1967).

14. Ballantine, D., Glines, A., Alder, G., and Metz, D.J. J. Polym. Aci. 34, 419 (1959).

15. Kennedy, J.P. and Smith, R.R. Polym. Prep., Amer. Chem. Soc., Div. Polym. Chem., 13(2), 710 (1973).

PHOTOASSISTED MODIFICATION OF AND GRAFTING TO POLYETHYLENE

James F. Kinstle and Stuart L. Watson, Jr.

Department of Chemistry, The University of Tennessee

Knoxville, Tennessee   37916

INTRODUCTION

Polyethylene (PE) is widely used in a number of applications, due to a combination of low cost, desirable mechanical properties, low permeability, low surface energy characteristics, etc. Yet in other potential applications, modification of the surface of the polymer - without significant alteration of bulk chemical or physical properties - would be advantageous. For example, the low surface energy of PE (1) is useful in several applications but contributes to low adhesion of inks, coatings, and adhesives. The PE surface can be roughened or micropitted in order to enhance physical interlock type adhesion (2), but the interfacial strength and permanence could be markedly enhanced by formation of chemical bonds or grafts to the PE substrate. This new grafted surface may be comprised of the coating, ink, or adhesive, or it may just exhibit physical or chemical properties different from the PE, thereby allowing better wetting or bonding in a subsequent step.

The studies related here concern incorporation of chemical functional groups at or near the surface of PE and use of these functional groups in grafting reactions. More specifically, the studies encompass (i) formation of C-N, C-O, and C-halogen bonds on the PE surface, utilizing both classical and photochemically assisted reactions, (ii) subsequent modification of these groups to photolabile functions, and (iii) photolysis of these groups in the presence of appropriate coreactants, including a monomer that undergoes free radical polymerization. The chemistry of the reactions in (i) and (ii) above was designed to maximize the formation of tertiary amine sites on the PE surface. In homogeneous solution,

it has been amply demonstrated that a tertiary amine function will
interact with an appropriate photochemically excited ketone to
form a pair of free radicals by an overall atom transfer reaction
(3), as illustrated schematically in equation 1.

The combination of benzophenone and a tertiary amine is an affec-
tive coinitiator system for photoinduced polymerization (4-6).
At least during low conversion polymerizations that have been
reported (6), such photoinduced polymerizations follow the expected mechanistic/kinetic pathways; i.e., the overall rate of polymerization is proportional to the square root of the concentration
of benzophenone (provided at least a stoichiometric quantity of
tertiary amine is present) and to the first power of the monomer
concentration. But the free radicals that are formed by this
coinitiator system are quite dissimilar in structure, and are not
equally effective with respect to their ability to initiate free
radical polymerization. The amine-derived radical is the more
effective initiator, as shown by the relatively selective and
efficient formation of polymeric grafts upon irradiation of a
homogeneous solution of benzophenone, a monomer (M_2), and an amine-
bearing substrate polymer (7), as shown schematically in reaction
scheme 2.

The present studies involve utilization of similar chemistry in a
similar process, with the important difference that incorporation
of amine coinitiator sites onto the polymer, photochemical interaction between ketone and amine, and the proposed grafting reaction
all are to be conducted under heterogeneous reaction conditions.
The monomers used in the present surface grafting studies were
methyl methacrylate (MMA) and 2-ethylhexyl methacrylate (2EHMA)
The polymer derived from MMA (PMMA) has a higher surface energy
than PE (1) so PE with a PMMA surface should exhibit increased
wettability, bondability, etc. The polymer derived from EHMA
(P2EHMA) has a surface energy similar to that of PE (8), but since
P2EHMA has a very low glass transition temperature, a P2EHMA surface

grafted onto PE may itself act as an adhesive.

It should be noted that the use of benzophenone to photochemically induce grafting to PE has been studied by Oster and his
coworkers (9, 10). They used specially prepared PE into which
benzophenone had been milled. Irradiation of this system led to
hydrogen abstraction from PE by excited benzophenone; the radical
site created on the PE chain was presumed to be the initiating
species for graft polymerization. The Oster system is quite different, in principle and in practice, from that of the present
study. The advantages and limitations of each system will be
discussed later.

EXPERIMENTAL

Materials

The high density PE used for most experiments was 0.038 mm
film of Marlex 6006 which was exhaustively extracted with 2-butanone
and dried to constant weight. In certain experiments, low density
polyethylene (which will be designated LDPE for differentiation)
of film thickness 0.032 mm was also used. The 2EHMA and MMA monomers were washed five times with dilute (0.5%) sodium hydroxide,
five times with distilled water, dried over magnesium sulfate, and
fractionally distilled under reduced pressure (11). Benzophenone
was recrystallized from pentane. Gaseous reagents, including
hydrogen chloride, chlorine, nitric oxide, nitrogen dioxide, and
nitrosyl chloride, were fractionally distilled three times on a
high vacuum line and stored on phosphorus pentoxide. All other
reagents and materials were appropriately purified to an exacting
degree, including bubbling of nitrogen through two solutions of
chromous ion and sulfuric acid (12).

Procedures

The vacuum system was constructed of glass, with a conventional vacuum pump and a two-stage mercury diffusion pump, with pressures monitored by a McLeod gauge and a calibrated Bourdon vacuum
gauge. All vacuum work was performed with illumination by a Kodak
Wratten Series 1 Safelight (13). Photochemical reactions were
performed in a Rayonet Type RS Preparative Photochemical Reactor
equipped with four 12.5 W RUL 3500 Å lamps. Photoreactions were
carried out in Pyrex vessels of appropriate dimensions. Isolations,
purifications, and identifications were carried out by conventional
techniques. The following examples illustrate these procedures.

For chlorination of PE surfaces, a portion of high density PE
film, 4 cm by 20 cm, was placed in a continuous extractor and washed for 24 hr with 2-butanone, followed by drying for 24 hr at room

temperature under vacuum. The dried film was weighed and placed in
the center of a photolysis cell, which was then attached to the
vacuum system and evacuated. Chlorine was admitted to the photolysis cell to a pressure of 380 mm Hg. The photolysis cell was
sealed and irradiated for 5.0 hr. The cell was then returned to
the vacuum system and the residual gases were removed. The film
was then placed in the continuous extractor and washed with
2-butanone for 24 hr, after which the film was dried to constant
weight at room temperature under vacuum before analysis.

    For conversion of Cl sites to amine sites, a procedure similar
to that of Adylov et. al. (14) was utilized. A 100 ml portion of
toluene and 10 ml diethylamine were placed in a 250 ml three-neck
round-bottom flask equipped with a condenser, magnetic stirrer, and
thermometer. The solution was heated to 50° and a weighed portion
of the chlorinated film was added. The mixture was stirred slowly
for 12 hr, after which the film was dried to constant weight at
room temperature under vacuum before analysis.

    Experimental details of the grafting procedure are related
later.

Analyses

    Instrumentation utilized in this study included Carey 14 and
17 Recording Spectrophotometers, Beckman IR5A and Digilab Fourier
Transform FTS-20 infrared spectrometers, Varian A60, T60, and HA100
nuclear magnetic resonance spectrometers, and an Hitachi Perkin-
Elmer RMU-6E Mass Spectrometer. ESCA (Electron Spectroscopy for
Chemical Applications) scans were performed on a modified AEI
ES100 spectrometer. Elemental analyses were performed by Galbraith
Laboratories, Inc. of Knoxville, Tennessee; all elemental analyses
reported here are averages of at least duplicate determinations.

RESULTS AND DISCUSSION

Initial Functionalization of PE

    LDPE was immersed in dilute or concentrated nitric acid or
subjected to the vapors over fuming nitric acid. Under the more
severe conditions the LDPE essentially disintegrated. Under less
severe conditions various C-N and C-O bonds were formed on the
polymer, including nitro, nitrate, and carbonyl functions. It was
obvious that surface modifications due to degradation, substitution,
and oxidation were being accomplished, and it was subsequently
shown that the concentrations and relative molar proportions of the
various groups could be altered by acid strength, time of exposure,
or by simultaneous uv irradiation. The desire to delineate the

course of these reactions prompted a change to a "simpler" system.

A strip of high density PE was suspended in the center of a Pyrex photolysis tube of about 500 ml volume. High vacuum conditions were established and maintained for one hour minimum, then a specific pure gas was admitted to the cell and its contents were subjected to uv irradiation. When the gas introduced was nitrogen, air, or nitrous oxide (N_2O) using final pressures from one-third to one atmosphere and irradiation times from fifteen to twenty four hours, no reaction involving the substrate was observed. This is the expected result, since none of these gases absorb uv radiation of the range employed. When a combination of PE plus nitric oxide (NO) or nitrogen dioxide (NO_2) gas was irradiated (one-third atmosphere of gas, nineteen to twenty hours irradiation time), nitro, nitrite, nitrate, and carbonyl functions were incorporated into or onto the PE, as shown in equation 3.

$$\{CH_2\text{-}CH_2\} \xrightarrow[h\nu]{NO \text{ or } NO_2} \underset{\sim CH \sim}{\overset{NO_2}{|}} \underset{\sim CH \sim}{\overset{ONO}{|}} \underset{\sim CH \sim}{\overset{ONO_2}{|}} \underset{\sim C \sim}{\overset{O}{\|}} \tag{3}$$

The same functional groups are introduced in each case, though the number of each type differs in the two systems. Rigorous exclusion of oxygen minimizes carbonyl formation; conversely, the amount of carbonyl can be enhanced by deliberately including some oxygen. Comparison of transmission and surface reflectance infrared spectra shown that reactions in the NO system occurred mainly on the PE surface, while those in the NO_2 system also yielded nitro functions in the interior of the PE film. The overall similarity in the product groups from irradiation of NO and NO_2 in the presence of PE is quite striking. This similarity is probably due in large part to equilibria between varios NO_x species, some of which are shown in reactions 4 and 5 (15).

$$3 \text{ NO} \longrightarrow N_2O + NO_2 \tag{4}$$

$$2 \text{ } NO_2 \rightleftharpoons N_2O_4 \tag{5}$$

N_2O_4 exists in three isomeric forms

planar or twisted

and the NO_2 formed by its dissociation, as well as NO, can behave as either N-centered or O-centered free radicals. Further NO_x interconversions are also likely, including those shown in equations 6-8.

$$\text{ONO} \;\text{---}\; \text{H abstraction} \to \text{HONO} \; \to \; \overset{\bullet}{\text{OH}} + \; \text{NO} \tag{6}$$

$$\text{NO} \;\text{---}\; \text{H abstraction} \to \text{H}_2\text{O} \; + \; \text{N}_2\text{O} \;\; [\text{ref 16}] \tag{7}$$

$$\text{NO}_2 \xrightarrow{\;h\nu\;} \text{NO} \; + \; \text{O} \qquad [\text{ref 17}] \tag{8}$$

These dynamic equilibria and secondary reactions are undoubtedly enhanced by the incident uv energy. While the presence of the various species does make it easy to rationalize the formation of the product functional groups, it is also a very complicating influence. In an attempt to simplify the reactions occurring in these systems, a mixture of Cl_2 and NO was evaluated. The idea was to use $Cl^\bullet$ as an efficient abstractor of H from PE, thereby rapidly forming radical sites on the PE that would in turn be rapidly consumed by addition of NO. Using a 2:1 molar ratio of NO:Cl, irradiation for sixteen hours in the presence of PE yielded all the products that had formed in the NO system plus an appreciable concentration of C-Cl bonds. While the desired selectivity may be exhibited at short reaction times, this has not yet been studied, since a relatively large number of modified sites was desired for the present investigation.

Another method of C-N bond formation in hydrocarbons is photochemical oximation (18), the usual course of which is shown in reaction 9.

$$\sim\!\!\!\curvearrowright\!\!\text{CH}_2\!\!\curvearrowright\!\!\sim + \;\; \text{NOCl} \xrightarrow{\;h\nu\;} \sim\!\!\curvearrowright\!\!\overset{\overset{\displaystyle \text{NO}}{|}}{\text{CH}}\!\!\curvearrowright\!\!\sim \longrightarrow \sim\!\!\curvearrowright\!\!\overset{\overset{\displaystyle \text{NOH}}{\|}}{\text{C}}\!\!\curvearrowright\!\!\sim \tag{9}$$

But when NOCl was photolyzed in the presence of PE, with or without added HCl, the oxime was not formed as the product. This is a striking example of a reaction that is well understood under homogeneous conditions that proceeds to a completely different set of products when carried out under heterogeneous conditions. The detailed course of this reaction has been investigated and will be reported elsewhere (19).

Chlorination of PE

The polyethylene film, after extensive extraction with butanone followed by drying to constant weight, was placed in the center of the cylindrical photolysis cell to insure equal amounts of chlorine and equal irradiation on both sides of the film. The cell was evacuated, and chlorine was distilled into the cell, typically to one-half atmosphere. The filled photolysis cell was then irradiated in the photoreactor for various times in the one to eight hour range. After the irradiation was complete, the residual gases were trapped with liquid nitrogen, and the photolysis cell was evacuated again under high vacuum conditions. The pressure of chlorine used

and the time of irradiation depended on the amount of chlorine
incorporation which was desired. It should be noted that the amount
of chlorine incorporation is very sensitive to trace amounts of
oxygen present in the reaction mixture. After evacuation of the
photolysis cell, the chlorinated polyethylene film was removed,
extracted with butanone, and dried to constant weight. Comparison
of transmission infrared spectra with the surface reflectance (ATR)
infrared spectra verified that chlorination occurred predominantly
on the surface of the PE, and that little or no elimination had
taken place. PE films of 0.038mm thickness containing 2.0, 3.5,
and 4.9 weight percent chlorine were prepared and characterized.

Reactions on Modified PE

PE containing the mixed functionalities from photochemical
interaction with NO or NO_2 was subjected to reduction with lithium
aluminum hydride (LAH). The LAH suspension reduced those functional
groups with C-N bonds to primary amines and those with C-O bonds
to secondary alcohols. Subsequent reaction with excess ethyl
bromide yielded the polymer with tertiary amine and alcohol sites,
as shown in reaction 10.

$$\text{(10)}$$

The conversion of chlorinated PE to N,N-diethylaminated PE was
studied in more detail. This reaction proceeds as shown in equation
11.

$$\text{(11)}$$

The chlorinated PE film sample was suspended in 100 cm^3 of 0.88 M
diethylamine in toluene. The mixture was slowly stirred and maintained at 20, 50, or 80°. As will be discussed in the grafting
reaction section, the 3.5% Cl film and 50° temperature was shown
to be the best combination, so this combination was used for an
evaluation of reaction times (6, 12, and 24 hours). In each case,
after the required reaction had been carried out the films were
extracted again with 2-butanone for 12 hr and were dried to constant weight under vacuum. The relative efficacy of each variant
in Cl content and diethylamination reaction temperature and time
is compared in the next section.

The film that had been chlorinated to 3.5% by weight and
reacted with diethylamine for 12 hr at 50° was analyzed by ESCA

(Electron Scattering for Chemical Applications). The surface of
the modified film contained about one atom percent nitrogen. While
this sounds like a low level of modification, note that triethylamine contains only 4.5 atom percent nitrogen. It is readily cal-
cuable that this sample (in the depth range of the ESCA spectrometer) contains one diethylamino group for approximately each 14
repeat units of the polyethylene substrate.

Photoinduced Grafting to Modified PE

The PE films were suspended in benzene solutions of benzophenone and monomer. The reaction mixtures were deoxygenated by three
freeze-pump-thaw cycles, were equilibrated to room temperature,
and attached in groups of three to Standco Stirrer (H. H. Sticht
Co.). The apparatus was then placed in the Rayonet Photochemical
Reactor and irradiated at $27\pm4°$. After irradiation the reaction
mixture was washed into a tared beaker, frozen, and placed under
room temperature vacuum for 24 hr followed by heating to approximately 50° for 48 hr under vacuum. The dried reaction mixture was
washed with methanol to remove any unreacted benzophenone and then
dried under vacuum for 48 hr at approximately 50°. The grafted
film was extracted for 12 hr with butanone to remove homopolymer
poly(methyl methacrylate) or poly(2-ethylhexyl methacrylate).

The PE samples subjected to these grafting conditions were
(i) unmodified high density polyethlene (PE), (ii) PE containing
Cl substituents on the surface, (iii) PE containing hydroxyl and
primary amine sites, (iv) PE containing hydroxyl and diethylamino
sites, and (v) PE containing only diethylamino sites. The PE
samples with the best defined structures are from classes (i),
(ii), and (v) and their use as grafting substrates is stressed in
succeeding discussions. PE of class (iii) was, as expected, not
very efficient as a graft substrate, since the radical formed by
abstraction at the hydroxyl site (reaction 12) is relatively stable
and probably leads to grafting only during the termination step,
and the radical formed by abstraction at the amine site easily
loses another H atom to form the imine (reaction 13).

PE of class (iv) is a somewhat better graft substrate than that of class (iii), but the concentration of the reactive sites was known with less certainty, and the more reactive site (the diethylamino group) could be studied better in PE of class (v).

TABLE I

PHOTOINITIATED GRAFTING OF 2EHMA TO POLYETHYLENE
AND MODIFIED POLYETHYLENES[a]

Polymer[b] Substrate	% Conversion of 2EHMA	Grafting Efficiency, %[e]	% P2EHMA[f]
PE	2.92	3.31	3.49
CLPE[c]	2.58	4.84	4.49
NEPE[d]	3.14	10.14	11.06

a All at 0.5 $\underline{M}$ 2EHMA, 0.02 $\underline{M}$ BP in 25 ml benzene; irradiation time 30 min

b Original film weights in 0.063 - 0.066g range; all weights to 0.0001

c 3.47% Cl by weight

d NEPE formed by reacting CLPE with 0.88 $\underline{M}$ diethylamine in toluene for 12 hr at 50°

e $\dfrac{\text{wt P2EHMA grafted}}{\text{total wt P2EHMA formed}}$ x 100 %

f $\dfrac{\text{wt P2EHMA grafted}}{\text{total wt grafted composite}}$ x 100 %

The results of the photoinitiated grafting of 2EHMA to high
density polyethylene (PE), chlorinated polyethylene (CLPE), and
diethylaminated polyethylene (NEPE) are shown in Table I. During
substrate preparation, it was difficult to prepare samples of
identical size and weight. Difference in sample size can cause
differences in per cent conversion of monomer to polymer and in
graft efficiency. Therefore, the most valid comparison between
substrates is a comparison between the percent grafted P2EHMA in
the final grafted samples. Comparison of the three substrates in
Table I with respect to % P2EHMA demonstrates the increased grafting which occurs upon amine incorporation. The HDPE and CLPE
substrates also have some P2EHMA grafted. This is probably due
to abstraction of hydrogen by triplet BP from unmodified methylene
units in these substrates. The diethylaminated film was the most
effective substrate for formation of graft polymer as expected.
This is due to efficient initiation of polymerization by the
substrate-bound amine-derived radical, as shown in equation 14.

$$\begin{array}{c}
CH_3CH_2 \quad CH_2CH_3 \\
\diagdown N \diagup \\
| \\
\sim\!CH\!\sim
\end{array}
\xrightarrow[h\nu]{\phi\!-\!\overset{\displaystyle O}{C}\!-\!\phi}
\quad
\phi\!-\!\overset{\displaystyle OH}{\underset{\displaystyle \cdot}{C}}\!-\!\phi
\quad
\begin{array}{c}
CH_3CH_2 \quad \overset{\displaystyle \cdot}{C}HCH_3 \\
\diagdown N \diagup \\
| \\
\sim\!CH\!\sim
\end{array}
\xrightarrow{M_2} \text{grafted PE}
\qquad (14)$$

To optimize the conditions for incorporation of the maximum
amount of grafted P2EHMA, the effects of per cent chlorine and the
temperature of the reaction of diethylamine with CLPE were studied.
The results are given in Table II, and inspection of these data
indicates that the grafting of 2EHMA maximizes at the diethylamination temperature of 50°C, while the highest per cent chlorine substrate shows a continual decrease in grafting of 2EHMA as the
amination reaction temperature is increased.

Reaction of the amine with CLPE results in both substitution
and dehydrohalogenation. The data in Table II show the effects of
at least three competing influences for the formation of the amine
functionality and the occurrence of the dehydrochlorination reaction. First the CLPE substrate is insoluble in the toluene solvent
used in the amination reaction at all temperatures employed.
However, some solvent swelling of the substrate may occur, and this
solvent swelling will increase as the temperature is raised, allowing more reaction of the amine. Second, the relative proportion
of dehydrochlorination increases as the percent chlorine increases
(14), and this effect is probably beginning to appear at the highest
per cent chlorine substrate. Third, the activation parameters for
the two competing processes favor an increase in dehydrochlorination
over amination as the temperature is increased (20).

TABLE II

PHOTOINITIATED GRAFTING OF 2EHMA TO MODIFIED POLYETHYLENE:
EFFECT OF ORIGINAL % CHLORINE AND DIETHYLAMINATION TEMPERATURE[a]

% Cl[b]	Reaction Temp.,[c] °C	% P2EHMA[d]
1.99	20	5.43
1.99	50	6.51
1.99	80	4.21
3.45	20	5.85
3.45	50	8.53
3.45	80	5.73
4.91	20	6.28
4.91	50	4.94
4.91	80	3.45

a   All at 0.5 $\underline{M}$ 2EHMA, 0.02 $\underline{M}$ BP in 25 ml benzene; irradiation
    time 30 min

b   Chlorine % by wt of CLPE before reaction with diethylamine

c   Temp of reaction of 0.88 $\underline{M}$ diethylamine in toluene with CLPE
    for 25 hr

d $\dfrac{\text{wt P2EHMA grafted}}{\text{total wt grafted composite}} \times 100\ \%$

The effect of the time of the reaction of diethylamine with
CLPE was also investigated, and the results are presented in Table
III. The data indicate that the greatest incorporation of grafted
P2EHMA is obtained at an amine reaction time of 12 hr. This effect
is unusual and may be due to increased dehydrochlorination at long
reaction times, resulting in less hydrogen abstraction from methylene units. Some of the chlorinated methylene units may be sterically constrained from undergoing amine substitution, but they may
eventually undergo dehydrochlorination.

TABLE III

PHOTOINITIATED GRAFTING OF 2EHMA TO MODIFIED POLYETHYLENE:
EFFECT ÒF REACTION TIME OF DIETHYLAMINE
WITH CHLORINATED POLYETHYLENE[a]

% Cl[b]	Reaction[c] Temp., °C	Reaction[d] Time, hr	% P2EHMA[e]
3.50	50	6	7.12
3.50	50	12	8.67
3.50	50	24	6.23

a   All at 0.5 $\underline{M}$ 2EHMA, 0.02 $\underline{M}$ BP in 25 ml benzene; irradiation
time 30 min

b   Chlorine % by wt of CLPE before reaction with diethylamine

c   Temp of reaction of 0.88 $\underline{M}$ diethylamine in toluene with CLPE

d   Time of reaction of diethylamine with CLPE

e $\dfrac{\text{wt P2EHMA grafted}}{\text{total wt grafted composite}} \times 100\ \%$

    The same series of reactions was accomplished using MMA in-
stead of 2EHMA, and the results are presented in Tables IV through
VI. The data are not as definitive as those of P2EHMA, but the
same trends are present. The decreased effectiveness for MMA as
compared to 2EHMA is probably due to the previously observed higher
rate of free radical polymerization for 2EHMA and because the mass
of a 2EHMA unit is almost twice that of MMA. The results in Table
IV demonstrate the increased effectiveness of NEPE over HDPE and
CLPE with respect to weight per cent grafting of MMA. The effect
of per cent chlorine and the temperature and time of the amination
reaction are given in Tables V and VI, and the trends are similar
to the 2EHMA results.

    The graft efficiencies (percentage of monomer consumed that is
chemically grafted to the substrate) is relatively low in each of
the experiments. The following analysis of the system explains
why this is so. The film substrate was suspended in a solution
of benzophenone (BP) and MMA or 2 EHMA in benzene. In view of the
short lifetime of triplet BP in benzene solution (1×10^{-5} sec.)
(12) an estimation of the fraction of BP triplets in solution which
will be capable of diffusing to the substrate is required. Assuming that the BP triplet can be approximated as a spherical particle

TABLE IV

PHOTOINITIATED GRAFTING OF MMA TO POLYETHYLENE
AND MODIFIED POLYETHYLENES[a]

Polymer[b] Substrate	% Conversion of MMA	Grafting[e] Efficiency, %	% MMA[f]
PE	3.32	3.14	3.49
CLPE[c]	2.20	3.82	3.18
NEPE[d]	2.18	5.15	4.31

a   All at 1.0 $\underline{M}$ MMA, 0.02 $\underline{M}$ BP in 25 ml benzene; irradiation time 30 min

b   Original film weights in 0.062 to 0.072 g range; all weights to 0.0001

c   3.47 % Cl by weight

d   NEPE formed by reacting CLPE with 0.88 $\underline{M}$ diethylamine in toluene for 12 hr at 50°

e $\dfrac{\text{wt PMMA grafted}}{\text{total wt PMMA formed}}$ x 100 %

f $\dfrac{\text{wt PMMA grafted}}{\text{total wt grafted composite}}$ x 100 %

of radius 0.4 nm, the Stokes-Einstein Equation may be used to calculate its diffusion coefficient (22). The Stokes-Einstein Equation is given in Equation 15, where D represents the diffusion coefficient, k is Boltzmann's constant, T is

$$D = kT \, (6\pi\nu r)^{-1} \tag{15}$$

temperature, ν is the viscosity of the medium, and r is the radius of·the spherical particle. Einstein related the diffusion coefficient to the distance that the particle will travel in a particular time (22), and his relationship is given in Equation 16, where X represents distance and τ is the time interval.

$$D = 0.5 \, X^2 \tau^{-1} \tag{16}$$

Combination of Equations 15 and 16 yields Equation 17 in which X represents the distance that the particle will travel in the

TABLE V

PHOTOINITIATED GRAFTING OF MMA TO MODIFIED POLYETHYLENE:
EFFECT OF ORIGINAL % CHLORINE AND DIETHYLAMINATION TEMPERATURE[a]

% Cl[b]	Reaction Temp.,[c] °C	% PMMA[d]
1.40	20	2.87
1.40	50	4.06
1.40	80	3.36
2.96	20	3.19
2.96	50	3.18
2.96	80	2.26
4.12	20	3.15
4.12	50	2.55
4.12	80	2.59

a All at 1.0 $\underline{M}$ MMA, 0.02 $\underline{M}$ BP in 25 ml benzene; irradiation time
   30 min

b Chlorine % by wt of CLPE before reaction with diethylamine

c Temp of reaction of 0.88 $\underline{M}$ diethylamine in toluene with CLPE
   for 25 hr

d $\dfrac{\text{wt PMMA grafted}}{\text{total wt grafted composite}} \times 100\ \%$

time interval τ. At a temperature of 30°C, using the lifetime of

$$X = (\tau kT)^{0.5}\ (3\pi\nu r)^{-0.5} \tag{17}$$

triplet BP to be 1×10^{-5} sec in a benzene solution (21) of viscosity 0.564 cp (23), equation 17 reveals that triplet BP will diffuse about 1.4×10^{-5} cm before deactivation. Comparison of the
distance of diffusion with the depth of the solution in which the
substrate is suspended (about 1 cm) shows that only a small fraction
of the BP triplets formed will be active in forming free radicals
on the substrate capable of initiating free radical polymerization.

TABLE VI

PHOTOINITIATED GRAFTING OF MMA TO MIDIFIED POLYETHYLENE:
EFFECT OF REACTION TIME OF DIETHYLAMINE
WITH CHLORINATED POLYETHYLENE[a]

% Cl[b]	Reaction[c] Temp, °C	Reaction[d] Time, hr	% PMMA[e]
3.13	50	6	2.34
3.13	50	12	3.82
3.13	50	24	3.52

a All at 1.0 $\underline{M}$ MMA, 0.02 $\underline{M}$ BP in 25 ml benzene; irradiation time
   30 min

b Chlorine % by wt of CLPE before reaction with diethylamine

c Temp of reaction of 0.88 $\underline{M}$ diethylamine in toluene with CLPE

d Time of reaction of diethylamine with CLPE

e $\dfrac{\text{wt PMMA grafted}}{\text{total wt grafted composite}} \times 100\ \%$

It has been demonstrated that solutions of either MMA or
2EHMA with BP in benzene undergo only a small amount of polymerization upon ultraviolet irradiation (6). However, comparison of
the reaction zone for graft polymer formation around the film
substrate (about 3×10^{-5} cm considering both top and bottom of the
film) with the total depth of solution (about 1 cm) allows the
prediction that the majority of polymer formation will not be due
to graft formation, but it will occur as a result of irradiation
of the solution away from the substrate. This process will then
yield inherently low graft efficiency.

This is a good point to compare our system, with its difficulties and limitations, to the previously mentioned graft system of
Oster et al (9, 10). In their PE samples that contained milled-in
BP, photoinduced BP abstraction and resulting polymerization can
occur away from the substrate only if the BP is extracted from the
substrate by the solvent action of the monomer. This extraction
may be a problem in some cases, but even when it is not, their
system suffers from an efficiency problem of a different type than
ours. Specifically, the benzophenone in the interior of the PE
substrate cannot interact directly with a monomer, nor can the
monomer get to any radical site produced by a secondary reaction

within the PE matrix. So minimization of extraction possibilities
means the monomer cannot penetrate; conversely maximization of
penetration into the substrate means the benzophenone near the
surface is likely to be leached out of the substrate. Additionally,
in the present studies it was desired to retain the bulk properties
of the PE substrate, without adulteration by additives or alteration
by "interior" reactions.

    The limitations of our system, as discussed above and as
verified by the experimental data, does not negate the value of
this approach or of the results. Remember that the surface of
the polyethylene film used in the present study was about 0.038 mm
in thickness. If the surface is considered to be the outer 20 Å
on both the top and bottom of the film (24), it constitutes only
about 1×10^{-2}% of the total thickness. If the density of the
surface region is the same as the density of the bulk of the film,
the weight of the surface is also only about 1×10^{-2}% of the total
weight of the film. Therefore, an increase in film weight of as
little as 1% due to polymer grafted to the surface can cause dramatic changes in the surface properties. This is verified very nice-
ly, albeit qualitatively to date, by inspection of the surface-
grafted PE samples prepared here. The PE that contains the grafted
P2EHMA surface is a material that permanently contains an adhesive.
It bonds very well to itself or to other smooth surfaces. The
composite so formed delaminates by adhesive failure between the
P2EHMA and the other component in all cases; this is quite different from the case in which PE is bonded to a substrate with "free"
P2EHMA. Further, the PE that contains the grafted PMMA surface
exhibits better wetting characteristics, and presumably enhanced
printability and bondability, when compared with unmodified PE.

                              CONCLUSIONS

    Polyethylene surfaces can be modified by photoassisted react-
ions with halogens (introducing -Br or -Cl sites), with various
NO_x species (introducing -NO, $-NO_2$, -ONO, $-ONO_2$, and sometimes
$C=O$ groups), or with NOCl. These reactive sites can be converted
to other functional groups, including 2° alcohols, 1° amines, or
3° amines. The most effective of these substituent groups with
respect to formation of free radical graft sites is the 3°amine
(diethylamino substituent) group, which can be used as part of the
benzophenone-amine photo-coinitiator system. Though these sites
markedly enhance photoinduced formation of free radicals that
initiate graft polymerization onto the polyethylene substrate, the
overall grafting efficiency is still low because much of the energy
is absorbed by benzophenone in the solution away from the polymer
surface. Graft efficiency might be improved by using a thin liquid
film of monomer and benzophenone instead of a large-volume solution

or, in cases where photochemically transparent films are the graft substrate, by irradiating through the film to the modified side so that absorption must occur at the interface. Modified polyethylenes (PE) have been prepared by grafting on methyl methacrylate (MMA) or 2-ethylhexyl methacrylate (2EHMA). The PE-g-MMA has different surface energy than the original PE, and the PE-g-2EHMA has a permanently bonded adhesive surface.

ACKNOWLEDGEMENTS

We thank the Phillips Petroleum Company, Bartlesville, Oklahoma, for the specially prepared films of high density polyethylene, Dr. R. W. Moddeman of the University of Dayton Research Institute for providing the ESCA scans, and Joseph B. Wicker for his contributions to the early stages of this work. We also thank the members of the UT Polymer Consortium and the UT non-service fellowship program for their financial support of SLW.

REFERENCES

1. D.K. Owens and R.C. Wendt, J. Appl. Polym. Sci., 13, 1741 (1969).

2. O.J. Sweeting, "The Science and Technology of Polymer Films," Vol. II, John Wiley and Sons, New York, N.Y., 1971, pp 131-211.

3. S.G. Cohen, A. Parola, G.H. Parsons, Jr., Chem. Rev., 73, 141 (1973).

4. M.R. Sandner, C.L. Osborn, D.J. Trecker, J. Polym. Sci., Polym. Chem. Ed., 10, 3173 (1972).

5. A. Ledwith and M.D. Purbrick, Polymer, 14, 521 (1973).

6. J.F. Kinstle and S.L. Watson, Jr., J. Radiat. Curing, 3(1), 2 (1976).

7. J.F. Kinstle and S.L. Watson, Jr., J. Radiat. Curing, 2(2), 7 (1975).

8. C.M. Hansen and P.E. Pierce, Ind. Eng. Chem., Prod. Res. Develop., 13, 218 (1974).

9. G. Oster, G.K. Oster, H. Moroson, J. Polym. Sci., 34, 671 (1959).

10. G. Oster, British Patent 856,884 (1960); Chem. Abstr., <u>55</u>,
     13907 f (1960).

11. S.R. Sandler and W. Karo, "Polymer Synthesis," Vol. 1,
     Academic Press, New York, N.Y., 1974, pp 271-272.

12. A.J. Gordon and R.A. Ford, "The Chemist's Companion," John
     Wiley and Sons, New York, N.Y., 1972, p 438.

13. "Kodak Filters for Scientific and Technical Uses," Kodak
     Publication B-3, Eastman Kodak Company, Rochester, N.Y.,
     1970, p 52.

14. S.A. Adylov, D.E. Il'ina, B.A. Krentsel, and M.V. Shishkina,
     Vysokomolekul. Soedin., <u>5</u>, 316 (1963); Chem. Abstr., <u>59</u>,
     2965f (1963). It must be noted that the abstract of this
     paper is incorrect in that it reports N,N-dibutylaniline in-
     stead of dibutylamine and 6.86% instead of 60.86% chlorine.

15. F.A. Cotton and G. Wilkinson, "Advanced Inorganic Chemistry,"
     Third Edition, Interscience Publishers, New York, N.Y., 1972,
     pp 354-360.

16. F.C. Kohout and F.W. Lampe, J. Chem. Phys., <u>46</u>, 4075 (1967);
     A.B. Callear, R.W. Carr, J. Chem. Soc.-Faraday II, <u>71</u>, 1603
     (1975).

17. F.E. Blacet, T.C. Hall, P.A. Leighton, J. Am. Chem. Soc., <u>84</u>,
     4011 (1962).

18. E.V. Lynn, J. Am. Chem. Soc., <u>41</u>, 368 (1919); E. Muller and
     H. Metzger, Chem. Ber., <u>87</u>, 1282 (1954); E. Muller, Pure
     Appl. Chem., <u>16</u>, 153 (1968).

19. J.F. Kinstle and S.L. Watson, Jr., manuscript in preparation.

20. R.W. Alder, R. Baker, J.M. Brown, "Mechanism in Organic
     Chemistry," Wiley-Interscience, London, 1971, p 230.

21. J.A. Bell and H. Linschitz, J. Am. Chem. Soc., <u>85</u>, 528 (1963).

22. F. Daniels and R.A. Alberty, "Physical Chemistry," 3rd ed,
     John Wiley and Sons, New York, N.Y., 1966, pp 404-406.

23. "Handbook of Chemistry and Physics," R.C. Weast and S.M. Selby,
     Eds., The Chemical Rubber Co., Cleveland, Ohio, 1967, p F-35.

24. W.E. Moddeman and C.R. Cothern, J. Environ. Sci., <u>28</u>, 27
     (1975).

Contributors

R. G. Bauer, Goodyear Tire & Rubber Co., Research Division
    142 Goodyear Blvd., Akron, OH   44316

C. Beslin, Ecole Superieure de Chimie de Mulhouse, 3, rue
    A. Werner 68093 MULHOUSE Cedex (France)

O. P. Bovkunenko, Donetsk State University, Donetsk, 340055
    (U.S.S.R.)

W. H. Buck, E. I. duPont de Nemours and Company, Inc.
    Wilmington, DE 19898

I. Cabasso, Gulf South Research Institute, P.O. Box 26518
    New Orleans,LA   70136

J. Cifaratti, Dept. of Chemistry, State University of
    New York at Albany, Albany, NY   12203

A. Conde, Universidad Industrial de Santander
    Bucaramanga, Columbia

N. Devia-Manjarres, Lehigh University, Bethlehem, PA   18015

J. Diamant, Dept. of Chemical Engineering, University of
    California, Berkeley, CA   94720

G. C. Eastmond, Department of Inorganic, Physical & Industrial
    Chemistry, Liverpool University, Liverpool L69 3BX, U.K.

R. Foreman, Department of Chemistry, State University of New
    York at Albany, Albany, NY   12203

R. P. Foss, Central Research & Development Department,
    Experimental Station, E. I. duPont de Nemours & Co., Inc.
    Wilmington, DE   19898

Emil M. Friedman, The Goodyear Tire & Rubber Company
    Research Division, 142 Goodyear Blvd., Akron, OH   44316

H. L. Frisch, Department of Chemistry, State University of
     New York at Albany, Albany, NY   12203

K. C. Frisch, Polymer Institute, University of Detroit
     Detroit, MI   48221

V. A. Gasin, Donetsk State University, Donetsk, 340055 (U.S.S.R.)

C. A. Glotfelter, Armstrong Cork Company, Research & Development
     Center, Lancaster, PA   17604

D. R. Hansen, Shell Development Company, Houston, TX   77001

J. F. Harris, Central Research & Dev. Department
     E. I. duPont de Nemours and Co., Inc.
     Wilmington, DE   19898

J. E. Herweh, Armstrong Cork Company, Research & Dev. Center
     Lancaster, PA   17604

S. Hohne, Donetsk State University, Donetsk, 340055(U.S.S.R.)

S. D. Hong, Dept. of Chemical Engineering, University of
     California, Berkeley, CA   94720

J. L. Illinger, Army Materials & Mechanics, Research Center
     ATTN: DRXMR-RA, Watertown, MA   02172

M. Iskander, Dept. of Chemistry, Rensselaer Polytechnic
     Institute, Troy, NY   12181

C. H. M. Jacques, General Motors Research Laboratories
     Warren, MI   48090

J. Jagur-Grodzinski, Weizmann Institute of Science,
     Rehovot, Israel

J. P. Kennedy, Institute of Polymer Science, University
     of Akron, Akron, OH   44325

J. F. Kenney, M & T Chemicals Inc., Rahway, NJ   07065

J. F. Kinstle, Dept. of Chemistry, The University of Tennessee
     Knoxville, TN   37916

D. Klempner, Polymer Institute, University of Detroit,
     Detroit, MI   48221

Sonja Krause, Dept. of Chemistry, Rensselaer Polytechnic
    Institute, Troy, NY   12181

B. V. Kravchenko, Donetsk State University, Donetsk
    340055 (U.S.S.R.)

D. Krevor, Dept. of Chemical Engineering, State University
    Of New York at Buffalo, Buffalo, NY   14214

R. V. Kucher, Institute of Physical Organic Chem. & Coal Chem.
    Academy of Sciences of the Ukrainian SSR 340048,
    Donetsk-48, R. Luxemburg Str., 70

A. A. Kuznetsov, Institute of Physical Organic Chem. & Coal Chem.
    Academy of Sciences of the Ukrainian SSR, 34008, Donetsk
    48, R. Luxemburg Str., 70

J. L. Locatelli, Ecole Superieure de Chimie de Mulhouse
    3, rue A. Werner, 68093 MULHOUSE Cedex (France)

C. E. Locke, University of Oklahoma, Norman, OK   73019

William A. Ludwico, Mobay Chemical Corp., Penn Lincoln Parkway W.
    Pittsburgh, PA   15205

J. A. Manson, Lehigh University, Bethlehem, PA 18015

S. Marti, Ecole Superieure de Chimie de Mulhouse,
    3, rue A. Werner, 68093 MULHOUSE Cedex (France)

V. I. Melnichenko, Donetsk State University, Donetsk, 340055
    (U.S.S.R.)

D. K. Metzler, Celanese Research Co., P.O. Box 1000,
    Summit, NJ   07901

O. M. Mezentseva, Donetsk State University, Donetsk,
    340055 (U.S.S.R.)

N. A. Noskova, Donetsk State University, Donetsk, 340055(U.S.S.R.)

J. Oziomek, The Central Research Laboratories
    The Firestone Tire & Rubber Co., Akron, OH   44317

R. Palma, Dept. of Chemistry, State University of New York
    At Albany, Albany, NY   12203

D. G. Phillips, CSIRO, Dept. of Textile Industries,
    Belmont, Victoria, Australia

P. J. Phillips, Dept. Of Chem. Eng., State University of
    New York at Buffalo, Buffalo, NY 14214

J. Pulido, Lehigh University, Bethlehem, PA   18015

J. L. Regregier, Ecole Superieure de Chimie de Mulhouse
    3, rue A. Werner, 68093 MULHOUSE Cedex (France)

G. Riess, Ecole Superieure de Chimie de Mulhouse, 3 rue
    A. Werner, 68093 MULHOUSE Cedex (France)

S. L. Rosen, Carnegie-Mellon University, Pittsburgh, PA   15213

Diego C. Rubio, The Goodyear Tire & Rubber Co.
    Research Division, 142 Goodyear Blvd., Akron, OH   44316

T. Russell, Dept. of Polymer Science and Engineering and
    Polymer Research Institute, University of Massachusetts
    Amherst, MA   01002

M. Schlienger, Ecole Superieure de Chimie de Mulhouse
    3, rue A. Werner, 68093 MULHOUSE Cedex (France)

N. S. Schneider, Polymer and Chemistry Division
    Army Materials and Mechanics Research Center
    Watertown, MA   02172

R. Schwartz, Dept. of Chemistry, State University of New York
    at Albany, Albany, NY   12203

R. B. Seymour, Dept. of Polymer Science, University of
    Southern Mississippi, Hattiesburg, MS   39401

W. H. Sharkey, Central Research and Development Dept.
    Experimental Station, E. I. duPont de Nemours & Co., Inc.
    Wilmington, DE   19898

M. Shen, Dept. of Chemical Engineering, University of California
    Berkeley, CA   94720

L. H. Sperling, Lehigh University, Bethlehem, PA   18015

G. A. Stahl, BF Goodrich Research & Development Center
    9921 Brecksville Road, Brecksville, OH   44141

R. S. Stein, Dept. of Polymer Science and Engineering and
    Polymer Research Institute, University of Massachusetts
    Amherst, MA   01002

S. A. Sundet, Plastic Products and Resins & Elastomer Chemicals
Department, E. I. duPont de Nemours & Co., Inc.
Experimental Station, Wilmington, DE 19898

Paik C. S. Sung, Dept. of Materials Science and Engineering
M.I.T., Cambridge, MA 02139

R. C. Thamm, Plastic Products & Resins & Elastomer Chemicals Dept.
E. I. duPont de Nemours & Co., Inc.

D. Vofsi, Weizmann Institute of Science, Rehovot, Israel

S. L. Watson(Jr.), Dept. of Chemistry, The University of
Tennessee, Knoxville, TN 37916

L. Watters, Halliburton Services, Duncan, OK 73533

W. Y. Whitmore, Armstrong Cork Co., Research & Dev. Center
Lancaster, PA 17604

Y. Yamashita, Dept. of Synthetic Chem., Faculty of Engineering
Nagoya University, Nagoya 464, Japan

V. D. Yenalyev, Donetsk State University, Donetsk,
340055 (U.S.S.R.)

G. M. Yenwo, Lehigh University, Bethlehem, PA 18015

H. Yoon, Polymer Institute, University of Detroit, Detroit, MI
48221

Ju. S. Zaitsev, Institute of Phys. Organic Chem. & Coal Chem.
Academy of Sci. of the Ukrainian SSR, 340048, Donetsk-48
R. Luxemburg, Str., 70

V. V. Zaitseva, Institute of Phys. Organic Chemistry and
Coal Chemistry, Academy of Sciences of the Ukrainian
SSR, 340048, Donetsk-48 R. Luxemburg Str., 70